Nanocomposites for the Desulfurization of Fuels

Tawfik Abdo Saleh
King Fahd University of Petroleum and Minerals, Saudi Arabia

A volume in the Advances in Chemical and Materials Engineering (ACME) Book Series

IGI Global
DISSEMINATOR OF KNOWLEDGE

Published in the United States of America by
IGI Global
Engineering Science Reference (an imprint of IGI Global)
701 E. Chocolate Avenue
Hershey PA, USA 17033
Tel: 717-533-8845
Fax: 717-533-8661
E-mail: cust@igi-global.com
Web site: http://www.igi-global.com

Library of Congress Cataloging-in-Publication Data

Names: Saleh, Tawfik A., editor.
Title: Nanocomposites for the desulfurization of fuels / Tawfik Abdo Saleh, editor.
Description: Hershey, PA : Engineering Science Reference, 2020. | Includes bibliographical references. | Summary: "This book explores the synthesis, properties, and technological developments of nanomaterials and their applications in petroleum"-- Provided by publisher.
Identifiers: LCCN 2019036918 (print) | LCCN 2019036919 (ebook) | ISBN 9781799821465 (h/c) | ISBN 9781799821472 (s/c) | ISBN 9781799821489 (eISBN)
Subjects: LCSH: Fossil fuels--Desulfurization--Materials. | Petroleum chemicals--Desulfurization--Materials. | Nanocomposites (Materials)--Industrial applications.
Classification: LCC TP156.D43 N36 2020 (print) | LCC TP156.D43 (ebook) | DDC 547/.82--dc23
LC record available at https://lccn.loc.gov/2019036918
LC ebook record available at https://lccn.loc.gov/2019036919

This book is published in the IGI Global book series Advances in Chemical and Materials Engineering (ACME) (ISSN: 2327-5448; eISSN: 2327-5456)

British Cataloguing in Publication Data
A Cataloguing in Publication record for this book is available from the British Library.

For electronic access to this publication, please contact: eresources@igi-global.com.

Advances in Chemical and Materials Engineering (ACME) Book Series

ISSN:2327-5448
EISSN:2327-5456

Editor-in-Chief: J. Paulo Davim, University of Aveiro, Portugal

MISSION

The cross disciplinary approach of chemical and materials engineering is rapidly growing as it applies to the study of educational, scientific and industrial research activities by solving complex chemical problems using computational techniques and statistical methods.

The **Advances in Chemical and Materials Engineering (ACME) Book Series** provides research on the recent advances throughout computational and statistical methods of analysis and modeling. This series brings together collaboration between chemists, engineers, statisticians, and computer scientists and offers a wealth of knowledge and useful tools to academics, practitioners, and professionals through high quality publications.

COVERAGE

- Chemical Engineering
- Heat Treatments
- Statistical methods
- Coatings and surface treatments
- Composites
- Biomaterials
- Computational methods
- Industrial chemistry
- Wear of Materials
- Fatigue and Creep

Titles in this Series

For a list of additional titles in this series, please visit:
https://www.igi-global.com/book-series/advances-chemical-materials-engineering/73687

Advanced Catalysis Processes in Petrochemicals and Petroleum Refining Emerging Research and Oportunities
Mohammed C. Al-Kinany (King Abdulaziz City for Science and Technology, Saudi Arabia) and Saud A. Aldrees (King Abdulaziz City for Science and Technology, Saudi Arabia)
Engineering Science Reference • ©2020 • 205pp • H/C (ISBN: 9781522580331) • US $195.00

Nanotechnology in Aerospace and Structural Mechanics
Noureddine Ramdani (Research and Development Institute of Industry and Defense Technologies, Algeria & Harbin Engineering University, China)
Engineering Science Reference • ©2019 • 356pp • H/C (ISBN: 9781522579212) • US $265.00

Polymer Nanocomposites for Advanced Engineering and Military Applications
Noureddine Ramdani (Research and Development Institute of Industry and Defense Technologies, Algeria & Harbin Engineering University, China)
Engineering Science Reference • ©2019 • 435pp • H/C (ISBN: 9781522578383) • US $225.00

Process Analysis, Design, and Intensification in Microfluidics and Chemical Engineering
Harrson Silva Santana (University of Campinas, Brazil) João Lameu da Silva Jr (Federal Institute of Education, Science, and Technology of South of Minas Gerais, Brazil) and Osvaldir Pereira Taranto (University of Campinas, Brazil)
Engineering Science Reference • ©2019 • 367pp • H/C (ISBN: 9781522571384) • US $215.00

Applications of Nanofluid Transportation and Heat Transfer Simulation
Mohsen Sheikholeslami (Babol Noshirvani University of Technology, Iran)
Engineering Science Reference • ©2019 • 692pp • H/C (ISBN: 9781522575955) • US $245.00

For an entire list of titles in this series, please visit:
https://www.igi-global.com/book-series/advances-chemical-materials-engineering/73687

IGI Global
DISSEMINATOR OF KNOWLEDGE

701 East Chocolate Avenue, Hershey, PA 17033, USA
Tel: 717-533-8845 x100 • Fax: 717-533-8661
E-Mail: cust@igi-global.com • www.igi-global.com

Table of Contents

Detailed Table of Contents

This chapter describes different desulfurization technologies used for the removal of sulfur from petroleum products or from refined products. These technologies include hydrodesulfurization and non-hydrodesulphurization such as extractive desulphurization, adsorptive desulfurization, precipitative desulphurization, oxidative desulphurization, and desulfurization by membranes. Types of reactors including batch and fixed bed reactors are discussed. The chapter also highlights some of the common mechanism to explain the desulphurization process.

Desulfurization (removal of S compounds) of fuels is an important research topic in recent years. Several techniques have been reported to remove the sulfur-containing compounds in fuels. One of these techniques is adsorptive desulfurization (removal based on chemisorption and physisorption), which has received much attention because of low energy consumption and facile operation condition. This chapter discusses the methods employed under this technique and the types of nanocomposites and hybrid materials (adsorbents) that have been investigated as potential adsorbents. The strategies to enhance sulfur adsorption capacity and main challenges will be discussed.

The ultra-low sulfur diesel (ULSD) is required to comply with stricter government policy on low sulfur content of transportation fuels. Better knowledge of the different factors that concern deep desulfurization of fuels is necessary to achieve ultra-low sulfur content and cheaper way of producing ULSD. Both the capital and operating cost of the adsorptive desulfurization process is cheaper compare to the conventional hydroprocessing. In the future, the need to produce more volume of fuels with very low sulphur content from low-grade feedstocks like heavy oil and light cycle oil in order to meet up with the global demand for sulphur-free fuels is pertinent. Several on-going researches are pointing to the use of adsorbents for removal of sulfur compounds from hydrocarbon refining stream. In this chapter, varieties of carbon nanomaterials suitable for adsorptive desulfurization are discussed. If the active lifetime, where the capacity of the adsorbents are adequate, the approach is practically feasible for commercial application.

Polyoxometalates have been demonstrated to be efficient catalysts for the activation of oxidants in desulfurization processes. Successful results on desulfurization using polyoxometalates and hydrogen peroxide to desulfurize model oils and liquid fuels were reported and can be found in the literature. The desulfurization is an actual subject with notable interest for refineries and fuel cost, and consequently, it is important to focus the scientific community to work in desulfurization technology in order to develop catalytic systems based on polyoxometalates capable to be reused, stable, efficient, and selective. Therefore, the main goal is the design of heterogeneous polyoxometalate based catalysts. This chapter pretends to inform the research society about the scientific directions that have been taken using heterogeneous polyoxometalate catalysts in oxidative desulfurization of simulated and real liquid fuels. In addition, future perspectives are proposed to cover the actual needs of this area.

Ni/ZnO nano-sorbent systems have been extensively used in the reactive adsorption desulfurization (RADS) of gasoline steams, especially in China, to meet the more rigorous regulation on the sulfur content. The apparent advantage of RADS is that most of the olefins are kept in the product with low consumption of hydrogen and little loss of octane. The authors discussed in this chapter the relation of catalytic properties with components and structures of Ni/ZnO sorbent. Based on detailed characterization and reaction results, they revealed the dynamic change of Ni/ZnO sorbents during RADS, the mechanisms of desulfurization, and the sulfur transfer and sulfur adsorption. Apart from the RADS of Ni/ZnO nano-sorbent for cleaner gasoline production, they also presented other potential applications.

Desulfurization of fuel oils is an essential process employed in petroleum refineries to reduce the sulfur content to levels mandated for environmental protection. Hydrodesulfurization (HDS), which is currently being employed, is limited in treating refractory organosulfur compounds and only reduces the sulfur content in fuels to a range of 200-500 ppmS. In this chapter, several scientific and technological advances reported in the literature for the desulfurization of fuels are reviewed and discussed. Amongst these techniques, oxidative desulfurization (ODS) and adsorptive desulfurization (ADS) are proposed as additional steps to complement HDS in meeting the mandated ultra-low sulfur levels (10 ppmS). In the ODS technique, refractory organosulfur compounds are oxidized to organosulfones, followed by solvent extraction or adsorption of the organosulfones. The chemistry involved in the development and fabrication of sulfur/sulfone responsive adsorbents is also discussed. The use of molecular imprinted polymers (MIPs) and coordination polymers (CPs) for the selective adsorption of organosulfone compounds (in ODS) and/or organosulfur (in ADS) offers various properties such as imprinting effect, hydrogen bonding, π-π interactions, van der Waals forces, π-complexation, and electrostatic interactions. CPs, in particular metal organic frameworks (MOFs), have been reported to possess suitable features to overcome most of these challenges associated with adsorptive ultra-deep desulfurization when design strategies to achieve good selectivity are strictly followed. Matching the sizes of the cavities to the critical dimensions of the sulfur containing compounds (SCCs), using suitable metal centres which allow for coordinative interaction with the SCCs and using linkers with suitable functionality as to enhance specific interaction (dispersion forces) with the SCCs were considered to be pivotal features to prioritize. The prospects for the use of MIPs and CPs for future industrial applications in desulfurization are envisaged.

The biodesulfurization (BDS) process consists of the use of microorganisms for the removal of sulfur from fossil fuels. Through BDS it is possible to treat most of the organosulfur compounds recalcitrant to the conventional hydrodesulfurization (HDS), the petroleum industry's solution, at mild operating conditions, without the need for molecular hydrogen or metal catalysts. This technique results in lower

Preface

Nanotechnology in desulfurization is a fascinating subject of recent origin. The expansion of nanomaterials is the focal point of research and technology that is mostly related to chemistry, physics, applied sciences, petroleum, and engineering. Research on nanotechnology has mainly focused on the aspects of synthesis of nanomaterials that have unique chemical, thermal and mechanical properties applicable for a wide range of applications. A variety of properties and phenomena has been investigated, and many of the studies have been directed toward understanding the properties and applications of nanomaterials.

Due to their enhanced chemical and mechanical properties, the nanomaterials play promising roles in enhancing the desulfurization. Nanomaterials have properties that are useful for enhancing surface-to-volume ratio, reactivity, strength, and durability. In pursuit of the same goal, this book offers detailed, up-to-date chapters on the synthesis, properties and technological developments of nanomaterials, and their applications in petroleum.

The first chapter of the book encompasses the occurrence of organosulfur compounds in petroleum and various techniques for the removal of these compounds. The next two chapters encompass the nanocomposites, hybrid materials and carbon-based nanomaterials for adsorptive desulfurization including preparation and evaluation.

The fourth chapter discussed the polyoxometalates as efficient catalysts for the activation of oxidants in desulfurization processes with the main goal to design heterogeneous polyoxometalate based catalysts. This chapter pretends to inform the research society about the scientific directions that have been taken using heterogeneous polyoxometalate catalysts in oxidative desulfurization of simulated and real liquid fuels. In Chapter 5, the relation of catalytic properties with components and structures of Ni/ZnO sorbent is discussed. Based on detailed characterization and reaction results, the dynamic change of Ni/ZnO sorbents during RADS is highlighted, the mechanisms of desulfurization, and the sulfur transfer and sulfur adsorption, etc.

In the next chapter, several scientific and technological advances reported in the literature for the desulfurization of fuels are reviewed and discussed. Amongst these techniques are oxidative desulfurization (ODS) and adsorptive desulfurization (ADS) which are proposed as additional steps to complement HDS in meeting the mandated ultra-low sulfur levels (10 ppmS).

Advances in the reduction of the costs inherent to fossil fuels' biodesulfurization towards its potential industrial applications are discussed in chapter seven. In the next chapter, environmental concerns and the importance of desulfurization is discussed.

In this book, we have tried to cover many aspects of nanomaterials for desulfurization, which is of current interest. This book is written for a large readership, including university students and researchers from diverse backgrounds such as chemistry, petroleum, materials science, physics, and engineering. It can be used not only as a textbook for both undergraduate and graduate students but also as a review and reference book for researchers in these fields. We hope that the chapters of this book will provide the readers with valuable insight into state-of-the-art advanced and functional nanomaterials and technologies.

However, it is possible that some topics have been left out owing to constraints on the size of the book and possible errors in judgment. We trust that the preface will be useful to students, teachers, and researchers.

Tawfik A. Saleh
King Fahd University of Petroleum and Minerals, Saudi Arabia

Chapter 1
Advanced Desulfurization Technologies and Mechanisms

Tawfik A. Saleh
King Fahd University of Petroleum and Minerals, Saudi Arabia

ABSTRACT

This chapter describes different desulfurization technologies used for the removal of sulfur from petroleum products or from refined products. These technologies include hydrodesulfurization and non-hydrodesulphurization such as extractive desulphurization, adsorptive desulfurization, precipitative desulphurization, oxidative desulphurization, and desulfurization by membranes. Types of reactors including batch and fixed bed reactors are discussed. The chapter also highlights some of the common mechanism to explain the desulphurization process.

INTRODUCTION

Overview of the Process of Crude Oil Refining

The oil and gas industry can be simply classified into three major divisions: upstream (or exploration and production- E&P), midstream and downstream.

DOI: 10.4018/978-1-7998-2146-5.ch001

1. Upstream division involves exploring for crude oil deposits and the production of crude oil. The upstream division includes searching for potential underground or underwater crude oil and natural gas fields, drilling exploratory wells, and subsequently drilling and operating the wells that recover and bring the crude oil or raw natural gas to the surface. Examples of firms that would belong in the upstream segment of the industry include companies that own rights to drill for oil and companies that provide support services to the drilling segment of the industry.
2. Midstream division involves the distribution of crude oil to refiners; the refining of crude oil into saleable products; and the distribution of products to wholesalers and retailers. The midstream operations are often taken to include some elements of the upstream and downstream sectors. For example, the midstream sector may include natural gas processing plants that purify the raw natural gas as well as removing and producing elemental sulfur and natural gas liquids (NGL) as finished end-products.
3. Downstream division involves the refining of petroleum crude oil and the processing and purifying of raw natural gas, as well as the marketing and distribution of products derived from crude oil and natural gas. The downstream division reaches consumers through products such as gasoline or petrol, kerosene, jet fuel, diesel oil, heating oil, fuel oils, lubricants, waxes, asphalt, natural gas, and liquefied petroleum gas (LPG) as well as hundreds of petrochemicals. Gasoline stations are perhaps the most visible downstream companies, but companies that deliver heating oil or propane would also fall into this category. Note that some of the midstream operations are often included in the downstream category and are considered to be a part of the downstream sector.

The overall well-to-consumer supply chain for petroleum products can be illustrated as shown in Figure 1.

Desulfurization

After the crude oil is being treated in the distillation unit followed by several refinery processing, various refined products are obtained, Figure 2. Several sulfur-based compounds are available in refined products or petroleum. Examples are listed in Figure 3. Desulfurization is a chemical process for the removal of sulfur from petroleum. It involves the removal of sulfur from a molecule and/or the removal of sulfur compounds from oil/gas refinery streams. Desulfurization provides the bulk of sulfur used in industry (Claus process and Contact process), sulfur-free compounds that could otherwise not be used in a great number of catalytic processes, and also

Figure 1. Well-to-consumer supply chain for petroleum products. Upstream midstream and downstream

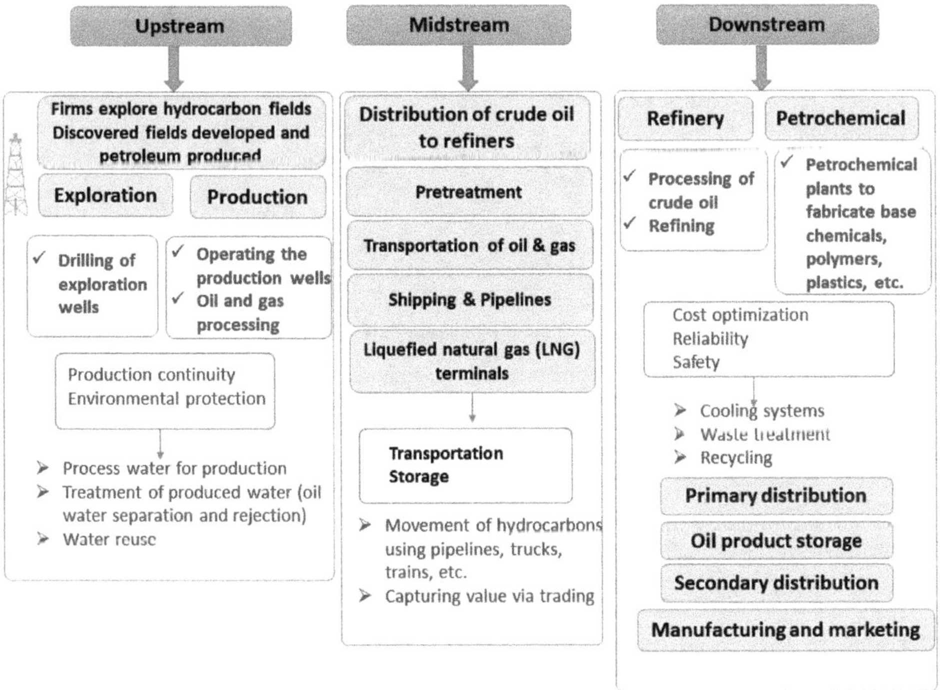

Figure 2. Schematic flow chart of a typical petroleum refinery before hydrodesulphurization

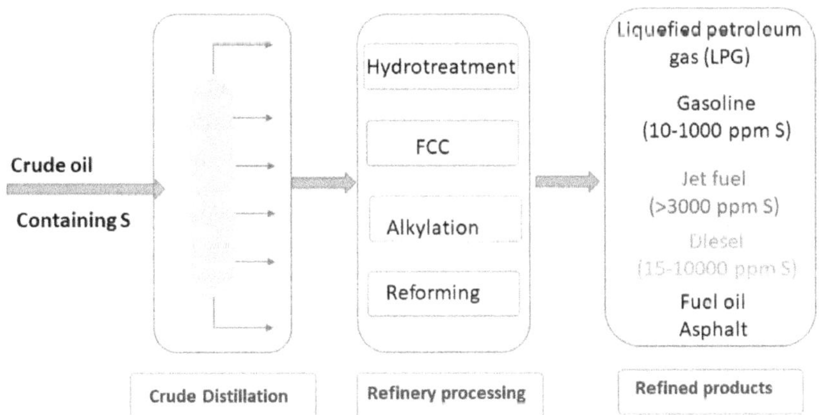

Figure 3. Reactivity of sulfur compounds with varying sizes commonly found in fossil-based fuels versus the difficulty in removing them via HDS
(Lee and Valla, 2019).

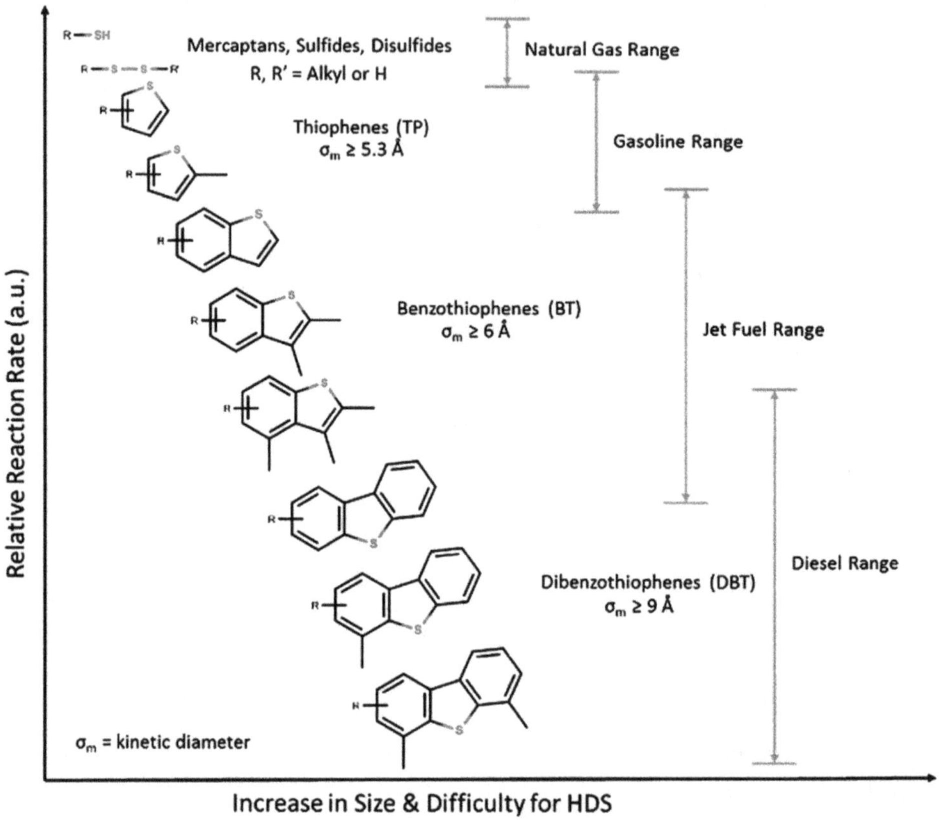

reduce the release of harmful sulfur compounds into the environment, such as sulfur dioxide which leads to acid rain. Processes used for desulfurization include hydrodesulfurization, SNOX process and the wet sulfuric acid process (WSA process).

Desulfurization Technologies

The demand for clean hydrocarbon fuels, with almost zero sulfur, has been on the rise as fossil-based fuel continues to be the primary source of energy. After hydrotreatment and processing, a typical gasoline fraction contains sulfur in a range from 10 to 1000 ppm, jet fuel contains around 3000 ppm, while diesel contains about 10000 ppm (ICCT 2011; Lee and Valla 2019), Figure 2. Therefore, liquid fuels are subjected to desulfurization to further achieve the sulfur standards for transportation fuels set

Figure 4. Types of Desulfurization Process, reactors, and mechanisms

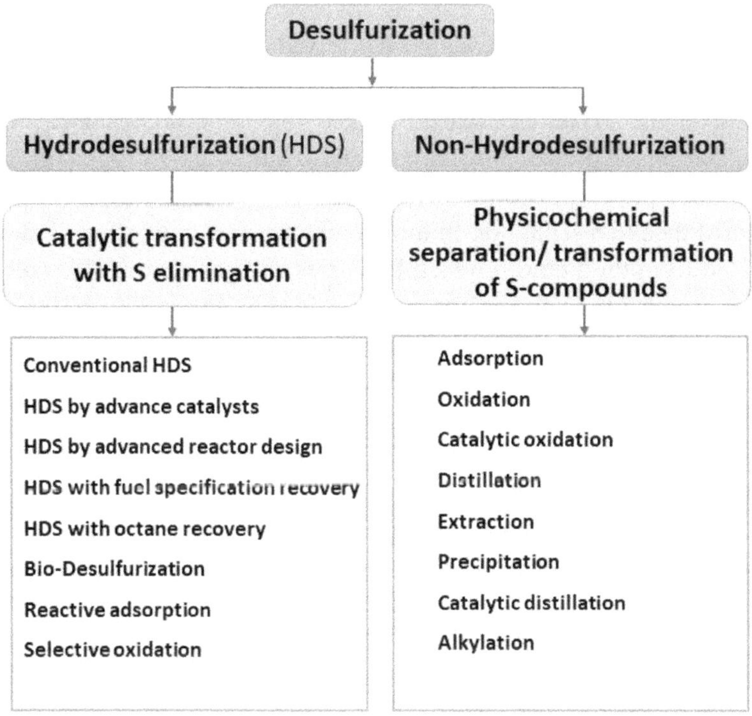

by the Environmental Protection Agency (EPA). In addition to hydrodesulfurization (HDS), there are several Alternative approaches to deep desulfurization. These methods are listed in Figure 4.

Hydrodesulphurization (HDS)

HDS is used as an industrial process for desulfurization of petroleum. The feed-in HDS process is treated with hydrogen gas in a specially designed reactor under high partial pressure and elevated temperature in the presence of a catalyst. The organically bonded sulphur is converted into H_2S during the reaction, which is separated from the feed in the stream of flue gases. The conventional HDS process commonly operates at a temperature ranging from 300 to 450 °C and pressure of 30 to 300 psi, however, the process conditions may vary depending on the nature of feed and its sulfur contents as well as the degree of desulphurization required (Javadli & Klerk, 2012). In conventional HDS the catalysts mostly used are sulfided CoMo type or NiMo and CoMo supported on Al_2O_3 (Furimsky, 1996; Shafi et al., 2000). The advanced HDS operation is used for deep desulfurization of high sulfur feed and carried out under severe conditions of temperature and pressure, i.e. at elevated temperature and high pressure, over the catalysts using advanced supports. In HDS unit, the feed which may be naphtha, kerosene or gas oil is mixed with hydrogen gas and heated in a fire-heated tubular furnace up to 315°to 425 °C temperature. The vaporized feed is then pumped into a fixed bed reactor packed with the catalyst under pressure up to 70 bars. During the process, the compounds of sulfur and nitrogen present in the feed react with H_2 to form H_2S and NH_3. The products from the reactor are passed to cooler, and then to a separator. In the separator, gases consisting of H_2S and H_2 are separated from the liquid product, high-pressure gas rich in H_2 is recycled to the HDS reactor and the low-pressure gas rich in H_2S is passed to amine contactor for removal of H_2S gas. The flue gases can be used for heating the furnaces. Liquid products are pumped to the stripping column, where remaining H_2S, gases, and other impurities are then removed as desulfurized product.

Example of HDS units is illustrated in Figure 5. This unit was reported to evaluate the hydrodesulfurization activity of the catalysts like AlMoCo and AlCNFMoCo, (AlHammadi et al. 2018). The HDS system is composed of a reactor, a hydrogen gas supplier, gas controller, and the valves for controlling the input and output samples. The HDS reaction occurred at T = 300 °C and 55 bar hydrogen partial pressure. The catalyst is inserted in the reactor vessel with 100 ml of the model fuel containing DBT at an initial concentration of 550 ppm-S in Decalin as a solvent model fuel. When the reaction temperature was reaching up to 300 °C, the first sample was collected and considered as the zero point. Then, after one hour of the reaction, a sample was collected by a manual valve and the reaction was monitored for 6h. Finally, the concentrations of sulfur in the collected samples are detected by gas chromatography employing sulfur chemiluminescence detector (GC-SCD). It should be noted that the identification of the products can be performed using gas chromatography mass spectrometry.

Extractive Desulphurization (EDS)

In EDS, sulfur can be removed from petroleum by extraction using an appropriate solvent. Extractive desulphurization (EDS) involves the removal of sulphur compounds from the petroleum by transferring into a solvent in which their solubility is higher than the petroleum hydrocarbons. The criteria of the solvent used for EDS process is that it should bear high solubility for sulphur compounds, should be least expensive and immiscible with petroleum and its boiling point should be different from sulphur compounds so that removal of sulfur compounds could be accomplished easily.

EDS functions at ambient conditions and conveniently applicable in industry. The solvent and feedstock are allowed to mix in the mixing tank, by virtue of their solubility in the solvents, the sulfur compounds transfer to solvent. The solvent

Figure 5. Illustration of the system used for testing the HDS reaction process

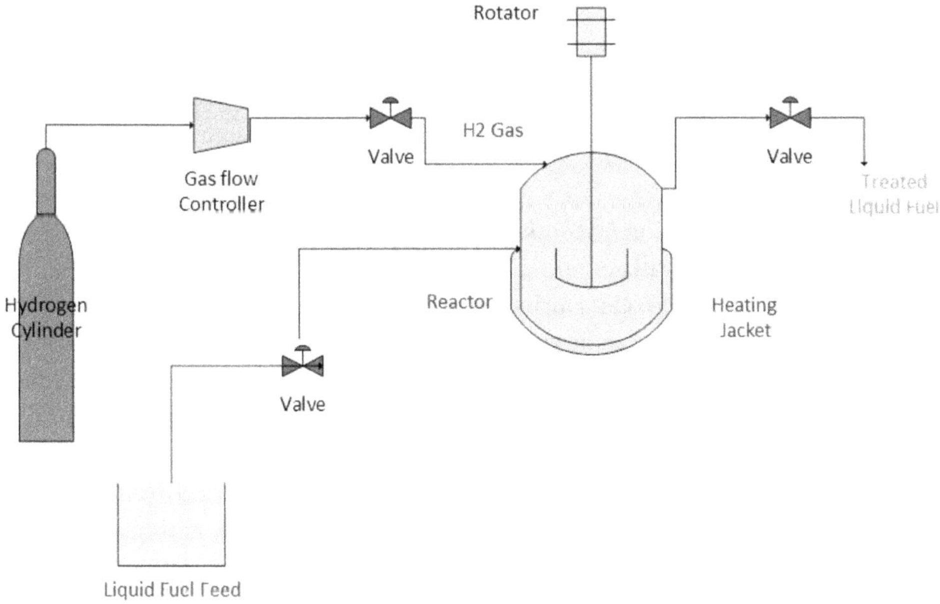

is then removed from the feedstock in the separator section and is recycled after removal of extracted sulfur compounds through distillation.

Commonly used solvents for EDS include ethanol, acetone, polyethylene glycols, various amines dimethyl formamide (DMF), misetyl oxide, furfural, dimethyl sulfoxide (DMSO) and N-methylpyrolidone (NMP). Commonly two or more solvents are blended together and used for EDS i.e. acetone:ethanol mixture or te traethyleneglycol:methoxytriglycol mixture, in order to enhance the extraction of organosulphur compounds. It was shown earlier that the EDS of various distillate fractions by polyethylene methylether, about 30% of sulfur removal is attained, however later on it was reported that using a blend of aqueous acetone, methanol, misetyl oxide and formic acid, about 86 to 96% of desulfurization is attained in different distillates fractions.

Out of its limitations, EDS cannot remove all classes of sulfur compounds, therefore in different distillates fractions the overall desulfurization efficiency of the EDS extractive desulfurization is very low, however literature reports exhibited that if employed in combination with other processes like hydrotreatment and oxidation, the efficiency of the process can be markedly enhanced (H. Rang et al., 2006). It was shown that in case of gas oil the sulphur content were reduced up to 71% by extraction with polyethylene glycol and dimethylether, but deep desulfurization of gas oil was attained by successive hydrotreatment and extraction (Pfüller, 2003). Similarly, in the case of kerosene, extraction by similar solvents followed by hydrotreatment, about 94% of sulfur was removed. The efficiency of the EDS is more enhanced when prior oxidation of the sulphur compounds is carried out followed by, extraction with some polar solvents, however such process will be discussed in detail in later sections.

Adsorptive Desulfurization

Adsorptive desulphurization involves the selective adsorption of the organosulphur compounds in the light distillates on a suitable adsorbent. The process may be operated in column mode or batch mode. During the process when the feed is contacted with solid adsorbent, sulphur compounds are adsorbed on the surface of the adsorbent and the sulphur free feed is allowed to pass. The process could be operated in cycles until the sulphur breakthrough is attained. Different characteristics of the adsorbent materials influence the efficiency of the adsorptive desulfurization; these include the capacity of adsorption, stability, selectivity, regeneration ability and durability, etc.

There are various forms of adsorptive desulfurization which will be discussed in this section and these include reactive adsorption, polar adsorption, selective adsorption, integrated adsorption, π-complexation. The schematic showing different forms of desulfurization are given in Figure 6.

Reactive Adsorption

Reactive adsorption is a form of adsorptive desulfurization whereby the sulfur compounds in fuel are removed by chemical interaction between the fuel and the sorbent material. The reaction involves removing or transferring the sulfur compounds from the fuel followed by attachment of the compounds to the sorbent material which will allow the sulfur free fuel to be collected in the main stream. The process uses metal-based sorbent for sulfur removal to form metal sulfide. The sorbent material can be regenerated by disposing off the sulfur compounds in the form of SO_2, H_2S, or elemental sulfur depending on the method employed. Reactive adsorption can

Figure 6. Types of adsorptive desulfurization

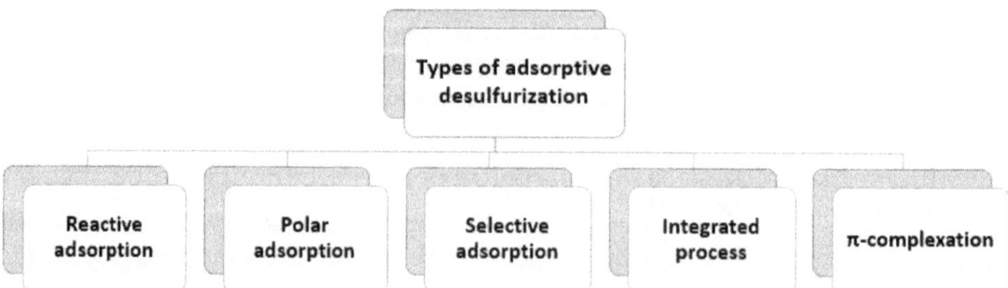

be carried out at ambient conditions, at elevated temperatures and with the aid of hydrogen at elevated temperatures. The efficiency of reactive adsorption depends on the adsorption capacity of the sorbent material, its affinity to the sulfur compounds, its thermal and mechanical stability and its ability to be regenerated. The overall mechanism of the process is outlined below (Babich *et al* 2003):

This process is widely used in the US as the S Zorb process developed by Philips Petroleum for efficient desulfurization at higher temperatures of between 340 -410°C and lower pressure of H_2 between 2-20 bars. Reactive adsorptive desulfurization combines the advantages of both the catalytic hydrodesulphurization and adsorption desulfurization and thus highly efficient for deep desulfurization. Research Triangle Institute (RTI) has developed a technology called TReND based on reactive adsorption desulfurization that utilizes metal oxides as sorbents materials to effectively remove sulfur compounds from fuels in the presence of hydrogen at a temperature of 426-535°C. Mercaptans are completely removed from fuels in this process in the absence of hydrogen but thiophenes require a considerable amount of hydrogen for its successful removal (B.S Turk *et al* 2001). Reactive adsorption can be aided with hydrogen at elevated temperatures using selective adsorbents. ZnO has been reported and industrially applied as a promising adsorbent for desulfurization processes. Addition of hydrogen in the process helps in the complete removal of thiophene in fuel, Figure 8.

Polar Adsorption

IRVAD process which is a low-cost method of providing low sulfur gasoline was developed and commercialized by Black and Veatch Pritchard and Alcoa Industrial chemicals to address even low concentrations of sulfur compounds in fuels. Activated alumina in particular Alcoa Selexsorb has been widely used as the sorbent material for the removal of polar compounds. The adsorption process is performed in a counter-current moving bed with alumina adsorbent in contact with liquid hydrocarbon in a multi-stage adsorber. The adsorbent is normally regenerated using hydrogen at various temperatures due to its heat capacity, thermal conductivity, and its availability. The regenerated adsorbent is then recycled for another round of the reaction. The process operates at lower pressures and does not consume a high amount of hydrogen. The

Figure 7.

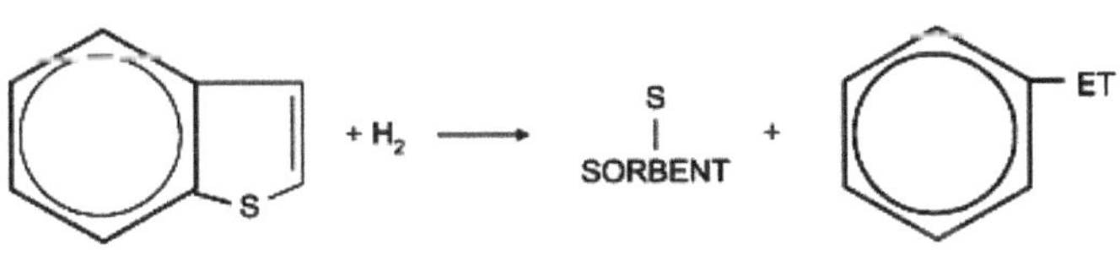

Figure 8. Reactive adsorption of benzothiophene (BT) on sorbent material in the presence of hydrogen

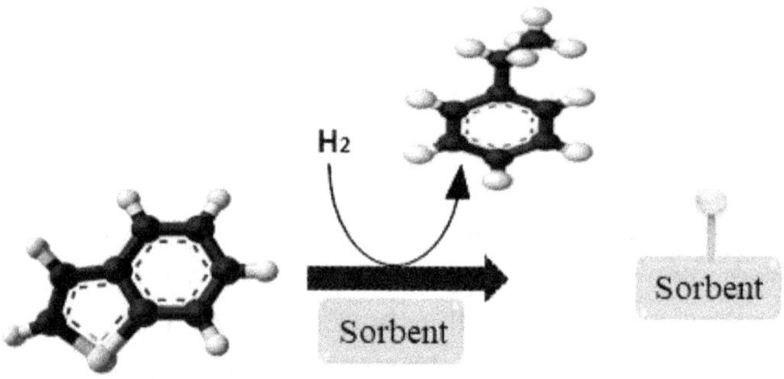

mechanism is based on the polarity of sulfur in gasoline and it is known to reduce sulfur from fuels to as low as 0.5ppmw.

Selective Adsorption for Removal of Sulfur Compounds (SARS)

The idea behind SARS is to remove sulfur compounds from fuels which constitute only less than 1% of the fuels at ambient conditions without hydrogen and to leave behind the remaining hydrocarbon contents of fuels which can be used as ultra-low sulfur fuels. This process was first demonstrated by Pennsylvania state university (C. Song *et al* 2003; Chunshan 2003) and if fully developed could be used in refinery operations and other mobile/stationary applications. The prime goal of SARS is to design appropriate adsorbents with surface sites having a high affinity for sulfur in the presence of aromatics. The adsorbents need to be effective, selective and apt for sulfur removal. Nickel-based sorbents and air regenerable metal oxide-based sorbents are the most common adsorbent materials used in this process. The basis for the SARS technology is that there is site-specific interaction between sulfur and metal species that is possible with selected organometallic complexes.

The likely adsorption configurations of thiophenic compounds on adsorbents used in SARS process could be explained from the known coordination geometries that thiophene exerted upon contact with organometallic complexes. Examples of coordination geometries are given in Figure 9 (C song *et al* 2003).

It is a known fact that both thiophenic sulfur compounds and non-sulfur aromatic compounds in fuels can interact with metal species by pi-electrons which is why the likely coordinating geometries that will best explain the interaction of sulfur atom of the thiophenic compounds with the adsorbent used in SARS are: ^{1}S (sulfur atom of thiophenes and a metal relation) and S-μ3 (sulfur atom of thiophenes and two metal species).

Figure 9. Coordination geometries of thiophene in organometallic complexes

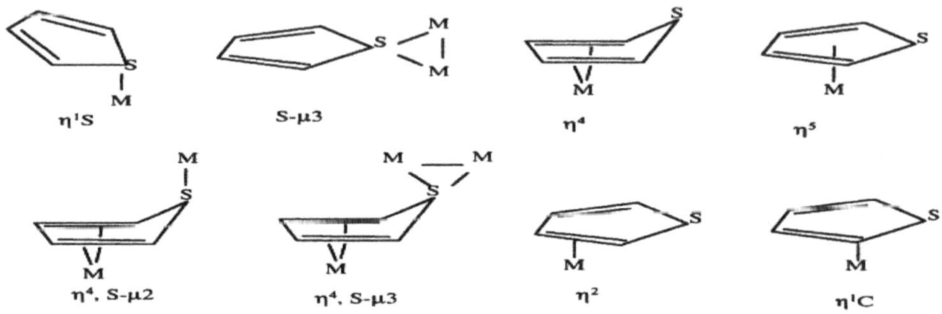

Integrated Process Adsorption

This is a recent technology that aids to combine selective adsorption with hydrodesulphurization technology using highly efficient catalysts for effective removal of sulfur compounds from fuels. It will confer added advantages like the reduction of cost, better efficiency, faster rates of desulfurization and reduction in reactor volumes than using single processes alone. The schematic diagram for the technology is given below (Chunsang *et al* 2003):

Precipitative Desulphurization (PD)

PD involves the conversion of organosulphur compounds into insoluble precipitates in the hydrocarbon feed with the help of different precipitating agents. The sulfur compounds precipitated to insoluble form can be easily removed from the feed stock through filtration or gravity settling. Different precipitating agents used to follow different reaction chemistry to accomplish the conversion of soluble sulfur compounds into insoluble precipitates present in the liquid hydrocarbon fuels e.g. formation of metal thiolates, S-alkyl sulfonium salts, and charge-transfer complex.

Thiols (mercaptans) in light distillate fractions of crude petroleum can be converted into metal thiolates by reacting with metal oxides or hydroxides, the metal thiolates formed are solid which can be easily recovered from the feed. In this process, the hydrocarbon feed is contacted with various active metal oxide or hydroxide; thiols are changed into metals thiolates through a hetrogeneous phase reaction which can be separated from the feed through filtration process. The metals thiolates can be treated with dilute acid to recover the parent thiolc and respective metal salts; the choice of the acid depends upon the metal salt desired. The process occurs under ambient condition and therefore no specialized reactor is required.

$$MO + 2R\text{-}SH \rightarrow M(SR)_2 + H_2O$$

$$M(SR)_2 + 2HNO_3 \rightarrow M(NO_3)_2 + 2R\text{-}SH$$

Precipitative desulfurization has been investigated with a number of metal oxides and hydroxides using model or real oils as feed. It was shown that PbO and $Hg(OH)_2$ are highly active in precipitating various types of thiols under ambient conditions in 1 h reaction time. In addition MnO_2 is also found efficient in removing thiols from crude oil as well as distillate fraction, however, its reaction requires a longer time to complete almost 3 h.

Desulfurization by Membranes

Polymer membranes are used to remove sulfur compounds from light distillate fractions by pervaporation (PV) technique, in which the sulfur compounds are selectively permeated from the feed through reduced pressure across the membrane. The polymer membrane in a fixed in a cell, to one portion of cell the preheated feed is introduced, the other portion of the cell, on downstream of the membrane is supplied with vacuum pump. A condenser or cold trap is also provided between the cell and the vacuum pump. Under reduced pressure, the sulfur compounds from the feed diffuse into the polymer membrane along with some hydrocarbons, which permeates through the membrane as vapor and is collected in the condenser as liquid permeate. The sulfur free feed is removed and sent to storage and sulfur rich permeate can be hydrotreated to remove sulfur. For continuous operation the feed is allowed to contact the membrane on its one side, and the vacuum is applied on its other side to build and continuous pressure gradient. The semipermeable membrane works as a separation medium in the process, the sulfur compounds are removed from the feed on the basis of affinity or relative diffusivity for the polymer membrane. The amount of feed components permeating through the unit area of the membrane in unit time is referred as flux, whereas the ratio of the amount of sulfur in feed to that in permeate is characterized as sulfur enrichment factor. These two parameters are of main concern in the operation of pervaporation process.

Oxidative Desulphurization (ODS)

ODS is based on selective oxidation of sulfur compounds in hydrocarbon fuel, followed by removal of the oxidized compounds from the feed through extraction or adsorption. In this process the nature of the organosulphur compounds is altered by oxidation with a suitable oxidizing agent, followed by separation. Hence the overall ODS process basically consists of two processes, namely the oxidation and extraction or adsorption. Appropriate oxidant is used to selectively oxidize the sulfur compounds in hydrocarbon stream into respective sulfoxides and then to sulfones, which possess relatively high polarity than the parent sulfur compound. Because of their high polarity the oxidized sulfur compounds can be removed from feed through liquid-liquid extraction or adsorption process. There are several types of ODS including (Saleh, 2016);

- ODS with Organic Peracids
- ODS with Air or molecular oxygen
- ODS with Organic Hydroperoxides
- ODS with H_2O_2- Polyoxometalates (POM)
- ODS with Ionic liquids (IL)
- Photo-oxidative desulphurization
- Ultrasound Assisted ODS

Reactors

Chemical reactions involving synthesis, precipitation, decomposition, isomerization, acid-base reaction, organic reactions, redox reactions, precipitation, displacement and many more deals with the conversion of chemical species known as reactants into chemical compounds known as products. A reactor (the main of chemical process) is a device designed for executing a chemical transformation from reactants to desired products. Most of the reactions carried out in chemical manufacturing, oil industries, metallurgical industries, pollution control plants and oil companies use reactors for their operations. In general, different types of reactors exist based on their operating conditions, chemical species involved, and the shapes and sizes. Various devices exist for carrying out chemical transformations e.g. tanks, columns, boilers, furnaces, mixers, drums, tubes etc. However, not all the devices have the characteristics of a reactor. A reactor needs to provide adequate time for successful completion of a chemical reaction; provide the necessary heat exchange; and finally, allow the interaction between phases involved for better product formation.

Reactors are classified based on their mode of operations, geometry of the reactors and types of phases involved.

- Based on their mode of operations, reactors can be classified as batch or continuous
- Based on their geometry, reactors can be classified as fluidized bed reactors, tubular reactors, stirred tank reactors, packed bed reactors, and plug flow reactors.

- Based on the types of phases, reactors can be classified as: single phase (involving gas or liquids) or heterogeneous phase (involving gas plus liquid or liquid plus liquid or gas plus solid or liquid plus solid or fluid plus solid).

Batch Systems

Batch system are otherwise called closed systems is an unsteady state system which involves carrying out a chemical transformation as a function of time where all the reactants are immersed in a reactor at the beginning of the process and the product will be collected after the completion of the reaction. In this process, all the reagents and reactants required for the transformation will be provided before the commencement of the reaction because there will be no room for addition or substitution after initiating the reaction. Batch reactions are normally conducted using tanks for the production of dyes, inks, in thermal cracking and polymerization. It is used mostly for small scale processes requiring long reaction times and for the production of different compounds in the same vessel. Batch reactors are the most common types of industrial reactors and their volume may exceed 100,000L. Batch reactors are heated or cooled using heat exchangers and internal or external coils. Reactants are usually preheated prior to the chemical reaction.

The batch adsorber is designed mainly to reduce adsorbent costs, which is highly important when dealing with expensive sorbent materials such as zeolites, silica, active carbon and resins. It also helps in reducing the contact time for a fixed percentage when cheap adsorbents are used meaning that it optimize cheap resource for even better efficiency. This will help in reducing capital cost in every industry, Figure 10 and 11.

Fixed Bed Systems

Fixed bed adsorption or otherwise known as percolation is a cyclic batch system that is universally used in chemical and other related industries. The process utilizes an

Figure 10. Illustration of the typical steps involved in batch system

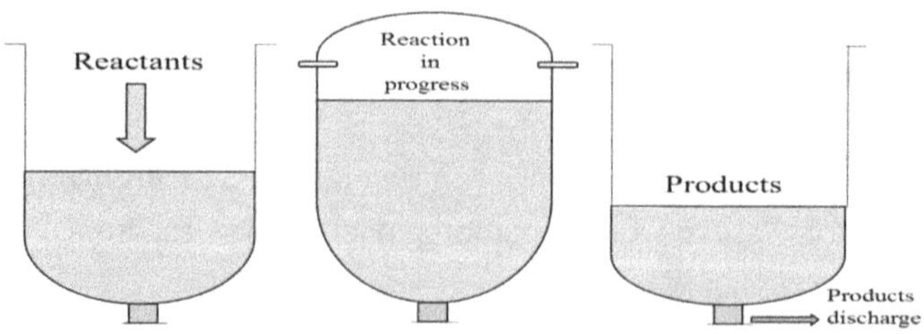

Figure 11. Other forms of typical batch systems

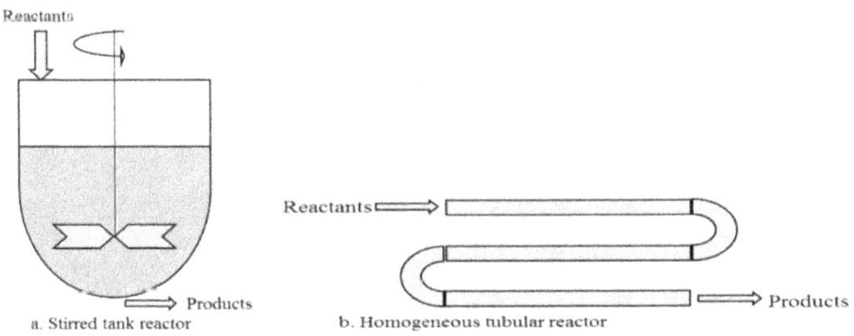

adsorbent material stacked inside a column for separation of one or more components from a mixture in a liquid or gas stream as it flows through the packed bed. The adsorbent bed is saturated and also regenerated in a periodic manner. This form of adsorption unit is ideal for small and medium installations because it is an unsteady state rate controlled process meaning conditions vary with time in the adsorber.

Adsorption occurs at a particular moving point in the bed which is called Mass Transfer Zone (MTZ). It is widely used in both gas and liquid treatments, it is perfect for the removal of pollutants and dissolved organics from water. There are certain factors that determine the number and arrangements of fixed beds which include: flow rate of the material needed to be treated, the allowable pressure drop, and length of the mass transfer zone, regeneration method employed for the adsorbent, available capital and energy demands. Steady flow of products is achieved when 2 beds are used such that one will aid in adsorption while the other will aid in the regeneration of the adsorbent material. High flow rate and short MTZ is achieved when the beds are set in parallel while longer MTZ and the low flow rate is achieved when the beds are set in series. For better flow rate and longer MTZ, the beds will be arranged both in parallel and series.

The fixed bed systems are inexpensive and easy to fabricate but their design is not always accurate due to the need for appropriate timing of the MTZ. Since adsorption is an exothermic process, the overall efficiency of the desorption process in fixed bed system may be affected due to poor heat transfer. A control system is required together with multiple beds piping and valve systems for efficiency of the adsorption and desorption process, Figure 12 and 13.

Mechanisms of Desulfurization

Adsorptive desulfurization is based on physisorption or chemisorption process of organosulfur compounds on solid sorbents. The mechanism of desulfurization, in general, has not been extensively studied in full details. Most of the reports on this

Figure 12. Schematic for the typical fixed bed system

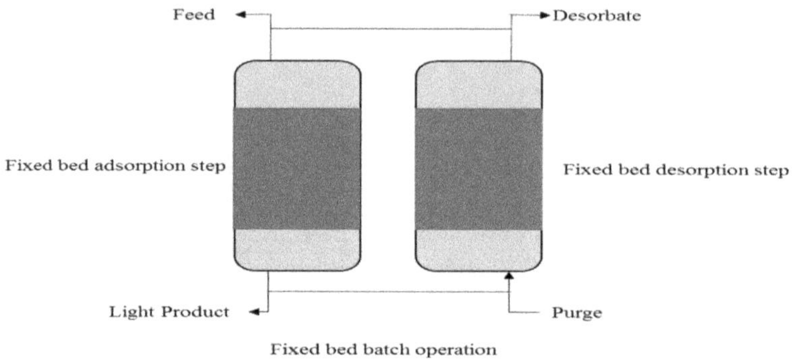

Figure 13. Schematic for the typical fixed bed system

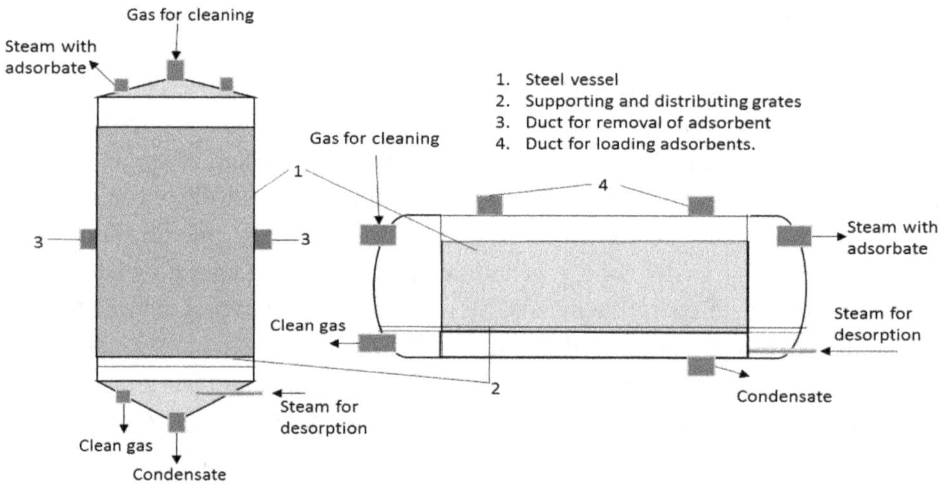

field proposed the mechanism theoretically or experimentally without proper insight into the subject. However, the studies of (Liang *et al* 2010; Duan *et al* 2014; Lichun *et al* 2011; and Xuan *et al* 2013) have outlined some of the mechanisms involved in adsorptive desulfurization.

Based on the idea to regenerate sulfur-poisoned Ni catalyst in hydrogen atmosphere, (Tawara *et al* 2000) used Ni/ZnO and Ni/Al_2O_3 as catalysts for adsorptive ultra-deep hydro desulfurization of kerosene using hydrogen at 600K. The idea was that sulfur poisoned nickel catalyst particles would be combined with ZnO or Al_2O_3 particles in the presence of hydrogen, then each poisoned catalyst would release a few ppb of H_2S which will be attached to the ZnO or Al_2O_3 particles. This process, in turn, will allow the effective regeneration of sulfur poisoned nickel catalyst. Residual sulfur was not detected in the treated kerosene with Ni/ZnO. The catalyst completely removed residual sulfur even after 800h of operation and no methane was produced in contrast to the other materials used.

ZnO/Al_2O_3 is used as the main component of the sorbent while Ni is used as the main site for the hydro desulfurization. Thiophene in the second step of the reaction is decomposed on the Ni site after reacting with hydrogen. Further reaction with hydrogen leads to the removal and transfer of hydrogen sulfide to ZnO surface. Subsequent reactions with thiophene further convert ZnO completely to ZnS which may later be regenerated or discarded.

(Lichun *et al* 2011) conducted a study on the reactive adsorption desulfurization of model oil containing DBT over a Ni/ZnO adsorbent. The reaction mechanism under different atmospheres was illustrated and the role of hydrogen in the process was investigated. The result showed different mechanisms when nitrogen and hydrogen were used respectively. Desulfurization over Ni/ZnO using nitrogen was achieved based on physical and chemical adsorption though the rate was very low. However, the desulfurization under hydrogen turns out to be based on reactive adsorption desulfurization and the rate was on the contrary fast. Hydrogen is very important in reactive adsorption desulfurization when Ni/ZnO adsorbent is used, it aids in the decomposition of DBT on Ni species and in the formation of Ni_3S_2 and the transfer of sulfur moieties to ZnO.

(Xuang *et al* 2013) further outlined the mechanism involved in the removal of thiophene by reactive adsorptive desulfurization from model fuels. Reduced $NiZnO/Al_2O_3$-diatomite was used as the adsorbent in the studies and the result indicated that S-M bonding of thiophene to Ni sites of the adsorbent was decomposed to form Ni_3S_2 and C_4 olefins formed were further saturated by hydrogen to CH_4 which was released back to the stream. The final step involves the transfer of sulfur from Ni_3S_2 to ZnO in a reduced atmosphere which will yield ZnS and a free Ni species ready for another round of the reaction.

The sulfur transfer mechanism in the presence of hydrogen involves three-step reactions: the first step involves the decomposition of DBT in the model oil on Ni surface of Ni/ZnO to form Ni_3S_2, the second step involves the reduction of Ni_3S_2 in the presence of hydrogen to form H_2S followed by the release of active nickel species from Ni_3S_2. The final step involves the storage of H_2S in the adsorbent followed by the final conversion of ZnO into ZnS, Figure 14.

The mechanism involved in the removal of thiophene by reactive ADS from model fuels is illustrated by (Xuan et al., 2013). Reduced $NiZnO/Al_2O_3$-diatomite was used as the adsorbent in the studies and the result indicated that S-M bonding of thiophene to Ni sites of the adsorbent was decomposed to form Ni_3S_2 and C_4 olefins formed were further saturated by hydrogen to CH_4 which was released back

Figure 14. Mechanism of reactive ADS of DBT containing model oil over adsorbent of Ni/ZnO in hydrogen.

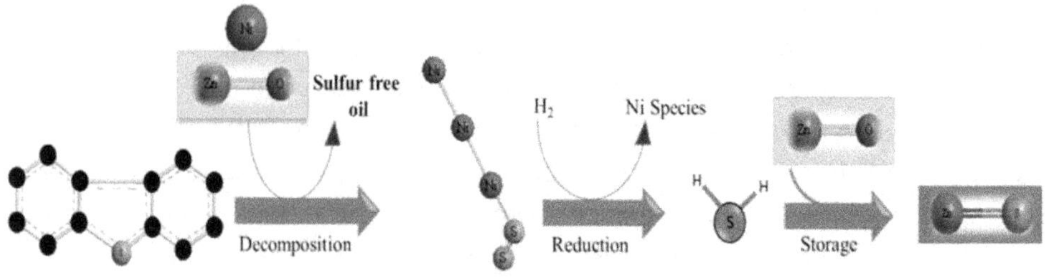

to the stream. The final step involves the transfer of sulfur from Ni_3S_2 to ZnO in a reduced atmosphere which will yield ZnS and a free Ni species ready for another round of the reaction.

Zeolites are acid catalyst used in olefin hydroforming and aromatic conversion. They are also used in HDS and as a bi-functional catalyst. Zeolites contain both acidic and a basic site. The lewis basic site on zeolites is responsible for the catalytic hydrogen desulphurization of most sulfur compounds in fuels; the bronsted acidic site has little or no role to play in the catalysis. Thiophenic ring cracking is the rate-determining step in HDS of thiophene and DBT (Rozanska et al., 2003). A periodic density functional theory study of the thiophenic derivative cracking catalyzed by proton or lithium exchanged modernite (zeolite) can be a good method to prove the cracking of thiophene derivatives; benzothiothiophene is cracked by a catalyst where one of the oxygen atoms in the catalyst is the catalytic center for the reaction. The mechanism of benzothiophene cracking is illustrated in Figure 15.

Mechanism of adsorption of DMDBT from solutions in hexadecane on three different nanopourous AC has been illustrated by (Kostas et al., 2014). Each of the nonporous carbons has microporous or combined micropores and mesopores

Figure 15. Mechanism of BT cracking by zeolites

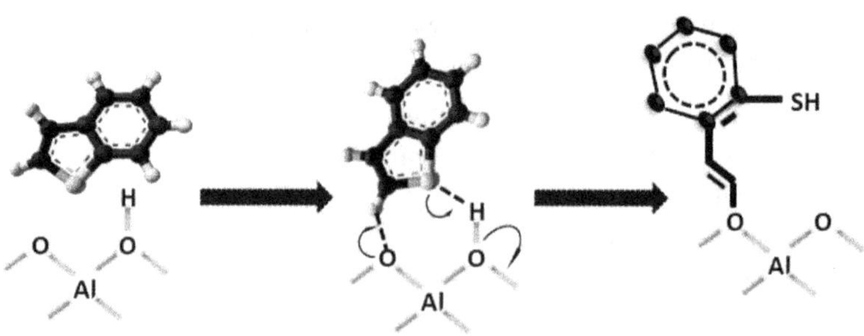

structures. The capacity for 4, 6-DMDBT adsorption increases with an increase in the volume of pores. The pores with a diameter lower than 10 A are said to adsorb better because the diameter is similar to that of 4, 6 DMDBT molecule. Acidic functional groups on the surfaces of the larger pores are said to contribute to the adsorption via polar interactions. Most noticeable functional groups are carboxylic acids because they interact with the oxidation products of 4, 6 DMDBT (sulfoxides, sulfones, and sulfonic acids) via hydrogen bond. Dispersive interactions between the delocalized π-electrons within the benzene rings of 4, 6 DMDBT and the electron-rich region on the nanoporous carbon aromatic ring also plays a major role in the adsorption. Figure 16 illustrates the mechanism.

HDS Reaction Mechanism

A clear example of HDS is illustrated in Figure 17 (AlHammadi et al. 2018). There are two main reaction mechanisms for the HDS process of DBT, namely, the hydrogenolysis pathway (DDS) and the hydrogenation desulfurization (HYD) pathway. In the DDS mechanism, the removal of the sulfur is due to hydrogen, without abetting the DBT aromatic rings. In the HYD mechanisms, the aromatic rings are hydrogenated and then desulfurized.

The possible reaction mechanism for the HDS of DBT over the AlCNFMoCo catalysts is illustrated in Fig. 17 (AlHammadi et al. 2017). In the DDS pathway, the removal of sulfur by direct C-S bond hydrogenolysis produces biphenyl (BiPh) which can be considered as the predominant organic product in this pathway. Subsequently, the BiPh is hydrogenated to produce cyclohexylbenzene (CHB). By contrast,

Figure 16. The possible adsorption mechanism of 4,6 DMDBT on AC

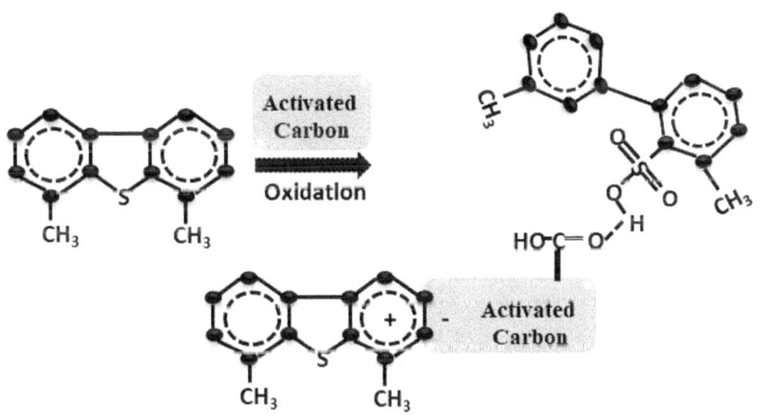

Figure 17. Possible pathways for the HDS of DBT over nano-catalysts

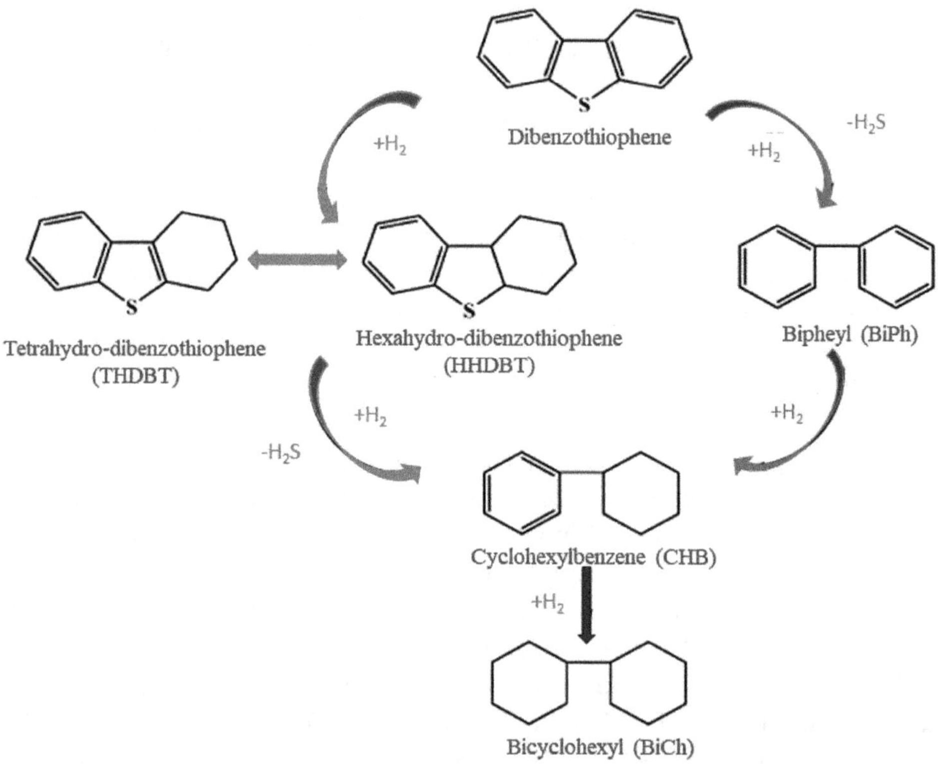

the hexahydrodibenzothiophene (HHDBT) or/and tetrahydrodibenzothiophene (THDBT) are the intermediate products of the main reaction in the HYD pathway mechanism. These intermediates are desulfurized to give a secondary product which is cyclohexylbenzene (CHB). In both mechanisms, bicyclohexyl (BiCh) is formed in trace amounts, which is the result of the hydrogenation of the CHB in a slow pattern.

REFERENCES

AL-Hammadi, S. A., Al-Amer, A. M., & Saleh, T. A. (2018). Alumina-carbon nanofiber composite as a support for MoCo catalysts in hydrodesulfurization reactions. *Chemical Engineering Journal*, *345*, 242–251. doi:10.1016/j.cej.2018.03.106

Babich, I. V., & Moulijin, J. A. (2003). Science and technology of novel processes for deep desulfurization of oil refinery streams: A review. *Fuel*, *82*(6), 607–631. doi:10.1016/S0016-2361(02)00324-1

Bansal, R. C., Donnet, J. B., & Stoeckli, F. (1988). *Active Carbon*. New York: Mercel Dekker.

Bej. (2004). Revamping of diesel hydro-desulfurizers: Options available and future research needs. *Fuel Process. Technol*, *85*(13), 1503–1517.

Bezeverkhyy, I., Ryzhikov, A., Gadacz, G., & Bellat, J. P. (2008). Kinetics of thiophene reactive adsorption on Ni/SiO_2 and Ni/ZnO. *Catalysis Today*, *130*(1), 199–205. doi:10.1016/j.cattod.2007.06.038

Cheng, P., Zhang, S., Wang, P., Huang, S., & Tian, H. (2013). First-principles investigation of thiophene adsorption on Ni13 and Zn@Ni12 nanoclusters. *Computational & Theoretical Chemistry*, *1020*, 136–142. doi:10.1016/j.comptc.2013.07.044

Duan, A., Peng, Z., Liang, Z., Ji, J., & Zhen, Z. (2014). *Computational simulation application in the research of desulfurization mechanism*. The 9th International conference on computer science & education (ICCSE 2014), Vancouver, Canada.

Furimsky, E. (1996). Spent refinery catalysts: Environment, safety and utilization. *Catalysis Today*, *30*(4), 223–286. doi:10.1016/0920-5861(96)00094-6

Georgii, H. W. (1970). Contribution to the atmospheric sulfur budget. *Journal of Geophysical Research*, *75*(2), 2365–2371. doi:10.1029/JC075i012p02365

Huang, L., Wang, G., Qin, Z., Dong, M., Du, M., Ge, H., ... Wang, J. (2011). In situ XAS study on the mechanism of reactive adsorption desulfurization of oil product over Ni/ZnO. *Applied Catalysis B: Environmental*, *106*, 26–38.

Javadli, R., & Klerk, A. (2012). Desulfurization of heavy oil. *Applied Petrochemical Research*, *1*(1-4), 3–19. doi:10.1007/13203-012-0006-6

Kulkarni & Afonso. (2010). *Deep desulfurization of diesel fuel using ionic liquids: current status and future challenges*. Academic Press.

Landau, M. V., Herskowitz, M., Agnihotri, R., & Kegerreis, J. E. (2009). Ultradeep hydrodesulfurization and adsorptive desulfurization of diesel fuel on metal-rich nickel phosphides. *Industrial & Engineering Chemistry Research*, *48*(11), 5239–5249. doi:10.1021/ie9000579

Lee & Valla. (2019). Adsorptive desulfurization of liquid hydrocarbons using zeolite-based sorbents: a comprehensive review. *React. Chem. Eng*. DOI: doi:10.1039/c9re00036d

Leh & Chan. (1973). Sulfur compounds: Pollution, health effects and biological function. *J. Chem. Edu*, *50*(4), 246.

Li, X., Nagaoka, K., Simon, L. J., Lercher, J. A., Wrabetz, S., Jentoft, F., ... Papp, H. (2005). Interaction between sulfated zirconia and alkanes: Prerequisite for active sites-formation and stability of reaction on intermediates. *Journal of Catalysis*, *230*(1), 214–225. doi:10.1016/j.jcat.2004.11.045

Li, Y.-H., Wang, S., Cao, A., Zhao, D., Zhang, X., Xu, C., ... Wei, B. (2001). Adsorption of fluoride fromwater by amorphous alumina supported on carbon nanotubes. *Chemical Physics Letters*, *350*(5-6), 412–416. doi:10.1016/S0009-2614(01)01351-3

Meille, V., Shulz, E., Vrinat, M., & Lemaire, M. (1997). Hydrodesulfurization of alkyldibenzothiophenes over a NiMo/Al2O3 catalyst: Kinetics and mechanism. *Journal of Catalysis*, *170*(1), 29–36. doi:10.1006/jcat.1997.1732

Meng, X., Huang, H., & Shi, L. (2013). Reactive Mechanism and Regeneration Performance of NiZnO/Al_2O_3-Diatomite Adsorbent by Reactive Adsorption Desulfurization. *Industrial & Engineering Chemistry Research*, *52*(18), 6092–6100. doi:10.1021/ie303514y

Nagai, Y., Shinjoh, H., & Yokota, K. (2002). Oxidation selectivity between n-hexane and sulfur dioxide in diesel simulated exhaust gas over platinum-supported zirconia catalyst. *Applied Catalysis B: Environmental*, *39*(2), 149–155. doi:10.1016/S0926-3373(02)00082-6

Orita, H., Uchida, K., & Itoh, N. (2004). A volcano-type relationship between the adsorption energy of thiophene on promoted MoS2 cluster-model catalysts and the experimental HDS activity: Ab initio density functional study. *Applied Catalysis A, General*, *258*(1), 115–120. doi:10.1016/j.apcata.2003.08.016

Reddy, B. M., Sreekanth, P. M., & Lakshmanan, P. (2005). Sulfated zirconia as an efficient catalyst for organic synthesis and transformation reactions. *Journal of Molecular Catalysis A Chemical*, *237*(1-2), 93–100. doi:10.1016/j.molcata.2005.04.039

Robeus, E., Roquerol, J., Rodriguez-Reinoso, F., Sing, K. S.W., & Under, K. K. (n.d.). *Characterization of porous solids III*. Elsevier Amsterdam.

Roquerol, F., Roquerol, J., Sing, K. S. W., Llewellyn, P., & Maurin, G. (2014). *Adsorption by powders and porous solids: Principles* (2nd ed.). Elsevier, UK: Methodology and Applications.

Rozanska, X., van Santen, R. A., Hutschka, F., & Hafner, J. (2003). A periodic density functional theory study of thiophenic derivative cracking catalyzed by mordenite. *Journal of Catalysis*, *215*(1), 20–29. doi:10.1016/S0021-9517(02)00148-3

Saleh, T.A. (2016). *Applying Nanotechnology to the Desulfurization Process in Petroleum Engineering*. IGI Global.

Shafi, R., & Hutchings, G. J. (2000). Hydrodesulfurization of hindered dibenzothiophenes:an overview. *Catalysis Today*, *59*(3-4), 423–442. doi:10.1016/S0920-5861(00)00308-4

Shafi, R., & Hutchings, G. J. (2000). Hydrodesulfurization of hindered dibenzothiophenes: An overview. *Catalysis Today*, *59*(3–4), 423–442. doi:10.1016/S0920-5861(00)00308-4

Song, C. (2003). An overview of new approaches to deep desulfurization for ultra-clean gasoline, diesel fuel and jet fuel. *Catalysis Today*, *86*(1-4), 211–263. doi:10.1016/S0920-5861(03)00412-7

Song, C. S. (2003). An overview of new approaches to deep desulfurization for ultra-clean gasoline, diesel fuel and jet fuel. *Catalysis Today*, *86*(1-4), 211–263. doi:10.1016/S0920-5861(03)00412-7

Tawara, K., Nishimura, T., & Iwanami, H. (2000). Ultra-deep hydrodesulfurization of kerosene for fuel cell system (Part 2) regeneration of sulfurpoisoned nickel catalyst in hydrogen and finding of autoregenerative nickel catalyst. Sekiyu Gakkaishi. *Journal of the Japan Petroleum Institute*, *43*(2), 114–120. doi:10.1627/jpi1958.43.114

The International Council On Clean Transportation (ICCT). (2011). *Energy Econ. Appl. Optim.*, 1–38.

Tian, W.-H., Sun, L.-B., Song, X.-L., Liu, X.-Q., Yin, Y., & He, G.-S. (2010). Adsorptive Desulfurization by Copper Species within Confined Space. *Langmuir*, *26*(22), 17398–17404. doi:10.1021/la101856d PMID:20882950

Triantafyllidis, K. S., & Deliyanni, E. A. (2014). Desulfurization of diesel fuels: Adsorption of 4, 6-DMDBT on different origin and surface chemistry nanoporous activated carbons. *Chemical Engineering Journal, 236*, 406–414. doi:10.1016/j.cej.2013.09.099

Turk, B. S., & Gupta, R. P. (2001). RTI's TReND Process for deep desulfurization of naphtha, Am. Chem. Soc. *Div. Fuel Chem. Prepr., 46*, 392.

Unger, K. K. (1979). *Porous Silica, its properties and use as support in column chromatography*. Elsevier.

Zhang, S., Zhang, Y., Huang, S., Wang, P., & Tian, H. (2012). Mechanistic investigations on the adsorption of thiophene over Zn3NiO4 bimetallic oxide cluster. *Applied Surface Science, 258*(24), 10148–10153. doi:10.1016/j.apsusc.2012.06.096

Chapter 2
Nanomaterials and Nanocomposites for Adsorptive Desulfurization:
From Synthesis to Application

Saddam A. Al-Hammadi
Melbourne University, Australia

ABSTRACT

Desulfurization (removal of S compounds) of fuels is an important research topic in recent years. Several techniques have been reported to remove the sulfur-containing compounds in fuels. One of these techniques is adsorptive desulfurization (removal based on chemisorption and physisorption), which has received much attention because of low energy consumption and facile operation condition. This chapter discusses the methods employed under this technique and the types of nanocomposites and hybrid materials (adsorbents) that have been investigated as potential adsorbents. The strategies to enhance sulfur adsorption capacity and main challenges will be discussed.

INTRODUCTION

Adsorptive Desulfurization

Various desulfurization technologies have been explored either to complement the current hydrodesulphurization (HDS) technology. The main alternative techniques include oxidative desulfurization, extractive desulfurization, biodesulfurization

DOI: 10.4018/978-1-7998-2146-5.ch002

Figure 1. Types of adsorptive desulfurization

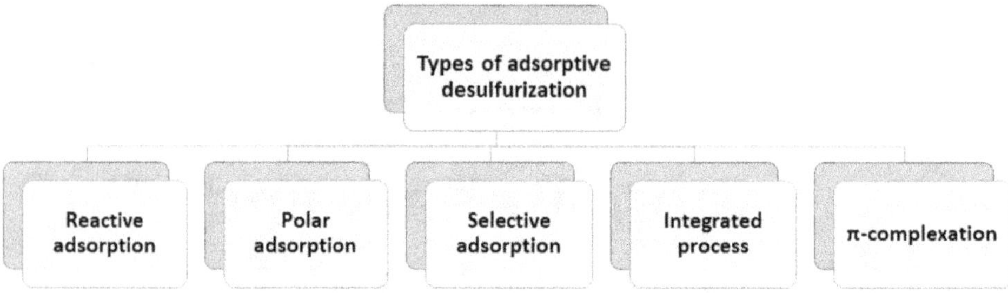

and adsorptive desulfurization (ADS). There are various forms of adsorptive desulfurization which will be discussed in this section and these include reactive adsorption, polar adsorption, selective adsorption, integrated adsorption, π-complexation. The schematic showing different forms of desulfurization is given in Figure 1. ADS is one of the most studied alternative technologies to replace or supplement conventional HDS. In ADS, sulfur compounds are removed via adsorption using a selective adsorbent. This means that the adsorbent plays an important role in the selectivity, capacity, and sustainability such as renewability of adsorbents, of the process.

Sulfur can be removed via two mechanisms, depending on the interaction between the adsorbent and the sulfur compounds:

1. Physical adsorption in which sulfur compounds and adsorbents are mainly bound by van der Waals forces, which is a relatively weak interaction. This enables the adsorbent to be regenerated relatively easily.
2. Chemical adsorption (such as reactive adsorption), which employs chemical bonding between the sulfur compounds and the adsorbents, which may alter the physical and chemical nature of the sulfur compounds.

ADS is a promising technique since it has the potential to be regenerative, cost-effective and environment-friendly while operating under ambient conditions. However, there are some challenges including selectivity and diffusion limitations which are the most important challenges in this type of desulfurization.

A reactive adsorption is a form of ADS whereby the sulfur compounds in fuel are removed by chemical interaction between the fuel and the sorbent material. The reaction involves removal or transfer of the sulfur compounds from the fuel followed by attachment of the compounds to the sorbent material which will allow the sulfur-free fuel to be collected in the mainstream. The process uses metal-based sorbent for sulfur removal to form metal sulfide, Figure 2.

Figure 2. Reactive adsorption of benzothiophene (BT) on sorbent material in the presence of hydrogen

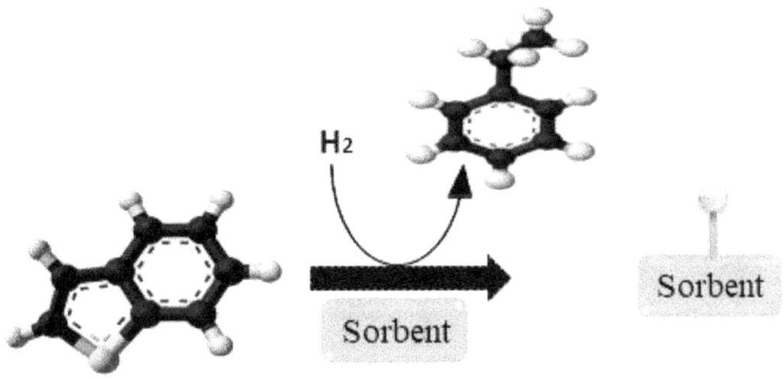

Polar adsorption (IRVAD process) is a less expensive method of providing low sulfur gasoline and it was developed and commercialized by Black and Veatch Pritchard and Alcoa Industrial chemicals to address even low concentrations of sulfur compounds in fuels. Activated alumina in particular Alcoa Selexsorb is widely used as the sorbent material for the removal of polar compounds. The adsorption process is performed in a counter-current moving bed with alumina adsorbent in contact with liquid hydrocarbon in a multi-stage adsorber. The adsorbent is normally regenerated using hydrogen at various temperatures due to its heat capacity, thermal conductivity, and its availability. The regenerated adsorbent is then recycled for another round of the reaction. The process operates at lower pressures and does not consume a high amount of hydrogen. The mechanism is based on the polarity of sulfur in gasoline and it is known to reduce sulfur from fuels to as low as 0.5ppmw.

Selective Adsorption for Removal of Sulfur Compounds (SARS) is to remove sulfur compounds from fuels which constitute only less than 1% of the fuels at ambient conditions without hydrogen and to leave behind the remaining hydrocarbon contents of fuels which can be used as ultra-low sulfur fuels. The prime goal of SARS is to design appropriate adsorbents with surface sites having high affinity for sulfur compounds in the presence of aromatics. The adsorbents need to be effective, selective and apt for sulfur removal. The likely adsorption configurations of thiophenic compounds on adsorbents used in SARS process can be explained from the known coordination geometries that thiophene exerted upon contact with organometallic complexes. Examples of coordination geometries are given in Figure 3. It is a known fact that both thiophenic sulfur compounds and non-sulfur aromatic compounds in fuels can interact with metal species by pi electrons which is why the likely coordinating geometries that will best explain the interaction of sulfur atom of the thiophenic compounds with the adsorbent used in SARS are: [1]S (sulfur

Figure 3. Coordination geometries of thiophene in organometallic complexes

atom of thiophenes and a metal relation) and S-μ^3 (sulfur atom of thiophenes and two metal species).

Integrated Process Adsorption is a recent technology that combines selective adsorption with HDS technology using highly efficient catalysts for effective removal of sulfur compounds from fuels. It has added advantages like reduction in cost, better efficiency, faster desulfurization rates and reduction in reactor volumes than using single processes alone.

Principles of Adsorption

Reactions carried out on the surfaces of different materials have a wide application since time immemorial. In the early days of existence, Greeks and Romans devised a method of reducing the tarnishing of metals, and the alchemist had a long quest of searching for a substance which will transform other substances in the name of "catalyst" which they later satisfied by adsorption principle (Richard 1996). Egyptians in the early days use the principle of adsorption to desalinate water and in the treatment of many diseases (Robens *et al* 1994).

The ability of porous solids to absorb substances in the vapor phase was recognized since in the eighteenth century (Douglas 1984), however, the utilization of adsorption technique to a large scale industrial application is relatively recent and pristine. (Hameed 2009 and Freitas *et al* 2008) in their work showed the application of adsorption in separation technologies which is relatively a new development compared to other technologies.

Adsorption is defined as the increase in the density of fluid within the vicinity of an interface (F. Rouquerol *et al* 2014). Adsorption occurs when a solid surface is exposed to a liquid or a gas, the gasses and liquids tend to adhere to the unoccupied spaces in solid particles due to the presence of unsaturated molecular forces within a solid material. (Khan 2012) explained the principle of adsorption as the removal of certain substances called adsorbate (liquid or gas) by an adsorbent (solid material). It can further be defined as the adherence of substances to solid materials. Nearly all substances may have trace capability for adsorption but depending on some favorable characteristics like surface area, active sites, pore size, and volume, some materials are far better than others in the way and manner they adsorb particles.

Adsorption involves two types of forces viz: physical forces which may include dispersive forces, polarization forces, or dipole and chemical forces due to the interaction of electrons between solid particles and the atoms of liquids or vapor phase (Roop C. B *et al* 2005). (Zeki 2013) indicated three principal forces responsible for adsorption which includes: van der Waal's forces, chemical affinity, and electrostatic attraction. There are two different types of adsorption mainly Physiosorption which is reversible and chemisorption that is irreversible.

The adsorption process is primarily studied based on certain types of equations and models. The most widely used equations are: Freundlich which is based on empirical equations that describe the adsorption characteristics on heterogeneous surface (Freundlich 1906), Langmuir equation which is based on monolayer adsorption process on energetically homogeneous surface (Langmuir 1918) and Brunauer-Emmett-Teller (BET) equation which is mostly used in gas phase adsorption and it is based on statistical analysis of adsorption sites occupied during a multi-layer adsorption of gasses (Brunauer *et al* 1938).

Some of the applications of adsorption include large scale removal of impurities from natural gas and organic pollutants from water (Douglas 1984), separation process in food engineering (Zeki 2013). Currently, more than 90% of all chemicals are produced based on catalyzed reactions on surfaces of substances that speed up chemical reactions (Richard 1996). Some of the applications of surface reactions involving adsorption as outlined by (Richard 1996) include integrated circuits production, production of fuels, in the disposal of noxious chemicals, corrosion and lubrication, processing of metals e.t.c.

The use of nanoparticles to adsorb vapor substances for practical application in the industries and purification process had been extensively studied over the past for many years (Tsuru, 2001). A known example is the utilization of aerosorb for the recovery of aromatic hydrocarbons when separating the cyclic shapes of aromatic compounds from the linear chains (Di Noto *et al* 2010).

Sulfur oxides (SOx) emissions produced mainly from human activities through combustions of fossil fuels in power plants, automobile exhaust, and refineries serve as a major source of atmospheric pollutions. They contribute to smog formation, acid rain, health-related problems and damage to the ecosystem (Mathieu *et al* 2012). The current methods in use for the removal of sulfur compounds in fuels are either pre-combustion techniques or post-combustion techniques. The pre-combustion techniques are the best methods to be utilized for the removal. Refineries currently rely on the hydro-desulfurization (HDS) for the removal of sulfur from petroleum products which is uneconomical (Ali *et al* 2006) considering the hydrogen consumption with a simultaneous release of carbon dioxide which is unfriendly to the environment. The technology and the mechanical set up is also not viable for long term goals (Hernandez *et al* 2004). These problems necessitate the developments of other friendly methods for the desulfurization of fuels.

Adsorbents

Adsorption on the commercial and industrial-scale depends solely on the type, quantity, economic cost and effectiveness of adsorbent material. This is why researches on adsorption studies currently focus keenly on the development of better adsorbent materials for large scale industrial applications. Most materials used as adsorbents are relatively porous and have a wide surface area with a pore diameter of tens of angstroms (Czaja *et al* 2009).

Activated carbon and silica gel were used in the early days of adsorption process as the prime adsorbent materials but the technology has improved currently to the utilization of molecular sieve adsorbents e.g. zeolites which hold great promise for more advanced separation methods. The most widely used adsorbents are: activated carbon, silica gel, zeolites, molecular sieve carbon, activated alumina, anhydrous calcium sulfate and the recent technology of combining polymers with adsorbent materials. These materials occur naturally in nature or they can be artificially manufactured. Each of these materials has distinct physical and chemical characteristics such as pore sizes and structure, porosity and nature of the adsorbing surface that are different from other materials.

Adsorbent materials are generally classified into organic sources (carbons, polymers and biomass) and inorganic (aluminas, silicas and zeolites) (Knaebel 2004). Some of the main required properties of these adsorbent materials are outlined as; (Buekens *et al* 2009)

- The material should be thermally and mechanically stable.
- It should be highly porous with balance between micro and macro-pores.
- It should have high surface area.
- It should have high affinity for the adsorbate.
- It should be cheap and be easily regenerated.

Table 1. The Characteristics of the isotherms

Isotherms	Characteristics
Type one Isotherms	They consist mostly of micro porous adsorbents. E.g. Silica gel and most carbons fall in this category
Type two Isotherms	They consist of non-porous surfaces and macro porous adsorbents. E.g. Graphitized carbon and some hardened powders of silica.
Type three Isotherms	They consist of non porous and macro porous solids. E.g. synthetic polymers used as adsorbent materials
Type four Isotherms	This class exhibit special attribute of hysteresis which means capillary condensation showed by the mesoporous adsorbents. E.g. Low density silica gel
Type five Isotherms	This class exhibit hysteresis as well but they show similar characteristics to Type three isotherms.
Type six isotherms	This isotherm share similar characteristics to type two isotherms but however, the adsorbent belonging to this class adsorb in a multilayered passion.

The adsorptive capacity of any adsorbent material depends on the number of pores in the internal area relative to the number of pores on the outer surface. The internal area is much higher than the outer surface area which is the major reason why adsorbent materials adsorb large quantity of the adsorbate. These materials are further classified based on their pore sizes which determines which material they can adsorb into three classes viz:

- Micro porous adsorbents (between 2A° to 20A°)
- Meso porous adsorbents (between 20 A° to 500 A°)
- Macro porous adsorbents (>500 A°)

A° = Angstrom which is equivalent to 10^{10} meters = 0.1 nm

Adsorption capacity or otherwise adsorption isotherms depend upon temperature and vapor pressure which help us to classify different forms of adsorbent (Sing 1982). The isotherms can be classified into different forms of the adsorption isotherms as shown in Figure 4. The characteristics of these isotherms is listed in Table 1.

Bulk Adsorbents

Adsorbents have different applications such as catalyst, desiccants, catalyst supports, storage of gasses, in purification of liquids, controlled drug delivery, and pollution control (F. Rouquerol et al 2014). This is why most industrial adsorbents have a large surface area, porous in nature and composed of fine particles (F. Rouquerol et al 2014). Table 2 below outlines some of the applications of the different adsorbent materials:

Figure 4. Graphical illustration of different isotherms models

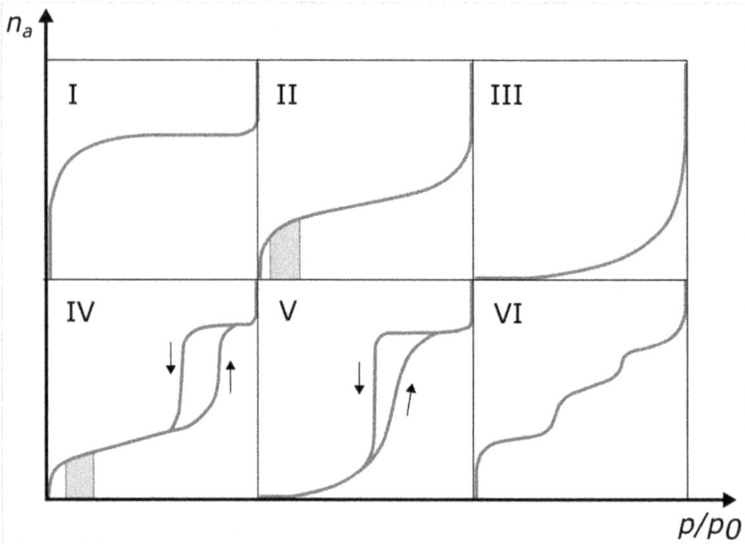

Zirconia

Zirconia is otherwise called zirconium di oxide (ZrO_2). It is a white crystalline solid found naturally in monoclinic crystalline structure. It is produced by calcination of zirconium. Upon addition of heat zirconia changes its monoclinic structure which

Table 2. Examples of the applications of the different adsorbents

Adsorbent	Applications
Carbons	Widely used in the removal of hydrogen, nitrogen, odors, vinyl chloride, desulfurization of SOx and denitrogenation of NOx from gasses, in water treatment, and in the treatment of nuclear waste.
Alumina	It is mostly used in drying of gasses (as a desiccant) and adsorption of fluoride, chromium, arsenic and selenium in water treatment. It is also used in desulfurization.
Silica	It is used as a desiccant in drying of gasses, and maintenance of relative humidity. Also used in desulfurization.
Zeolites Clay Polymers	Helps in drying of gasses, water treatment, sweetening of gasses and liquids, aids in pollution control of SOx and NOx through desulfurization and denitrogenation, recovery of CO_2, and in the separation of aromatics from paraffins etc. Helps in the removal of pigments, treatment of oils, and refining of mineral oils. Aids in color removal, water treatment, purification of proteins, steroids, amino acids and in the separation of some macromolecules.

Figure 5. (a) Appearance of zirconia (b) Structural changes of zirconia upon addition of heat

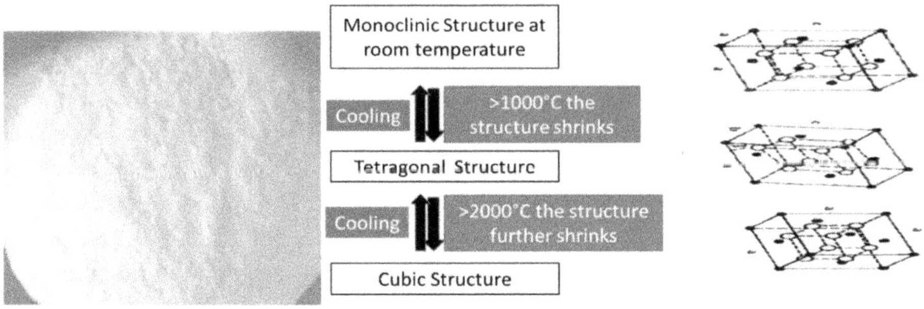

is temperature and corrosion resistant to tetragonal structure and later when the temperature rises the tetragonal structure changes to cubic structure. It does not dissolve in water and it is amphoteric in nature (Tanabe, 1985); monoclinic and tetragonal zirconia have received attention due to their wide spread applications as a catalyst, catalyst support and ceramic production (G. Cerrato *et al* 1997[a]). Zirconia has greater surface area, high dielectric constant and has the potential to replace silicon di oxide in fabrication technology, Figure 5.

(G. Cerrato *et al* 1997[a, b]) studied and characterize the surface of monoclinic zirconia, its morphology and its structure with fourier transmission infrared spectroscopy (FTIR), computer graphics, X-ray diffraction (XRD) and High resolution transmission electron microscope (HRTEM). XRD and SEM studies of some commercial zirconia powders was carried out by (G. Dercz et al 2008).

Zirconia nanoparticles has been synthesized using sol-gel process using metal organic precursors like zirconium alkoxide or inorganic precursors such as zirconyl chloride ($ZrOCl_2$) in a basic medium. The particle size from the study conducted by (I. Flavia *et al* 2014) was found within the range of 50-80nm, while the crystallite size was between 2-10nm compared to normal coarse zirconia which has particle size between 900nm-2.13μm and the crystallite size of >5nm to ≥20nm (I. Flavia *et al* 2014). (Tajudeen *et al* 2011) also used zirconyl chloride in a basic medium but by co-precipitation with bismuth oxide for the synthesis of zirconia.

Figure 6.

$$ZrOCl_2 + 2H_2O + 2NH_3 \xrightarrow[\text{Basic medium}]{\text{Benzyl alcohol surfactant}} ZrO_2 + H_2O + 2NH_4Cl$$

$$ZrO_2 \text{ minute particles} \xrightarrow{\text{Sodium dodecyl sulfate}} ZrO_2 \text{ nanoparticles}$$

Alumina

Alumina (Al_2O_3) otherwise called aluminum (III) oxide is an amphoteric chemical compound that is white and crystalline in appearance. It is odorless; it has high melting point, high surface area for adsorption, high thermal conductivity and highly insoluble in water. The most common alumina crystal structure is the trigonal corundum. Like zirconia, alumina is used in ceramic industries, in dentistry due to their chemical stability and tissue compatibility. Alumina is also used in metal production, as a catalyst and widely used as an abrasive due to its hardness, Figure 7.

Alumina occur in many metastable structures which completely transform into alpha-alumina upon the addition of heat >1150°C. Alumina exist in its natural form as corundum or emery but its gemstone possess different forms of colors e.g. ruby is red in appearance due to the presence chromium in its crystal lattice while sapphire is blue in color due to the presence of titanium or iron in its crystal lattice (Corrado *et al* 2014). Alumina powders are usually obtained from corundum, bauxite or hydrated aluminum oxide. The most common method in use for the conversion of bauxite to aluminum powder is the bayer process. Each aluminum atom is surrounded by oxygen atoms which account for the chemical stability of the compound (Corrado *et al* 2014).

The above mentioned properties of alumina has made researchers to focus more on its applications and the benefits it can offer to humanity. So many studies have been carried on the adsorption aspect of alumina among which are: the work of

Figure 7. (a) Appearance of alumina (b) Crystal structure of alumina

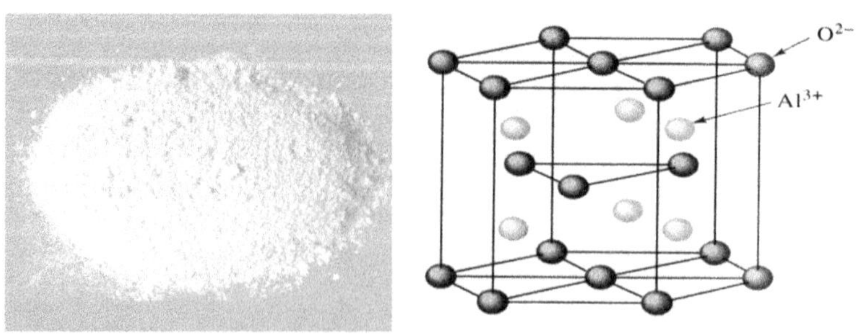

Figure 8. (a) Silica appearance (b) Arrangement of atoms of silica

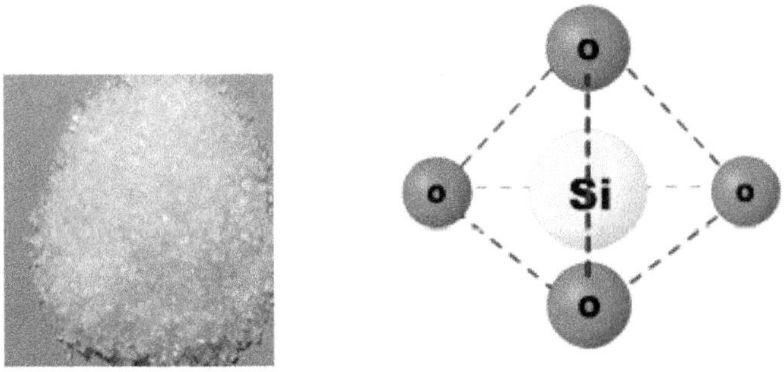

(Sushree *et al* 2008) and (Tony *et al* 2004) that utilizes alum impregnated with activated alumina and activated alumina for the removal of arsenic (V) and arsenic (III) from water respectively. The first study lowered the concentration of arsenic from 10mg/l to 40ppb while the second study revealed that activated alumina can be used for the effective removal of arsenic (III) from water. The study of (Yan-Hui et al 2001) successfully adsorb fluoride from water using amorphous alumina on carbon nanotubes; it was found that this combination was 13.5 times better than using activated carbon for the adsorption. Similar work conducted by (Subhashini *et al* 2005) and (Young *et al* 2002) on the removal of fluoride from water using activated alumina at neutral and close to neutral pH revealed that better fluoride removal occurs at lower flow rate and lower concentration and that sulfate ions affects the adsorption of fluoride through the formation of aluminum sulfate. Activated alumina proved to be an effective adsorbent material for the removal of chromium (VI) from water at optimum pH of 4 by the studies of (Suman *et al* 2007) and (Narsi *et al* 2004).

Silica

Silica also known as silicon di oxide (SiO_2) is found naturally in the crystalline form of quartz. It can also be found in some plants but it is usually synthetic in terms of chemical applications, Figure 8. Silica also exist in different phases depending on temperature, pressure, and the content of water. Silica can be crystalline or amorphous in appearance; it has a high surface area sometimes viscous, possess high porosity for adsorption and poorly soluble in water. Different types of silica exist which can be colloidal/sols silica, pyrogenic silica (aerosils, arc silicas, and plasma silicas), precipitated silica, and silica gels (hydrogels, aerogels, and aerogels). It is widely used as a filler, in the production of silicon and glass, as an adsorbent and desiccant, as a lubricant, and used as additive in the food and pharmaceutical industries.

The four parameters responsible for the chemical and physical properties of amorphous silica are: surface area, pore volume, pore size and particle size. The most widely known method for the preparation of silica is the sol-gel pathway (E.F. Vansant *et al* 1995).

Soluble Silica

This is the form of silica that is formed when amorphous or crystalline form of silica is disposed in liquid solution. The main constituent of the soluble silica is an acidic solution (monosilicic acid low in concentration).

Silica Sols

Sodium silicate is used to form silicilic acid which condenses to form a siloxane network for the formation of silica sols. By alteration of the pH between 8-9 and addition of some electrolyte to the mixture some colloidal particles are usually formed which later changes as the viscosity increases to form gels if the conditions are not stabilize or controlled. The colloidal particles which are usually spherical in shape and non-porous usually range in size from 1-100nm.

Silica Gels

Hydrogels are usually first to be formed from sols. As the viscosity increases in sols, the structure tends to bind together forming aggregates and colloids that contain mostly liquid substances within the pores. Hydrogels are usually determined by the starting material usually salt or alkoxyde, the solvent used for the production sometimes alcohols are used in that case the hydrogels are termed alcogel, the temperature of the medium, the pH of the medium and the electrolyte to be added.

Xerogel (compressed hydrogels) are usually formed after subjecting hydrogel to drying by thermal or other conventional methods. The hydrogel will lose an appreciable quantity of water which will compact the structure and reduce the pore size and the porosity. The xerogel that will be formed solely depend on the temperature that the hydrogel will be subjected to, the pH of the medium and concentration of the hydrogel particles.

Aerogel (alcogel extensively dried) usually has a particle size between 5-50nm; it is formed by subjecting silica compounds to high temperature hydrolysis till fine powdered substance is obtained. The pores of aerogel are stable and do not collapse upon contact with water. Grinding quartz to fine particle size (<1μm) produces aerogels.

Precipitated Silica

This form of silica are usually formed by coagulation and precipitation of silica solution. They have distinct structural characteristics though they are similar to gels however, in precipitated silica the gel is not allowed to be formed. The precipitation conditions are maintained and the formation of gel is avoided.

Polymeric Silica

Partial hydrolysis of silica compounds leads to the production of siloxane chains that are less viscous compared to gels. The viscous compounds formed are less soluble in organic solvents however, they have a carbon linked bond to their structure in form of alkoxy group which gives them a characteristics different from gels. Gel will be formed upon complete hydrolysis of polymeric silica solution.

Pyrogenic Silica

Silicas are formed from interaction of sand with high temperature usually in a furnace, plasma or arc. Pure silica is formed when silicon tetrachloride reacts with hydrogen and oxygen. Particle size silicas produced on this process are non-porous and it condenses to form a linear chain pattern, Figure 9.

Silica is one of the most used chemical adsorbent in chemical technology and chromatographic techniques (Eugene 2000), used as catalyst support and in the immobilization of enzymes. Recently the work of (Hajar *et al* 2014) combined organic polymer with silica aerogels to improve their mesoporosity, mechanical stability, and thermal conductivity for space applications. The study of (Saboktakin *et al* 2014) illustrated the applications of silica aerogel nanocomposites in architecture, (Yuhui *et al* 2009) demonstrated the use of hybridized nanosilica in combination with other nanomaterials for variety of applications. The newly formed nanomaterials

Figure 9. Chart showing family tree of silica

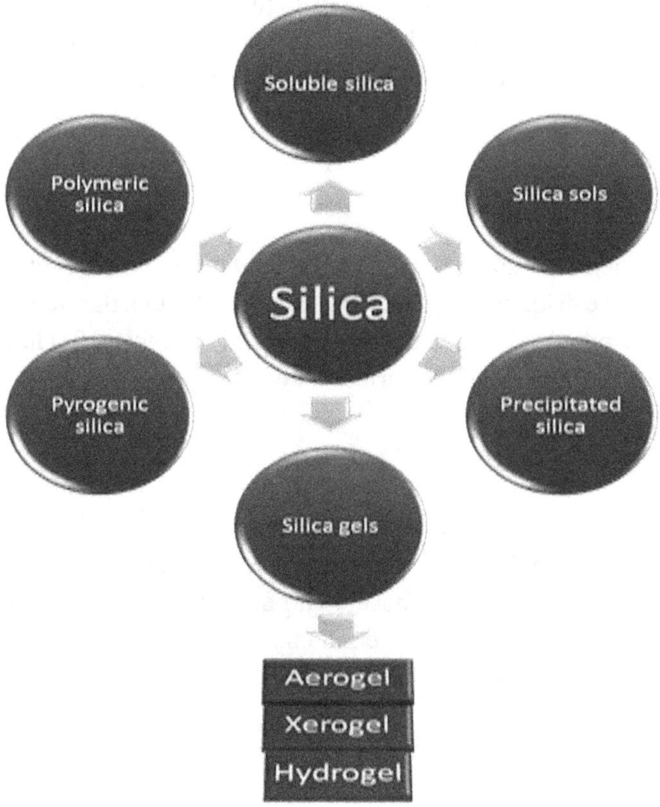

possess the characteristics of all the parent materials and their application in drug delivery, diagnosis and therapy had been outlined. Silica based adsorbent has not been extensively used in desulfurization technology compared to other adsorbent materials. The study of the mechanism of adsorption by silica is therefore immensely important.

Zeolites

Zeolites are termed aluminosilicates which are a combination of alumina and silica. They are micropourous, crystalline, white, and chalk like in appearance sometimes called molecular sieves. When alumina predominates in zeolites it becomes hydrophilic and when silica dominates the zeolites becomes hydrophobic. They have good thermal and mechanical properties qualifies them to be good catalyst and catalyst supports. The empirical formula for a typical zeolite containing a single cation "M" of valence "n " containing oxides of alkali or alkali earth metals (Na,

Figure 10. Chart illustrating different classes of carbon

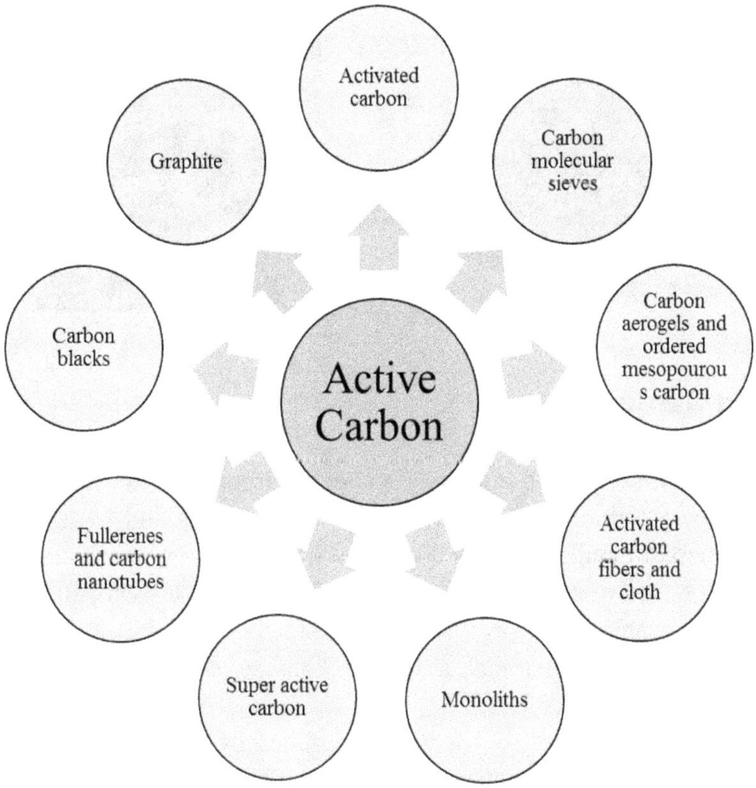

K, Ca etc.) is: ($M_{2/n}O$. Al_2O_3 . x SiO_2. yH_2O) x is the silica to alumina ratio ranging from ≥1 to 5 and y is the molar water of hydration. Each Al atom introduces one negative charge to the zeolites which needs to be balanced with exchangeable cations. The catalytic activity of zeolites depends on the nature of their cations or anions employed to neutralize charges. Lewis and/or Brønsted basic and/or acidic sites are then generated. Most zeolites are held by a binder which confer to them an adsorption advantage. Most commercial zeolites are normally produced using an autoclave under controlled conditions. Some examples of zeolites include: (A, X, Y, ZSM5, modernite and silicalite). They are widely used in drying, in catalytic membrane reactors, separation of oxygen from air, paraffins from naptha, and widely used in hydrodesulphurization.

Figure 11. (a) Appearance of activated carbon (b) SEM image of activated carbon from coconut shell

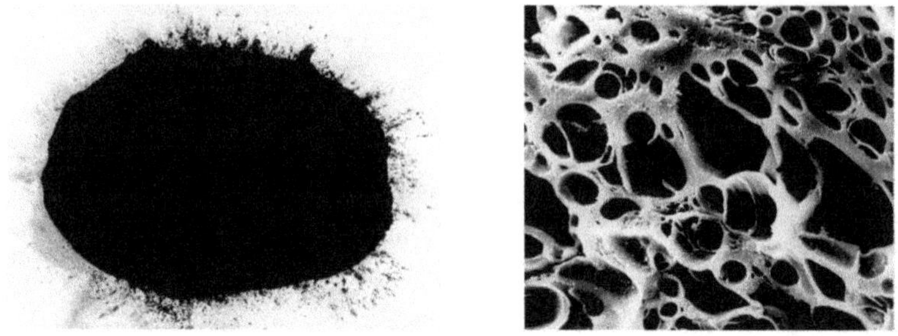

Activated Carbon

Activated carbon is a special class of organic adsorbent that is armophorous and possess microcrystalline structure with large surface area and high porosity. Its pores are dispersed having different sizes and shapes. It is sometimes referred to as activated coal or active charcoal; it is produced from carbon-rich parent materials e.g. charcoal, biochar, wood, peat, sewage sludge, animal bones, hair etc. that are subjected to physical or chemical activation. Classes of active carbons are given in Figure 10.

The physical production of activated carbon involves two processes viz: carbonization of the parent material and activation of the material formed. The chemical methods of producing activated carbon involves treatment of the parent material with acid, base or salt followed by contact with controlled atmosphere to aid in the activation process. It requires less amount of heat compared to the physical process of producing activated carbon.

Carbonization

In carbonization, the parent materials are subjected to pyrolysis at a temperature ≤ 800°C under controlled conditions of atmosphere and heat. At this stage, all elemental composition of the parent material used will be decomposed leaving behind only the aggregates of carbon that are randomly linked together forming residual pores which are used for adsorption. Activated carbon is widely known for its monumental adsorptive capacity; a gram of activated carbon contains an area of around 500m^2 while a pound of activated carbon contains an area equal to six football pitches.

Activation

The material formed after pyrolysis will be further modified, enhanced and developed to form great number of pores for tremendous applications using temperature between 800°C-900°C and a controlled atmosphere, Figure 11. This will improve the surface area of the activated carbon produced depending on the parent material. The activated material that will be produced after this process will have the potential to bind with different elements.

Activated carbon is divided into three classes based on their sizes viz: powdered activated carbon, granular activated carbon and extruded activated carbon. Activated carbon shows little selectivity in the adsorption of molecules of different sizes which is why it is sometimes impregnated or coated to desired material or molecular sieves for selective adsorption of persistent pollutants, air seperation and clean-up of off gasses in nuclear facilities. The modification of activated carbon to molecular sieves allows the development of effective micropore diameter for enhanced applications. It is widely used in purification of water, waste water and sewage and air; clean-up of oil spill, in groundwater remediation, widely used as a catalyst and in medical applications like treatment of poisoning, food and pharmaceutical industries e.t.c.

Nanomaterials

A nanomaterial is a broad name given to all types of materials found at the nanoscale; it is a material that has a unit size between the ranges of 1-100nm. They can be naturally occurring or chemically/mechanically synthesized with zero dimension (0-D), one dimension (1-D), two dimensions (2-D) or three dimensions (3-D). They have a wide range of both structural and non-structural industrial applications due to their sizes, shapes, chemistry and high surface area. They are applied in most industrial applications to improve performance because their surface properties far exceed their bulk properties.

Nano-materials such as alumina, zirconia, silica, and activated carbon have been the subject of research recently due to their promise in adsorption desulfurization. Some of these materials have a high affinity for organic sulfur compounds which are considered the third most abundant substances in petroleum aside from hydrocarbons (Soleimani *et al* 2007). Fuels with higher boiling point and higher density are considered to have higher sulfur compounds. These sulfur compounds in fuels can be either in organic or inorganic form. The inorganic compounds can be elemental sulfur, hydrogen disulfide or pyrite and they are not persistent in fuel compared to their organic counterparts. The organic sulfur compounds are divided into saturated (thiols, sulfides, and disulfides) or aromatic thiophene, Figure 12. The aromatics and their derivatives e.g. thiophene, benzothiophenes (BT) dibenzothiophene

Figure 12. Types of sulfur compounds found in crude oil

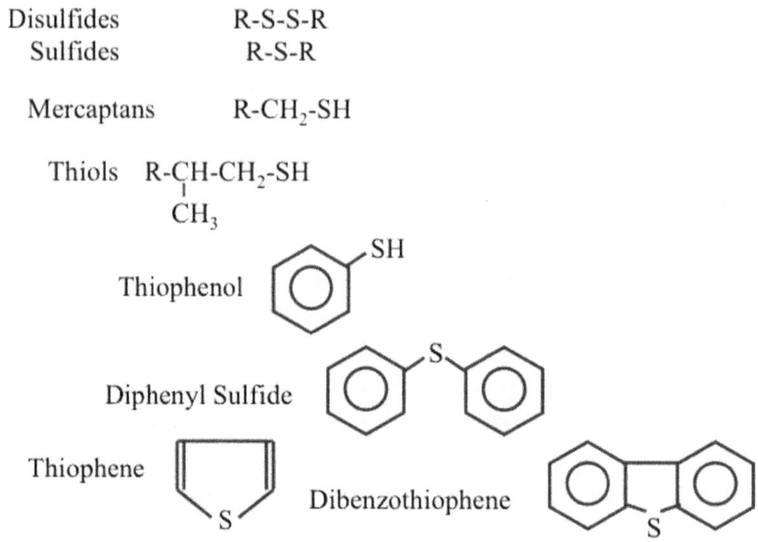

(DBT), 4-methylbenzothiophene (4MDBT), and 4, 6-dimethyl-dibenzothiophene (4, 6-DMDBT) are found in large quantities in fuel matrix. They are the prime targets in all desulfurization process and they are most persistent.

Synthesis of Nanomaterials

Nanomaterials are synthesized using two different approaches viz; bottom-up approach and top-down approach. In the bottom-up approach, the nanoparticles are first obtained at the atomic level and later integrated into the desired material e.g. is the formation of nanoparticle from colloidal dispersion or the formation of powders from sol-gel method followed by a later stage of integration. The top-down approach starts with bulk material at the macroscopic level followed by trimming of the material to the desired nanoparticles e.g. etching and ball milling, Figure 13.

The general outline for the production of nanomaterials involves three different processes; grinding, wet chemical processes otherwise called liquid phase processes

Figure 13. Diagram depicting the process for the synthesis of nanomaterial

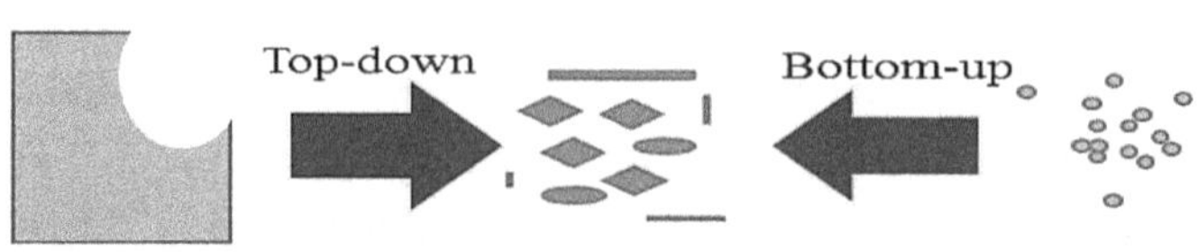

Figure 14. Diagram showing different synthesis routes of nanomaterials

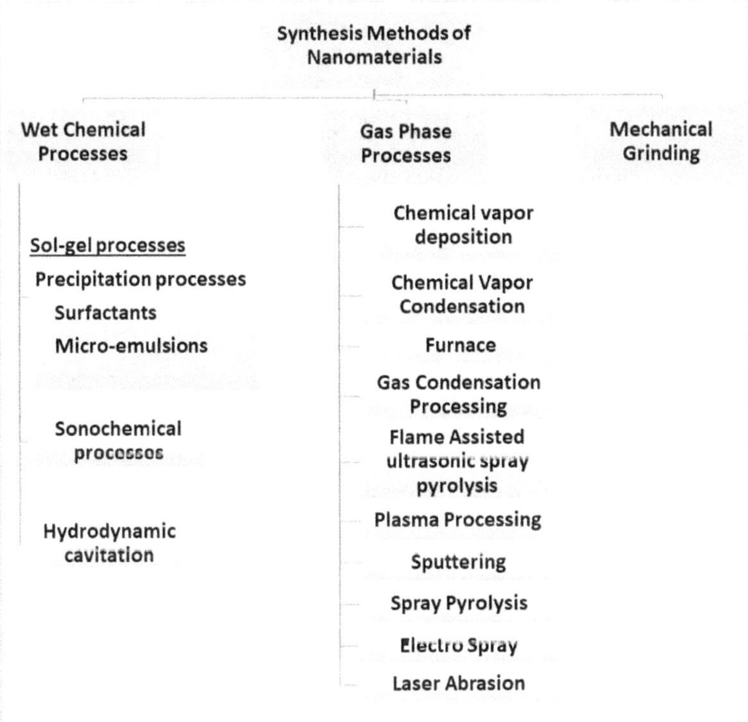

and lastly the gas phase processes of producing nanomaterials. These processes are outlined in Figure 14.

Synthesis of Nanocomposites

Nanocomposites are special classes of materials formed by combining two phases of different materials and they tend to retain the properties of each material used in their formation. A matrix (filler and resin) and a reinforcement (mostly fibers) are required for the formation of a composite. Properties of composites depend on the properties of each material, their relative amounts, and overall geometry. In essence, nanocomposites are materials having one or more of their phases in the nanoscale size embedded in a polymer matrix, ceramic matrix or metal matrix. They have nanoparticle properties, high thermal and mechanical stability, multifunctional capabilities, chemical functionalization, and huge interphase zone. Figure 15 depicts some classes and subclasses of nanocomposites.

Figure 15. Diagram depicting classes and subclasses of nanocomposites

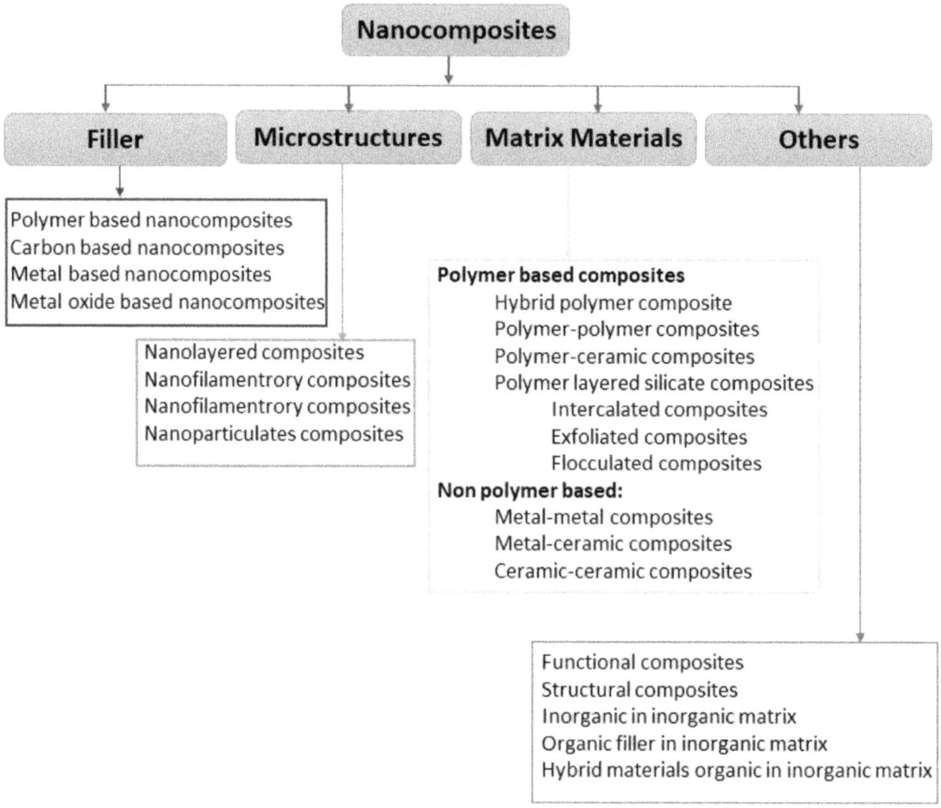

The combination confers added advantage to them which is why they are currently utilized in various fields of science and technology. They are widely used in catalysis, nanosensors and nanoprobes production, sorption process, chemical, and biological applications, fuel cells, non-linear optics, bio-ceramics, batteries with greater power output, environmental protection, anti-corrosion agents, drug delivery, UV protection gels, as fire-retardant material, in the production of lubricant and scratch-free paints and in adsorptive studies with clear application in desulfurization. There are various ways of synthesizing nanocomposites which will be outlined in this section. The processes do not take a specific method; the synthesis may involve combining two different methods or even more for the production. The most widely used methods are: Intercalation, sol-gel method, molecular composite formation method, high energy sonication, nanofiller direct dispersion method, hydrothermal synthesis, polymerized complex method, surfactant-assisted processing, solution-evaporation methods, electrochemical synthesis of polymers, chemical vapor deposition, microwave synthesis, and ball milling process. Figure 16 outlines the methods involved in the synthesis.

Sol-Gel Synthesis

This is the most effective and cost reliable method of producing nanomaterial, nanocomposites, and powders recently. Earlier it was widely used in glass and ceramic industries. It involves hydrolysis of the alkoxides by water, alcohol, ammonia, or acid followed by condensation of the products formed to produce a glass-like material. It is utilized when forming inorganic or hybrid composites at low temperatures and pressures. The factors affecting this method include pH, starting material, and solvent to be used.

Hydrothermal Synthesis

This involves combining the starting materials intended for composites synthesis with a certain amount of water followed by acid digestion in a reactor at high temperature and pressure. The reactants may not dissolve completely using this method.

Chemical Vapor Synthesis

This is a modification of the chemical vapor deposition where the process is directed towards the synthesis of nanomaterials instead of films. The whole idea of chemical vapor deposition is the attachment of solids produced from chemical reactions in the vapor phase to a heated surface. The energy needed for complete conversion of starting materials to nanoparticles is provided by hot walls, plasma, flame and laser reactors. The availability of appropriate starting material is the major limitation of this technology. It is currently used in the production of monoliths, powders, and films.

Microwave Synthesis

Microwave synthesis technique has been used in organic synthesis since time immemorial. More than 2000 papers have been published relating to the application of this technique (Kappe 2004). It is in use in pharmaceutical, biochemical, medical, food, ceramic industries, and academic institutions. Use of microwave synthesis in the production of nanocomposites entails combining starting materials by microwave irradiation to yield the desired nanomaterial. It is fast, economical and generates less by-products.

Solution-Evaporation Method

This is a method of composites formation which involves the mixing of nanotubes dispersion with a solution solely made up polymers that dissolve in solvents or thermoplastic polymers followed by evaporating the solvent. This will allow the nanotubes to move freely within the polymer matrix. The mixture can then be subjected to mixing and molding until the desired shape is achieved. The mixing intensity and its duration will determine the distribution of the nanoparticles within the polymer. The major drawbacks of this technique are that the polymer has to be soluble in the same solvent used to disperse nanoparticles and the problem of solvent removal limits the application of this technique in major industries.

Ball Milling Process

This is a technique that has been used in almost all industries for size reduction of materials. Nanocomposites production is no exception considering the recent utilization of high energy ball milling process for the production. Powdered materials are normally immersed for ball milling in the appropriate machine through the movement of the balls, collision to the supporting disc and centrifugal force exerted nanomaterials are easily formed. Mechanical alloying, mechanical milling and mechanochemical synthesis are the terms mostly used relating to the production of materials by ball milling. This method is perfect for the production of nanomaterials because it is efficient and cost-effective.

Intercalation Method

This is the most widely known method for the synthesis of polymer nanocomposites. The desired outcome depends on whether an intercalated or exfoliated hybrid is required. The process involves combining a starting material (a polymer) within layers of clay of material. When intercalate is required the organic material is immersed within the layers of the clay such that there will be expansion within the component mixtures. While in the exfoliated mixture, the whole layers of the clay material will be separated from each other but tied within the matrix of the organic component.

Figure 16. Synthesis routes of nanocomposites

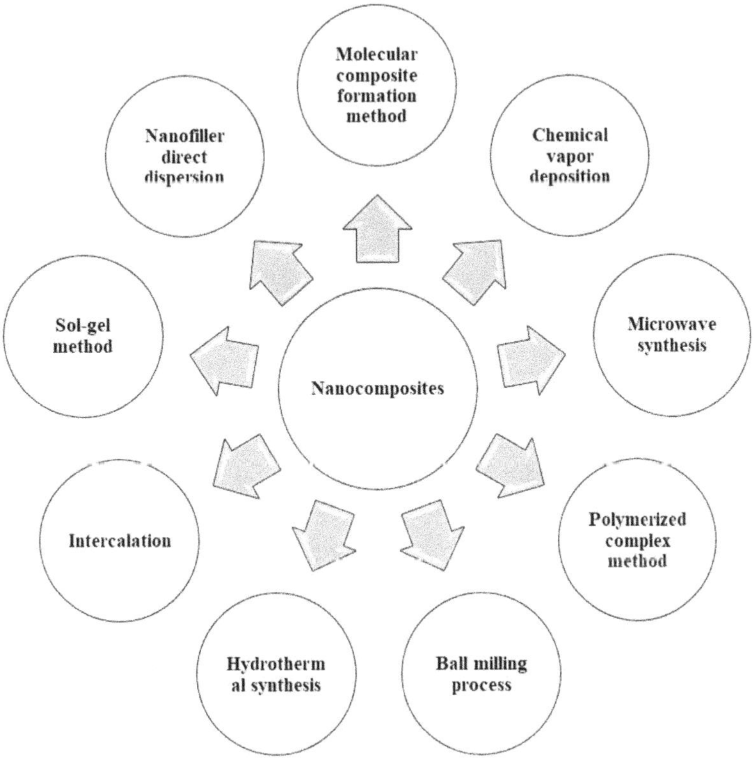

Nanoparticle Direct Dispersion Method

This is a method of producing nanocomposites by chemically modifying nanoparticle to increase compatibility with polymers. The advantage of using this method is that total homogeneity is achieved without compaction. Zinc oxide nanoparticles were prepared by combining zinc sulfate and ammonium bicarbonate in the study conducted by (Yujun et al., 2010).

Molecular Composite Formation

This is a method of producing nanocomposites that have extra mechanical stability due to the reinforcement provided by molecular rods. A strong and durable polymer is normally combined with a malleable polymer matrix at a molecular dimension with a solvent that will later be precipitated. An equilibrium compound is required as the third component for this method which will aid in the formation and dispersing of fibrils within the composite.

Figure 17. (a) SEM image of coarse zirconia (b) SEM image of synthesized zirconia and (c) SEM image of calcined zirconia at 3000x

a. b. c.

Applications of Nanomaterials and Nanocomposites in Desulfurization

Zirconia in Desulfurization

Zirconia (chemically unreactive) has a great promise in adsorption studies because of its properties. It has been used in oxidative desulfurization of thiophene by (Wang *et al* 2009). (Zubair *et al* 2012) in their study utilized calcined WOx/ZrO_2 at different temperatures as a catalyst for the oxidative removal of benzothiophenes and thiophenes from model fuels. It was found that the optimum calcination temperature of the catalyst was around 700°C and when the catalyst is equally reusable when used properly. (Yatsuka *et al* 2002) showed low selectivity of using Pt/ZrO_2 catalyst for oxidizing SO_2 from the exhaust gas. Copper was aided by zirconia in the study carried out by (P. Baeza *et al* 2008) for the removal of thiophene from n-octane at room temperature and pressure; the result showed the capacity of adsorption increases simultaneously with the increase in copper concentration. Chromium promoted sulfated zirconia prepared by wetness impregnation was used by (Sachin et al 2012) for oxidative desulfurization of model oil. The optimum dose was found to be 5g/l and the oxidation of DBT was found to be a slow process achieved after 6h. (Sachin et al 2011) extensively studied three different zirconia-based adsorbents viz: zirconia, calcined zirconia and sulfated zirconia that is calcined for the removal of dibenzothiophene in n-octane. The desulfurization process using zirconia was found to be exothermic and the data obtained best fit the BET isotherm model. The equilibrium between the adsorbents and dibenzothiophene was reached after 22hours of interaction.

Some of the SEM images of coarse zirconia, synthesized zirconia and calcined zirconia adopted from some of the researches carried out above are shown in Figure 17.

Alumina in Desulfurization

The principle of adsorptive desulfurization which allows an adsorbent material that is porous and has a high surface area to be placed in contact with the aromatic sulfur compounds in fuels as the substrates have received considerable attention in recent years. Researchers have tried to utilize alumina for adsorptive desulfurization due to its better adsorptive characteristics.

(Jae *et al* 2006) utilizes activated alumina for adsorptive desulfurization and denitrogenation though the result was not so successful in removing the sulfur compounds it revealed that the selectivity of the adsorption depends on the acidic-basic interaction and the molecular electrostatic potential. A study conducted by (Maria *et al* 2002) successfully showed the adsorption of benzothiophenes, dibenzothiophene, 4, 6-dimethyldibenzothiophene, indole, dibenzofuran and carbazole on zirconia, alumina, and magnesia using IR spectroscopy. They confirmed the adsorption to be due to the presence of Lewis sites on the adsorbent and that alumina was the strongest adsorbent within the adsorbent used. The study of (Jevanandam *et al* 2005) successfully utilized silver-impregnated nanocrystalline aluminum oxide and nanocrystalline metal oxides to adsorb thiophenes out of hydrocarbons. Nanocrystalline aluminum oxide was found to be more effective compared to the nanocrystalline metal oxides in the desulfurization. The adsorbent was found to have regeneration ability after heating at a moderate temperature. (Mansouri et al 2014) reduced the concentration of dibenzothiophene in n-hexadecane from 250ppmw sulfur concentration down to 0ppmw in the absence of hydrogen using a novel adsorbent consisting of well-dispersed nanoparticles prepared by impregnation on γ-Al_2O_3. (T. Zaki *et al* 2013) prepared alumina sorbents using cationic surfactants and microwave irradiation technology for the removal of aromatic sulfur compounds from petroleum wax found in Suez crude petroleum. The results from the study indicated the efficiency of the nano-alumina sorbents on desulfurization. (Srivastav et al 2009) uses aluminum oxide for the removal of dibenzothiophene in n-hexane. The study used X-ray diffractometer, FTIR and SEM to understand the mechanism of adsorption of DBT on alumina. The study found that the carbon-oxygen functional group on alumina was responsible for the adsorption of DBT. Langmuir isotherm best fit the adsorption data obtained from the studies and the adsorbent dose was optimum at 20g/l. The particle size of alumina from the study was found to be 0.074mm and the SEM image showed the small crystalline aggregate nature of alumina and its porosity. The d-spacing on the XRD carried out on the alumina in the study revealed the presence of aluminum oxide and boehmite as the major components.

Silica in Desulfurization

(Jun-Mi *et al* 2008) used mesoporous silica for adsorptive desulfurization and denitrogenation of refinery fuels. The study used plain mesoporous silica which has higher surface area and zirconia combined with silica (lower surface area) for the experiment. SEM images of the samples revealed both samples have a diameter between 2-4 μm. In addition, both adsorbents showed regenerative ability and were effective in the removal of nitrogen and sulfur compounds from the fuels. However, the structure of the plain mesoporous silica collapses after treatment with deionized water.

(Paolo et al 2008) conducted a study on the application of mesoporous silica in oxidative desulfurization for the purpose of producing ultra-low sulfur fuels; the adsorbent showed high affinity for sulfur compounds and high regeneration potential. The work of (Jessica *et al* 2014) successfully utilized mesoporous silica nanoparticles for efficient removal of dimethyldibenzothiophene and it proved that silica could perform better than other adsorbents used for desulfurization. The study on the adsorption/desorption desulfurization conducted by (Anne *et al* 2014) used cheap adsorbent prepared by lead loaded in mesoporous silica to reduce 1000ppm sulfur-containing gasoline to 50ppm in three rounds of cycles. (Anton *et al* 2012) the study used lithium modified mesoporous silica for the desulfurization of natural gas.

Activated Carbon in Desulfurization

Activated carbon is widely known as a versatile absorbent due to its surface area, ease of access, ease of preparation and its wide applications in almost every sector of the economy. It is said to have high selectivity for organosulfur compounds and nitrogen compounds removal due to the presence of hydrogen bonding interactions in their surface functional groups (Jae *at al* 2006). In addition, activated carbons are said to have surface functional groups that aid in chemisorption (Roop *et al* 1988) (R. N Fallah *et al* 2012). (Isam *et al* 2013) carried out a research on the adsorptive desulfurization properties of activated carbon; local diesel fuel of 450ppm sulfur content was treated using activated carbon at room temperature and the sulfur concentration was reduced to 210ppm using 5% adsorbent material. The study concluded that activated carbon has great potential to be used as adsorbent due to its low cost and high capacity for sulfur compounds.

A study by (Jintawat *et al* 2010) compared the activity of removal of dimethyl sulfide impurity from bioethanol using virgin activated carbon and activated carbon loaded with zinc oxide. The result showed loaded activated carbon has twice efficiency in comparison with the virgin activated carbon. The study equally concluded that activated carbon with larger mesopores will have better efficiency in adsorptive desulfurization. (Thitiwan *et al* 2013) utilized activated carbon derived from sewage sludge for adsorptive desulfurization of dibenzothiophene from n-octane at room temperature. Time (30mins-2hrs), carbonization temperature (400-800°C) and different activating agents were varied in the study to fully understand the adsorption process. The result indicated that DBT adsorption increases with increase in carbonyl functional groups and around 70.6% DBT removal were achieved using potassium hydroxide activating agent, carbonized at 600°C for 1hr which is by far higher than the percentage removed by commercial activated carbon. The adsorption of DBT on the sewage derived activated carbon followed a Langmuir isotherm. (Marko *et al* 2010) used batch adsorber to study the adsorptive desulfurization power of activated carbon and zeolite. The experimental results conformed to the Langmuir and Freundlich isotherms and activated carbon was found to have much better adsorption characteristics.

The study of (V. Selvavathi *et al* 2009) reported the use of modified activated carbons by nitric acid and nickel to have better efficiency compared to non-modified commercial activated carbon in the adsorptive desulfurization of refractory sulfur compounds from diesel oil. (Jae *et al* 2006) used activated carbon, activated alumina, and nickel-based adsorbent for denitrogenation and desulfurization of aromatic compounds from model diesel fuels in a fixed bed system; the result indicated that activated carbon has greater affinity and selectivity in the removal of both the most notorious refractory sulfur compounds and the nitrogen compounds compared to the other adsorbents used. The work of (Celia *et al* 2010) used activated carbon in adsorptive desulfurization of diesel fuel containing 72ppmw sulfur content in a batch system at 303.15K at atmospheric pressure and magnetic stirring for 18hrs. The experimental result best-fit Sips and BET models and the sulfur content was reduced to 15ppmw.

(Jie *et al* 2011) studied the adsorption affinity of polyaromatic sulfur heterocycles (PASHs) and polyaromatic hydrocarbon (PAHs) on activated carbon and how PAHs affects the adsorption of PASHs in both batch and fixed-bed systems. The result concluded that the electron-donor acceptor mechanism plays an important role in the adsorption of sulfur molecules and the presence of PAHs reduce the adsorption of PASHs. In addition, the adsorbent should possess pore size higher than the adsorbate for effective adsorption of large molecules and reducing diffusional resistance. Similar study conducted by (Jing *et al* 2012) on the side effects of aromatics, oxygen-containing additives, nitrogen containing compounds and moisture on the efficiency of adsorptive desulfurization of sulfur compounds from diesel fuel by activated carbon revealed that sulfur adsorption by activated carbon is due to the interaction of conjugated π electrons with the carbon surface and all the compounds have inhibitory effects on the adsorption which is due to the hydrogen bonding interaction between polar groups.

(S.P Hernandez *et al* 2010) used different sorbents loaded with nickel for desulfurization of diesel oil containing all the refractory sulfur compounds (250ppmw). The study confirmed the inhibitory effects of aromatics on adsorption. Activated carbon loaded with nickel in the study gave the higher breakthrough and uptake capacities compared to other sorbents due to the high selectivity of nickel on sulfur compounds and high surface area of activated carbon. (Mykola *et al* 2009) studied the textural and chemical factors affecting the adsorption capacity of activated carbon in diesel fuel. The study used polymer derived carbons and commercial activated carbon for the experiment. The polymer-derived carbon performed better in adsorption and selectivity. The result revealed that the presence of aromatic hydrocarbons did not affect the adsorption capacity of the activated carbon. The study concludes that adsorption efficiency was attributed to the level of micropores present and the presence of acidic groups in macropores that aid in adsorbing and oxidizing sulfur compounds.

The major concern about the use of adsorptive desulfurization is the regeneration of the adsorbent used. (Xue *et al* 2014) carried out a study on the regeneration methods to restore carbon adsorptive capacity of dibenzothiophene using thermal, solvent and ultrasound approaches. The result revealed that maintaining the integrity of micropores is cardinal to regeneration. Thermal regeneration method results in the loss of micropores and the important surface functional groups of activated carbon. Regeneration of activated carbon by toluene extraction restored its efficiency to about 70% fresh capacity after nine cycles of regeneration.

Nanocomposites in Desulfurization

Various nanocomposites have been reported as having potentials in adsorptive desulfurization. A study conducted by (Faghihian *et al* 2013) used zeolite-TiO_2 for the photodegradation of dibenzothiophene in fuels, TiO_2 was used as the photocatalyst and the removal by adsorption of the degradation product was carried out using β-zeolite. The result showed 88% degradation of dibenzothiophene and complete removal by the zeolite used for adsorption. The work of (Xinhai et al 2014) used a high-performance sulfur adsorbent ($NiO–CeO_2/Al_2O_3–SiO_2$) for the desulfurization of jet A fuels at ambient conditions. The composite reduced the sulfur content from 1140ppmw to as low as 10ppmw; it approached sulfur adsorption capacity of 1.4 and 3.22 mg-S/g-ads at breakthrough points of 10 and 50 ppmw, respectively and the treated fuel volumes reached 3.02 and 3.58 ml/g-ads at same breakthrough points respectively. The adsorbent was regenerated four times by calcinating helium gas for a short time before it reached 50% of its activity. At 50 ppmw S in effluent fuel, about 75%, 66%, 59% and 51% sulfur adsorption capabilities were recovered in the regenerated adsorbent after 1-4 cycles respectively.

(Shahadat *et al* 2013) in their study used Ag/TiOx–Al_2O_3 and Ag/TiOx–SiO_2 adsorbents for the adsorptive desulfurization of jet and diesel fuels at ambient conditions in a fixed bed continuous adsorption test. It was observed that dispersing TiOx on alumina and silica increased their adsorption and subsequent loading of Ag on the nanomaterials increased their adsorption. Ag/TiOx–Al_2O_3 was found to reduce sulfur content of ULSD from 1170ppmw to 75ppmw and it was regenerable. It equally showed the greatest affinity for benzothiophenes and the least affinity for dimethyldibenzothiophene. (Soumen *et al* 2013) used vapor phase adsorptive desulfurization of diesel on NiMCM-41 and NiY adsorbents under controlled condition. 1g of NiMCM-41 treated 20ml of diesel and the sulfur concentration was reduced from 450ppm to 50ppm while 1g of NiY treated 25ml of diesel and the sulfur concentration was reduce down to <50ppm. Both composites were regenerable under controlled oxidation at approximately 450°C.

π-complexation a form of a adsorptive desulfurization using copper species on nanomaterials such as zeolites, activated carbon, silica and alumina or nanocomposites to selectively remove thiophenic compounds in fuels at room temperature and pressure has been reported by (Yang *et al* 2003; Hernandez *et al* 2004; Hernandez *et al* 2005; Yang *et al* 2006; Wang-Hang *et al* 2010). Complexes are formed between the bulk materials and the cuprous ions which is an essential step in the removal of sulfur compounds from fuels. The extent of removal depends on the amount of the copper species and the limit of dispersion of the species on the nanomaterials. The cuprous species for adsorptive desulfurization can be directly introduced or formed from the reduction of cupric species.

The adsorptive desulfurization efficiency of Ni/Alumina and Ni/Aluminosilicate-5 is improved by varying the nickel/copper loadings on the sorbent materials (Sarda *et al*. 2012). The starting sulfur concentration was 350ppmw and hydrogen peroxide was used as the oxidizing agent in the process. The result showed that sulfur removal was dependent on the nature and amount of the metal used and the type of supporting material. More than 90% sulfur removal was achieved by the best sorbent Ni-Alumina in the presence of hydrogen peroxide. The use of membranes for desulfurization has been reported by (Ben *et al* 2012). The study used oleophilic polymer-inorganic nanocomposites for the pervaporative removal of organosulfur compounds from gasoline. The inorganic component used (silica nanoparticles) was in-situ synthesized through biomimetic mineralization in confined space using an alkaline inducer catalyst. The synthesized membrane demonstrated a good performance permeation flux of 7.36 kg/ (m^2h) and a selectivity of 4.98 to thiophene in gasoline.

Summary and Outlook

The chapter discusses the synthesis and applications of nanomaterials and nanocomposites in ADS. Major effects of sulfur on the environment, health, economy, and refining equipment have been discussed. The regulation set by USEPA on the level of sulfur in diesel and gasoline is currently 15ppm and various technologies are currently in use to achieve the set standard viz: HDS, ODS, BDS, ILE, and ADS. ADS is the most promising approach because it requires no hydrogen, oxidants or solvents and it operates at ambient conditions. Its major drawback is lack of selective and regenerable adsorbents. The principle of adsorption that relies on chemisorption and physisorption is based on the increase in the density of fluid within the vicinity of an interface. The types of ADS have been discussed which include: reactive, polar, selective, integrated process adsorptions and adsorption by π-complexation. Various classes of the adsorbents (AC, alumina, zirconia, zeolite, silica, etc.) and their applications have been discussed however for efficient adsorption, the material should possess the following properties: high surface area, porosity, affinity to the adsorbate and they should be cheap with thermal and mechanical stability. The synthesis routes of nanomaterials and nanocomposites with their applications in desulfurization have been extensively discussed. The chapter concludes with possible mechanisms of ADS on various adsorbents.

ACKNOWLEDGMENT

The author would like to thank the editor, Dr. Saleh for their assistance in the writing of this chapter.

REFERENCES

Al Zubaidy, I. A. H., Tarsh, F. B., Darwish, N. N., Majeed, B. S. S. A., Al Sharafi, A., & Abu Chacra, L. (2013). *Adsorption Process of Sulfur Removal from Diesel Oil Using Sorbent Materials*. Academic Press.

Anne Rodrigues, K. O., Ramos, J. E. T., Cavalcante, C. L., Rodriguez-Castellon, E., & Azevedo, D. C. S. (2014). Pd-loaded mesoporous silica as a robust adsorbent in adsorption/desorption desulfurization cycles. *Fuel*, *126*, 96–103. doi:10.1016/j.fuel.2014.02.019

Baeza, P., Aguila, G., Gracia, F., & Araya, P. (2008). Desulfurization by adsorption with copper supported on zirconia. *Catalysis Communications*, *9*(9), 751–755. doi:10.1016/j.catcom.2007.08.020

Ben Li, W. L., Wu, H., Yu, S., Cao, R., & Jiang, Z. (2012). Desulfurization of model gasoline by bioinspired oleophilic nanocomposite membranes. *Journal ofMembrane Science*, *415–416*, 278–287.

Brunauer, S., Emmett, P. H., & Teller, E. (1938). Adsorption of gasses in multimolecular layer. *Journal of the American Chemical Society*, *60*(2), 309–319. doi:10.1021/ja01269a023

Bu, J., Loh, G., Gwie, C. G., Dewiyanti, S., Tasrif, M., & Borgna, A. (2011). Desulfurization of diesel fuels by selective adsorption on activated carbons: Competitive adsorption of polycyclic aromatic sulfur heterocycles and polycyclic aromatic hydrocarbons. *Chemical Engineering Journal*, *166*(1), 207–217. doi:10.1016/j.cej.2010.10.063

Buekens, A., & Zyaykina, N. N. (2009). *Adsorbents and adsorption processes for pollution control* (Vol. 2). Pollution Control Technologies.

Cerrato, G., Bordiga, S., Barbera, S., & Morterra, C. (1997). A surface study of monoclinic zirconia (m-ZrO_2). *Surface Science*, *377-379*(50-55).

Cerrato, G., Bordiga, S., Barbera, S., & Morterra, C. (1997). Surface Characterization of monoclinic ZrO_2 I morphology, FTIR spectral features and computer modelling. *Applied Surface Science*, *115*(1), 53–65. doi:10.1016/S0169-4332(96)00586-7

Chaichanawong, J., Yamamoto, T., Ohmori, T., & Endo, A. (2010). Adsorptive desulfurization of bioethanol using activated carbon loaded with zinc oxide. *Chemical Engineering Journal*, *165*(1), 218–224. doi:10.1016/j.cej.2010.09.020

Czaja, A. U., Trkhan, N., & Muller, U. (2009). Industrial applications of metal organic frameworks. *Chemical Society Reviews, 38*(5), 1284. doi:10.1039/b804680h PMID:19384438

Dasgupta, S., Gupta, P., Nanoti, A., Goswami, A. N., Garg, M. O., Tangstad, E., & Stocker, M. (2013). Adsorptive desulfurization of diesel by regenerable nickel based adsorbents. *Fuel, 108*, 184–189. doi:10.1016/j.fuel.2012.12.060

Dercz, G., Prusik, K., & Pajak, L. (2008). X-ray and SEM studies on zirconia powders. *Journal of Mat & Manufacturing Eng, 31*(2).

Di Noto, V., Lavina, S., Wintersgill, M. C., & Fontanella, J. J. (2010). A formalism relating the conductivity of functionalized nanoparticles to constituent ligand molecules and application to water-containing silica. *Physical Chemistry Chemical Physics, 12*(23), 5993–5997. doi:10.1039/b924811k PMID:20396820

Douglas, R. M. (1984). Principle of adsorption and adsorption processes. J. Wiley & Sons.

Faghihian, H., & Naeemi, S. (2013). Application of a Novel Nanocomposite for Desulfurization of a Typical Organo Sulfur Compound. *Iran. J. Chem. Chem. Eng., 32*(3).

Fallah, R. N., & Azizian, S. (2012). Removal of thiophenic compounds from liquid fuel by different modified activated carbon cloths. *Fuel Processing Technology, 93*(1), 45–52. doi:10.1016/j.fuproc.2011.09.012

Flavia, I. P. N., Lakshmi Prabha, V., & Aswathy Paul, D. (2014). Nirmal: Synthesis and dielectric studies of monoclinic nanosized zirconia. Advances in Condensed Matter Physics.

Freitas, P. A. M., Iha, K., Felinto, M. C. F. C., & Suárez-Iha, M. E. V. (2008). Adsorption of di-2-pyridyl ketone salicyloylhydrazone on Amberlite XAD-2 and XAD-7 resins: Characteristics and isotherms. *J. Colloid Inter. Sci., 323*(1), 1–5. doi:10.1016/j.jcis.2008.04.001 PMID:18466914

Freundlich, H. M. F. (1906). Over the adsorption in solution. *Journal of Physical Chemistry, 57*, 385–471.

Hameed, B. H. (2009). Grass waste: A novel sorbent for the removal of basic dye from aqueous solution. *Journal of Hazardous Materials, 166*(1), 233–238. doi:10.1016/j.jhazmat.2008.11.019 PMID:19111987

Hernandez, S. P., Fino, D., & Russo, N. (2010). High performance sorbents for diesel oil desulfurization. *Chemical Engineering Science*, *65*(1), 603–609. doi:10.1016/j.ces.2009.06.050

Hernandez-Maldonado, A. J., & Yang, R. T. (2004). Desulfurization of transportation fuels by adsorption. *Catalysis Reviews. Science and Engineering*, *46*(2), 111–150. doi:10.1081/CR-200032697

Hernández-Maldonado, A. J., & Yang, R. T. (2004). Desulfurization of Diesel Fuels by Adsorption via p-Complexation with Vapor Phase Exchanged (VPIE) Cu (I)-Y Zeolites. *Journal of the American Chemical Society*, *126*(4), 992–993. doi:10.1021/ja039304m PMID:14746446

Hernandez-Maldonado Arturo, J. (2005). Desulfurization of commercial fuels by p-complexation: Monolayer CuCl/g-Al_2O_3. *Applied Catalysis B: Environmental*, *61*, 212–218. doi:10.1016/j.apcatb.2005.05.003

Jeevanandam, P., Klabunde, K. J., & Tetzler, S. H. (2005). Adsorption of thiophenes out of hydrocarbons using metal impregnated nanocrystalline aluminum oxide. *Microporous and Mesoporous Materials*, *79*(1-3), 101–110. doi:10.1016/j.micromeso.2004.10.029

Kappe, C. O. (2004). Controlled Microwave Heating in Modern Organic Synthesis. *Angewandte Chemie International Edition*, *43*(46), 6250–6284. doi:10.1002/anie.200400655 PMID:15558676

Khan, A. (2012). The theory of adsorption equilibria analysis based on general equilibrium constant expression. *Turkish Journal of Chemistry*, *36*, 219–231.

Kim, J. H., Ma, X., Zhou, A., & Song, C. (2006). Ultra-deep desulfurization and denitrogenation of diesel fuel by selective adsorption over three different adsorbents: A study on adsorptive selectivity and mechanism. *Catalysis Today*, *111*(1-2), 74–83. doi:10.1016/j.cattod.2005.10.017

Knaebel Kent, S. (2004). *Adsorbent selection*. Dublin, OH: Adsorption Research, Inc.

Koriakin, A., Kim, Y.-H., & Lee, C.-H. (2012). Adsorptive Desulfurization of Natural Gas Using Lithium-Modified Mesoporous Silica. *Industrial & Engineering Chemistry Research*, *51*(44), 14489–14495. doi:10.1021/ie301066n

Kwon, J.-M., Moon, J.-H., Bae, Y. S., Lee, D.-G., Sohn, H.-C., & Lee, C.-H. (2008). Adsorptive desulfurization and denitrogenation of refinery fuels using mesopourous silica adsorbents. *ChemSusChem*, *1*(4), 307–309. doi:10.1002/cssc.200700011 PMID:18605095

Langmuir, I. (1918). The adsorption of gases on plane surfaces of glass, mica and platinum. *Journal of the American Chemical Society*, *40*(9), 1361–1403. doi:10.1021/ja02242a004

Li, Yue, Su, & Ga. (2011). Equilibrium and a two-stage batch adsorber design for reactive or disperse dye removal to minimize adsorbent amount. *Bioresource Technology*, *102*(9), 5290–5296.

Maleki, H., Duraes, L., & Portugal, A. (2014). Synthesis of lightweight polymers reinforced silica aerogels with improved mechanical and thermal insulation properties for space applications. *Microporous and Mesoporous Materials*, *197*, 116–129. doi:10.1016/j.micromeso.2014.06.003

Mansouri, A. (2014). Ultra-deep adsorptive desulfurization of a model diesel fuel on regenerable Ni–Cu/-Al2O3 at low temperatures in absence of hydrogen. *Journal of Hazardous Materials*, *271*, 120–130. doi:10.1016/j.jhazmat.2014.02.006 PMID:24632365

Marı'n-Rosas, C., Ramı'rez-Verduzco, L. F., Murrieta-Guevara, F. R., Herna'ndez-Tapia, G., & Rodrı'guez-Otal, L. M. (2010). Desulfurization of Low Sulfur Diesel by Adsorption Using Activated Carbon: Adsorption Isotherms. *Industrial & Engineering Chemistry Research*, *49*(9), 4372–4376. doi:10.1021/ie901756b

Maria Angeles Larrubia, A. (2002). A FT-IR study of the adsorption of indole, carbazole, benzothiophene, dibenzothiophene and 4,6-dibenzothiophene over solid adsorbents and catalysts. *Applied Catalysis A, General*, *224*(1-2), 167–178. doi:10.1016/S0926-860X(01)00769-4

Mathieu, Y., Soulard, M., Patarin, J., & Moliere, M. (2012). Mesopourous materials for the removal of SO_2 from gas streams. *Fuel Processing Technology*, *99*, 35–42. doi:10.1016/j.fuproc.2012.02.005

Muzic, M., Sertic-Bionda, K., Gomzi, Z., Podolski, S., & Telen, S. (2010). Study of diesel fuel desulfurization by adsorption. *Chemical Engineering Research & Design*, *8*(8), 487–495. doi:10.1016/j.cherd.2009.08.016

Mykola, S. (2009). Textural and chemical factors affecting adsorption capacity of activated carbon in highly efficient desulfurization of diesel fuel. *Carbon*, *47*(10), 2491–2500. doi:10.1016/j.carbon.2009.05.001

Narsi, R. B. (2004). Adsorption of Cr (VI) on activated rice husk carbon and activated alumina. *Bioresource Technology*, *91*(3), 305–307. doi:10.1016/S0960-8524(03)00204-9 PMID:14607491

Palomino, J. M., Tran, D. T., Hausser, J. L., Dong, H., & Oliver, S. R. J. (2014). Mesopourous Silica nanoparticles for high Capacity adsorptive desulfurization. *J. Matter. Chem A*, *2*(36), 14890–14895. doi:10.1039/C4TA02570A

Paolo, D. F., & Scarsella, M. (2008). Functionalized hexagonal mesoporous silica as an oxidizing agent for the oxidative desulfurization of organosulfur compounds. *Industrial & Engineering Chemistry Research*, *47*(3), 973–975. doi:10.1021/ie071057y

Papirer, E. (2000). *Adsorption on silica surfaces*. CRC. doi:10.1201/9781482269703

Picconi, C. (2014). *Saverio Giovanni Condo, Tomaz Kosmae: Alumina and Zirconia based ceramics for load bearing applications*. Butterworth-Heinemann.

Richard, I. M. (1996). Principles of adsorption and reaction on solid surfaces. J. Wiley & Sons.

Roop, C. B. (2005). Activated Carbon Adsorption. Taylor and Francis Group.

Saboktakin, A. (2014). Improvements of reinforced silica aerogel nanocomposites thermal properties for architecture applications. *International Journal of Biological Macromolecules*, *72C*, 230–234. PMID:25172161

Sachin, K., & Vimal Chandra Srivastava, R. P. (2011). Studies on adsorptive desulfurization by zirconia based adsorbents. *Fuel*, *90*(11), 3209–3216. doi:10.1016/j.fuel.2011.06.029

Sachin, K., & Vimal Chandra Srivastava, R. P. (2012). Oxidative desulfurization by chromium promoted sulfated zirconia. *Fuel Processing Technology*, *93*(1), 18–25. doi:10.1016/j.fuproc.2011.08.017

Sarda, K. K., Bhandari, A., Pant, K. K., & Jain, S. (2012). Deep desulfurization of diesel fuel by selective adsorption over Ni/Al_2O_3 and Ni/ZSM-5 extrudates. *Fuel*, *93*, 86–91. doi:10.1016/j.fuel.2011.10.020

Selvavathi, V., Chidambaram, V., Meenakshisundaram, A., Sairam, B., & Sivasankar, B. (2009). Adsorptive desulfurization of diesel on activated carbon and nickel supported systems. *Catalysis Today*, *141*(1-2), 99–102. doi:10.1016/j.cattod.2008.05.009

Shahadat, H. A. M. H., & Bruce, J. (2013). Adsorptive desulfurization of jet and diesel fuels using Ag/TiOx–Al_2O_3 and Ag/TiOx–SiO_2 adsorbents. *Fuel*, *107*, 465–473. doi:10.1016/j.fuel.2012.11.030

Sing, K. S. W. (1982). Reporting Physiosorption Data Gas/Solid systems. *Pure and Applied Chemistry*, *54*(11), 2201–2218. doi:10.1351/pac198254112201

Soleimani, M. M., Bassi, A., & Margaritis, A. (2007). Biodesulfurization of refractory organic sulfur compounds in fossil fuels. *Biotechnology Advances*, *25*(6), 570–596. doi:10.1016/j.biotechadv.2007.07.003 PMID:17716849

Song, C. (2003). An overview of new approaches to deep desulfurization for ultra-clean gasoline, diesel fuel and jet fuel. *Catalysis Today*, *86*(1-4), 211–263. doi:10.1016/S0920-5861(03)00412-7

Song, C., & Ma, X. (2003). New design approaches to ultra-clean diesel fuels by deep desulfurization and deep dearomatization. *Applied Catalysis B: Environmental*, *41*(1-2), 207–238. doi:10.1016/S0926-3373(02)00212-6

Song, S. C. (2003). An Overview of new approaches to deep desulfurization for ultra-clean gasoline, diesel fuel and jet fuel. *Catalysis Today*, *86*(1-4), 211–263. doi:10.1016/S0920-5861(03)00412-7

Soumen, D. (2013). Adsorptive desulfurization of diesel by regenerable nickel based adsorbents. *Fuel*, *108*, 184–189.

Srivastav, A., & Srivastava, V. C. (2009). Adsorptive desulfurization by activated alumina. *Journal of Hazardous Materials*, *170*(2-3), 1133–1140. doi:10.1016/j.jhazmat.2009.05.088 PMID:19523762

Subhashini, G., & Pant, K. K. (2005). Equilibrium Kinetics and breakthrough studies for adsorption of fluoride on activated alumina. *Separation and Purification Technology*, *42*(3), 265–271. doi:10.1016/j.seppur.2004.09.001

Suman, M., & Khaiwal Ravinda, N. R. (2007). Adsorption of chromium from aqueous solution by activated alumina and activated charcoal. *Bioresource Technology*, *98*(4), 954–957. doi:10.1016/j.biortech.2006.03.018 PMID:16725320

Sushree, S. T., & Ashok, M. (2008). Enhanced adsorption capacity of activated alumina by impregnation with alum for removal of As (V) from water. *Chemical Engineering Journal*, *138*(1-3), 179–186. doi:10.1016/j.cej.2007.06.028

Tajudeen, O. A. (2011). Mophology and composition of nanocrystalline stabilized zirconia using SEM-EDS system. *Leonardo. Journal of Science*, (19): 81–92.

Tanabe, K. (1985). Surface and catalytic properties of $Zr0_2$. *Materials Chemistry and Physics*, *13*(3-4), 347–364. doi:10.1016/0254-0584(85)90064-1

Thitiwan, N. (2013). Adsorptive desulfurization of dibenzothiophene by sewage sludge-derived activated carbon. *Chemical Engineering Journal*, *228*, 263–271. doi:10.1016/j.cej.2013.04.067

Tony, S. S., & Pant, K. K. (2004). Equilibrium, Kinetics and thermodynamics studies for adsorption of As (III) on activated alumina. *Separation and Purification Technology*, *36*(2), 139–147. doi:10.1016/S1383-5866(03)00209-0

Tsuru, T. (2001). Inorganic porous membranes for liquid phase separation. *Separation and Purification Reviews*, *30*(2), 191–220. doi:10.1081/SPM-100108159

Unites States Environmental Protection Agency (USEPA). (2014). *EPA sets tier 3 motor vehicle emission and fuel standards*. EPA-420-F-14-009. Retrieved from www.epa.gov/otaq/documents/tier3/420f14009.pdf

Vansant, E. F., Vander Voort, P., & Vrancken, K. C. (1995). *Characterization and chemical modification of the silica surface*. Elsevier.

Wang, B., Zhu, J., & Ma, H. (2009). Desulfurization from thiophene by SO $^{(4)}$ $^{(2-)}$/ ZrO_2 catalytic oxidation at room temperature & atmospheric pressure. *Journal of Hazardous Materials*, *164*(1), 256–264. doi:10.1016/j.jhazmat.2008.08.003 PMID:18799257

Xiao, J., Song, C., Ma, X., & Li, Z. (2012). Effects of Aromatics, Diesel Additives, Nitrogen Compounds, and Moisture on Adsorptive Desulfurization of Diesel Fuel over Activated Carbon. *Industrial & Engineering Chemistry Research*, *51*(8), 3436–3443. doi:10.1021/ie202440t

Xinhai, X., Zhang, S., Li, P., & Shen, Y. (2014). Adsorptive desulfurization of liquid Jet-A fuel at ambient conditions with an improved adsorbent for on-board fuel treatment for SOFC applications. *Fuel Processing Technology*, *124*, 140–146. doi:10.1016/j.fuproc.2014.03.001

Xue, H., Lin, H., & Zheng, Y. (2014). Regeneration methods to restore carbon adsorptive capacity of dibenzothiophene and neutral nitrogen heteroaromatic compounds. *Chemical Engineering Journal*, *243*, 315–325. doi:10.1016/j.cej.2013.12.074

Yang, R. T., & Arturo, J. (2003). Desulfurization of Transportation Fuels with Zeolites under Ambient Conditions. *Science*, *301*(5629), 79–81. doi:10.1126cience.1085088 PMID:12843388

Yang, X., Erickson, L. E., Hohn, K. L., Jeevanandam, P., & Klabunde, K. J. (2006). Sol-Gell Cu-Al_2O_3 Adsorbents for Selective Adsorption of Thiophene out of Hydro carbon. *Industrial & Engineering Chemistry Research*, *45*(18), 6169–6174. doi:10.1021/ie060559t

Young, K. (2002). The adsorption of fluoride ion from aqueous solution by activated alumina. *Water, Air, and Soil Pollution*, *133*(1/4), 349–360. doi:10.1023/A:1012929900113

Yuhui, J., Li, A., & Sandra, G. (2009). Amorphous Silica nanohybrids: Synthesis, properties & Applications. *Coordination Chemistry Reviews*, *253*, 2998–3014.

Yujun, W., Zhang, C., Bi, S., & Luo, G. (2010). Preparation of ZnO nanoparticles using the direct precipitation method in a membrane dispersion micro-structured reactor. *Powder Technology*, *202*(1-3), 130–136. doi:10.1016/j.powtec.2010.04.027

Zaki, T., & Nermen, H. (2013). Characterization and application of nano-alumina sorbents for desulfurization and dearomatization of Suez crude petrolatum. *Fuel Processing Technology*, *106*, 625–630. doi:10.1016/j.fuproc.2012.09.055

Zeki, B. (2013). Food Process Engineering and Technology. In Adsorption and Ion Exchange (2nd ed.). Food Science and Technology.

Zubair, Jeon, & Hwa Jhung. (2012). Oxidative desulfurization of benzothiophene and thiophene with WOx/ZrO2 catalysts: Effect of calcination temperature of catalysts. *Journal of Hazardous Materials*, (205–206), 216–221.

Chapter 3
Advances in Carbon-Based Nanocomposites for Deep Adsorptive Desulfurization

Saddam A. AL Hammadi
Chemical Engineering Department, The University of Melbourne, Australia

ABSTRACT

The ultra-low sulfur diesel (ULSD) is required to comply with stricter government policy on low sulfur content of transportation fuels. Better knowledge of the different factors that concern deep desulfurization of fuels is necessary to achieve ultra-low sulfur content and cheaper way of producing ULSD. Both the capital and operating cost of the adsorptive desulfurization process is cheaper compare to the conventional hydroprocessing. In the future, the need to produce more volume of fuels with very low sulphur content from low-grade feedstocks like heavy oil and light cycle oil in order to meet up with the global demand for sulphur-free fuels is pertinent. Several on-going researches are pointing to the use of adsorbents for removal of sulfur compounds from hydrocarbon refining stream. In this chapter, varieties of carbon nanomaterials suitable for adsorptive desulfurization are discussed. If the active lifetime, where the capacity of the adsorbents are adequate, the approach is practically feasible for commercial application.

INTRODUCTION

Removal of sulfur compounds (desulfurization) is a vital unit operation in petroleum refining since the combustion products of sulfur compounds is the main reason for acid rain and environmental pollution. In addition, sulfur is also a catalyst poison

DOI: 10.4018/978-1-7998-2146-5.ch003

during industrial processes. Compounds with sulfur are removed catalytically at high temperature and pressure. Desulfurization is gaining a lot of attention and efforts have been channeled towards investigating several methods that are effective and economically viable. The attention is warranted by stricter environmental regulations on the amount of sulfur that should be present in transportation fuels (Song, 2003; Yang et al., 2004). Sulfur limit (mass-percent) of 0.015, 0.035, and 0.2 for gasoline, diesel fuel, and light fuel oil, respectively has been set. A new and more stringent limit of 0.003-0.005 mass-percent (30-50 ppm) is imminent for transportation fuels in Europe and United States of America (Song, 2003; State Announcer, 2001; Babich and Monlijn, 2003). It's worthy to note also that; desulfurization processes have found applications in converting used tyres and shale oil to fuel oils. Using calcium oxide in binding up sulfur oxide in emissions has been achieved for stationary applications in the desulfurization of non transportation fuels but the use of the harmful compounds still a challenge (Svobodal et al., 1994). Therefore, the need to innovate effective technologies for desulfurization processes is very paramount.

Numerous methods of desulfurization processes have been investigated. The need to achieve a lower level of sulfur in fuel oils has also called for different innovative ways of achieving deep desulfurization where synergy of methods yielding better results (Agarwal and Sharma, 2010; Sundaraman et al., 2009). Hydro-desulfurization (HDS) is a popular process, but there is wide variation in the reactivity of sulfur-containing heterocyclic compounds. Alkyl-substituted derivatives of dibenzothiophene like 4-methyldibenzothiophene, 4, 6-dimethyldibenzothiophene e.t.c from fuel oil have been reported to be relatively unreactive towards hydrotreating (Gate and Tropsoe, 1997). In order to achieve deep-desulfurization and take care of attending challenges of hydro-desulfurization which include high hydrogen consumption, energy (heat) cost, catalyst volume and many other methods are combined with HDS for better results (Rana et al., 2007). In recent years, efforts are being directed to other methods, including; adsorptive desulfurization, oxidative desulfurization where different catalysts are used (Kumar et al., 2012) extractive desulfurization involving ionic liquids, photochemical activation, bio-adsorption, bio-desulfurization, ultrasonic-desulfurization, microwave desulfurization and electrochemical approach (Bhatia and Sharma, 2006; Lam et al., 2012).

Adsorptive desulfurization where adsorbents are used to remove sulfur-containing compounds in fuel oils is holding the future promise of ultra-clean sulfur-free fuel oils. Adsorption processes can be performed at ambient temperature and pressure, thus saving a lot of energy as compared to other methods of desulfurization (Seredych et al., 2009). Therefore, intensive research is on-going to produce new adsorbents with great emphasis on good selectivity, high capacity for adsorption and regeneration of the adsorbents with special attention to the mechanism of adsorption (Yang et al., 2007).

Various materials have been reported to be good adsorbents for different adsorption processes. These materials include; different types of zeolites and metal-impregnated zeolites such as MCM-22 zeolites (Delitata et al., 2008), MCM-41 (lanthanum loaded, sorbents for diesel fuel) (Subhan et al., 2012), Y-zeolites with exchanged cation (NaY) and CNT as template, Ag(1)- and Cu(1)-beta zeolites among others (Cong et al., 2009). A deep desulfurization method was presented by Daimler-Chrysler AG (2000) where adsorption process is used for engine fuel with TiO_2, MgO, Al_2O_3 or SiO_2 (with metal additives) as adsorption materials.

Zhou et al., (2009) reported dibenzothiophene and 4,6-dimethyldibenzothiophene removal from diesel by using mesoporous carbon synthesized from silica HMS (template) and phenolic resins (source of carbon). Due to high thermal stability, large surface area, and well-arranged mesoporous structure of CMK-1, 3 and 5 carbons, they have been implicated as good adsorbents of DBT in petroleum fuels (Arbia and Parvin, 2011). Removal of organosulfur from commercial jet fuel like JP-5 (847 ppm sulfur) has been reported using metallic halides like; $PdCl_2$/SBA-15 and $PdCl_2$/MCM-41 with activated carbon lending excellent support (Wang et al., 2006 and 2008). Chemisorptions, as well as production and regeneration of novel sorbents, have attracted much attention and a lot of patents and patent applications are on record for desulfurization.

Classification of Carbon-Based Nanostructures as Adsorbents

Structure indeed determines properties. Different allotropes of low dimension can be formed from carbon, such as; graphene, CNT and C-60. The superiority in properties exhibited by CNT among other 1-D nanomaterials can be traced to its special stereogenic structure. Several attentions have been paid to CNTs over the past two decades; this is due to nothing but unique mechanical, thermal, electrical properties among other key applications in many fields of science and technology. The simple category of Carbon-based nanostructures as adsorbents can be illustrated as in Figure 1.

Carbon Nanostructures as Substrates

Over the years, carbon has been used as catalyst support employed for different applications in the chemical industry owing to its large surface area. The orientation of graphene layers can be varied with respect to the axes of nanofibres giving different morphologies of the carbon nanofibre which make available different pores for catalysts' nanoparticles (Salernitano et al., 2014). Carbon paper substrate with 3-D fibre network has been found appropriate in fuel cells with Silicon-Carbon composite deposited alongside the fibre network on the electrode (Matsiu et al.,2013).

Figure 1. Classification of carbon-based materials as adsorbents

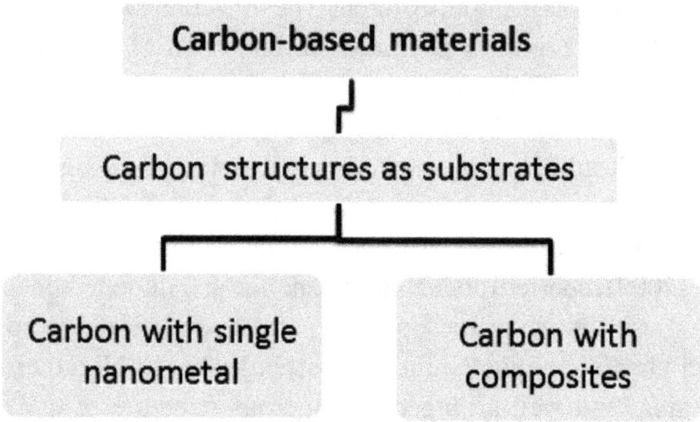

Structure indeed determines properties. Different allotropes of low dimension can be formed from carbon, such as; graphene, carbon nano-tube (CNT) and C-60. The superiority in properties exhibited by CNT among other 1-D nanomaterials can be traced to its special stereogenic structure. Several attentions have been paid to CNTs over the past two decades, This is due to nothing but unique mechanical, thermal, electrical properties among other key applications in many fields of science and technology (DeVolder, et al.,2013; McEuen, et al. 2002; Terrones, 2003; Zhang, 2013).

Carbon blacks and typically, acetylene black, have been deeply investigated electrochemically (Kinoshita and Bett, 1973) and gained applications in many areas, like catalysis, energy storage and fuel cell technology (Zhang, et al., 2010). An effective interaction (oxidation-reduction active species) is owned to a large surface area of carbon (particle size 1-50 nm, commercially available). Although carbon is not well defined structurally like CNT, graphene materials or nano-onions but their activities in carbon-based nano-sensor, polymer nanocomposites (selective shape sensing), fuel cell and energy storage are very excellent (Iijima, 1991; Molina-Ontoria, et al.,2013; Novoselov, et al., 2014 ; Loo, et al., 2013; Li, 2009; Blanco-L_pez, et al., 2004). The correct choice of nanocarbon substrate is very vital in electrocatalysis and fuel cells, the effects of this have been critically reviewed (Dribinskii, et al., 1989). Pt– carbon nanomaterial composites have found application in fuel cells that are characterized by CO_2 reduction (Perathoner, et al., 2007). With adequate optimization for stable operation conditions, carbon blacks give allowance for gas flow into catalysts' reactive layers.

Carbon With Single Nanometal

Carbon possesses ordered structure which provides its framework with good porosity. However, their uses are limited due to poor solubility. Thus, functionalized nanocomposites of metals and carbons are achieved to improve the properties of the materials. For example, the increasing development of CNTs chemistry makes possible the preparation of hybrid materials of metals and CNT with enhanced performance (Ebbesen et al., 1996; Ding et al., 2012; Wildgoose et al., 2006; Wu et al., 2011; Li et al., 2011; Eder, 2010). These hybrid materials can function as chemical sensors due to their high conductive (electrical) power and selective chemical nature. Another vital property they possess is a reasonably large surface area which provides them with good catalyst sites in reaction catalysis (Ding et al., 2012). Nanomaterials of low-dimensions can now be used for electronic applications, chemical sensors, adsorption (desulphurization) e.t.c (Geim A K, 2009). In the same manner, graphene supported metal nanoparticles also perform well (Huang et al., 2012; Bai and Shen, 2012).

Metal-carbon interactions play critical roles in catalytically growing carbon nanotubes and graphene via chemical vapor deposition (CVD) as well as in controlling the structures of these carbon allotropes and, consequently, have been intensively investigated in order to elucidate the catalytic mechanisms(Jourdain and Bichara, 2013; Mattevi, 2011; Li, 2010; Ding, 2008; Edwards, 2013; Seah, 2014; Cheng and Guo, 2002). Fe and Cu are two metal systems exhibiting distinct interactions with carbon (Naidich, 1981). With partially filled 3d orbitals, Fe exhibits a high affinity for carbon. Carbon has a finite solubility in Fe at high temperatures (0.022% carbon by weight in α-Fe at 727 °C and 2.14% carbon by weight in γ-Fe at 1147 °C) and forms metastable iron carbide (Fe_3C, 6.67% carbon by weight) with Fe (Jourdain and Bichara, 2013; Mattevi, 2011).

Due to its strong interactions with carbon, Fe was among the first catalysts and has been most intensively used for CVD growth of single-wall carbon nanotubes (SWCNTs). In contrast, with completely filled d-orbitals, Cu displays a low affinity for carbon. The carbon solubility in Cu is much lower than in Fe (~ 0.008% carbon in Cu by weight at 1085 °C) and there is no carbide formation (Jourdain and Bichara, 2013; Mattevi, 2011). Due to the weak Cu-C interactions, Cu was once considered to be unsuitable for CVD growth of carbon nanotubes (Deck and Vecchio, 2006).

However, this notion did not stop the exploration of using Cu as the catalyst. In fact, it has been demonstrated that Cu can catalyze the growth of SWCNTs (Yoshida et al., 2009; Zhou, 2006) and, furthermore, it is a better catalyst for developing SWCNT system using quartz and silicon as substrates (Li et al., 2010 and Cui et al., 2010). In terms of the metal-catalyzed CVD growth of graphene, the different interactions with carbon between Fe and Cu result in the formation of few-layer graphene (FLG) on Fe foil and monolayer graphene on Cu foil (Li et al., 2009). With the formation of methane, carbon hydrogenation reverses the reaction for CVD growth of carbon nanotubes and graphene using methane as the carbon source. It was demonstrated that Fe is an active catalyst in producing methane and in etching channels in graphite. Though an initial study showed that Cu was inactive in catalyzing the hydrogenation of graphite, Cu was later found to be active in catalytically etching channels in graphite (Baker, 1981 and 1995)

Metals of transition series are good options for growing graphene or CNTs. The metal-carbon system can be built by performing the growth and contacting in a single step with the same metal. But a pressing challenge here is oxide layer formation on the exposure of nanoparticles of metal to air as it tarnishes and form oxide coating. The formation of the oxide coatings is largely due to the large surface-volume ratio which translates to the high oxidation process. The oxidation process at the metal surface can, however, be prevented in the procedure (Geim AK, 2009).

Carbon-Based Nanocomposites

The good surface chemistry, stable structure, and very large surface area are factors enabling the fine surface templating property of graphitic supports (CNT, CNS and graphene) and their decoration with metal nanoparticles (Iwamoto and Grimblot, 1997). Figure 2 depicts an example of SEM image for nano-structured CNTs. The composites formed show enhanced or even synergistic activity in their applications (Salem and Hamid, 1997). The surface chemistry of the composites, however, is not fully established and still open to further studies (Nagai et al., 2000).

Carbon nanomaterial composites made by decorating carbon (graphene, carbon nanospheres, and carbon nanotubes) with transition metals (oxides or salts) is receiving great attention. Carbon-supported metal nanoparticles (MNPs) for example; cobalt, iron, or nickel based nanoparticles (CoOx–CoP/C or FeOx–CoP/C) have been confirmed to be catalytically active for oxygen reduction reaction (ORR); this is due to the nature of the transition metals oxide nanoparticles (Stephanie, et al., 2010). The confirmation of this great activity is shown in the less activity of cobalt porphyrin (CoP/C) when it's adsorbed on unmodified carbon. Sometimes a synergistic activity is observed (Salem, 1997).

In the investigation of the potency of activated carbon materials for desulfurization and denitrogenation gas oils, MAXSORB-II adsorbent, is an effective adsorbent for organosulfur compounds' adsorption at ambient temperature (Mochida et al, 2004a and 2003c). This material can be prepared by activating carbon prepared from petroleum coke with KOH. Oxidizing agents like HNO_3 and H_2O_2 were also used to further enhance the adsorptive power of commercially available activated carbon which is then heat-treated (Mochida et al., 2004b, 2003d). These oxidizing

Figure 2. Examples of SEM image of nano-structured CNTs

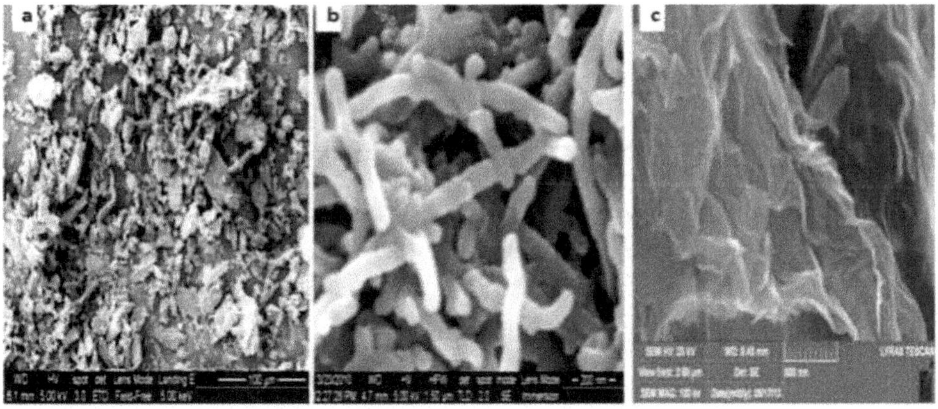

agents aid the release of CO (600-800 ^{0}C) at the carbon surface which thus enhances the absorption power.

Methods for the Carbon Nanostructures production

There are some advanced techniques, Figure 3, to prepare carbon nanomaterials such as CNT, CNS, CNF, graphene (Iijima,1991;Yudasaka et al.,1997; Li et al.,1996). These methods include:

- Mechanical methods such as ball milling
- Chemical methods such as sol gel method, thermal methods
- Electrochemical methods such as electro reduction, electro oxidation, etc.
- Physical methods such as laser ablation
- Arc discharge
- Chemical vapor deposition (CVD).

Other techniques that are commercially less developed compared to the above-mentioned three, which involve the use of flame, solar, electrolysis in their syntheses (Yuan et al., 2001;Hsu et al.,1996; Laplaze et al., 1998).

In the arc discharge technique, with a space of less than 1 mm between graphite electrodes, a direct current is passed under the inert environment argon at low pressure. Figure 4 shows an illustration of arc discharge techniques.

Figure 3. Techniques used in preparation of carbon structures

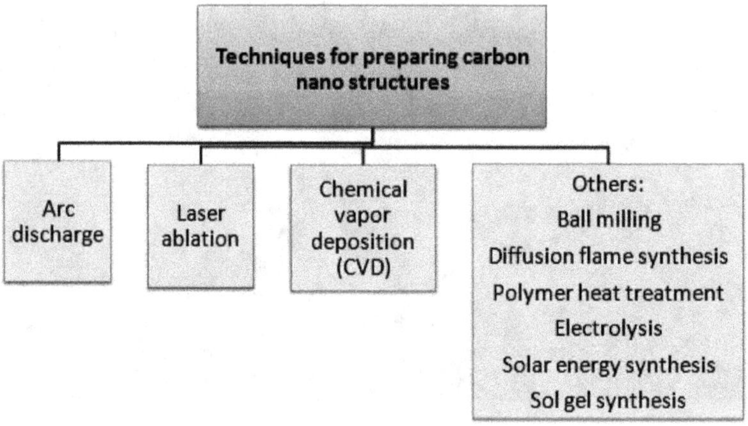

The carbon anode vaporizes on the generation of plasma of very high temperature on the application of current and the carbon material is deposited alongside other by-products of carbon at the cathode (Popov et al., 2004; Journet et al., 1997). The production of MWNTs with diameters ranging from 2 to 20 nm and micrometer-scaled length can be achieved (Ebbesen and Ajayan, 1992). Similarly, SWNTs of above 70% yield and diameter of close to 1.4 nm can be prepared. In the laser ablation technique, the principle is not so different from the arc discharge. The heating is done in a furnace with the aid of pulsed laser source under inert surrounding. In this

Figure 4. Schematic representation of arc discharge technique for the synthesis of carbon nanomaterials

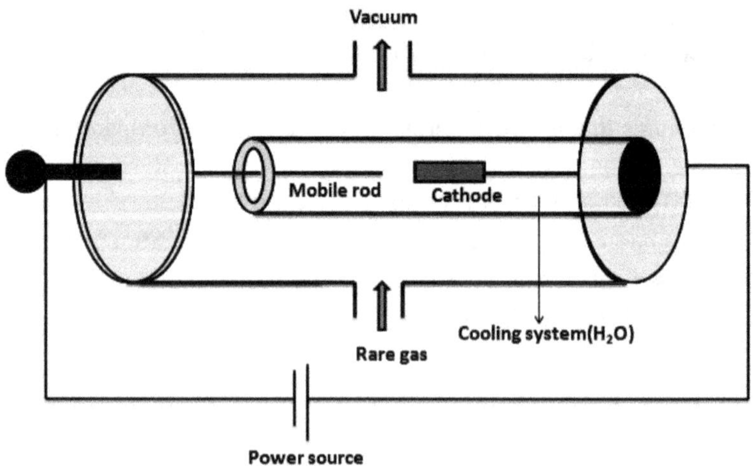

Figure 5. Schematic representation of Oven laser vaporization apparatus for carbon production

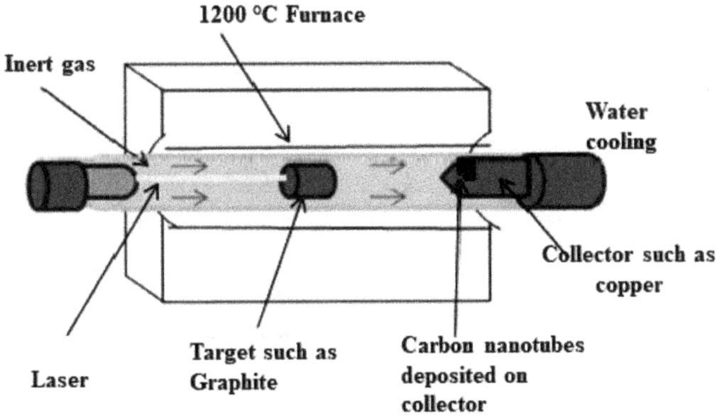

technique, high-temperature carbon vapor is generated from the surface of the graphite as illustrated in figure 5 (Guo et al. 1995; Journet et al., 1997). The carbon vapor is transported by the flowing helium or argon and then collected on copper-collector.

These first two techniques produce high-quality carbon nanomaterials in a large amount but yet with some shortcomings, however, have been identified; the reliance on high-temperature graphitic surface evaporation of carbon which is not cost-effective as this leads to high energy consumption and the adhering impurities which will require further purification of the carbon nanomaterials produced (Popov et al., 2004).

In view of the above challenges of the two techniques mentioned previously, chemical vapor deposition is the most commonly used technique. Chemical vapor deposition (CVD), Figure 6, ensure better control of carbon (e.g. CNT) growth than laser ablation and arc discharge, this makes it a more feasible method (Dai, 2002). It makes use of hydrocarbons as precursors in the presence of metal catalysts at a temperature ranging from 500 to 1000 ^{0}C. The hydrocarbons get decomposed at this relatively low temperature and thus the growth of CNTs as the system cools. This technique is simpler in that lower temperature is involved because the precursor is not solid thus consumes less energy (Journet, 1997).

Of a vital advantage of CVD worthy of mentioning is that it's very versatile. Virtually all the major components involved in the growth can be switched; catalyst and precursor can be in liquid or solid form, the particular precursor to being used, the possibility of incorporating other materials like nitrogen, temperature and pressure of the reaction, etc. (Nikolav et al., 1999). The choice of catalyst plays an important role in decomposition and dispersion of liquid hydrocarbon as a precursor to allowing in situ decomposition and possible better dispersion if liquid hydrocarbon is used

Figure 6. Schematic illustration Chemical vapor deposition (CVD)

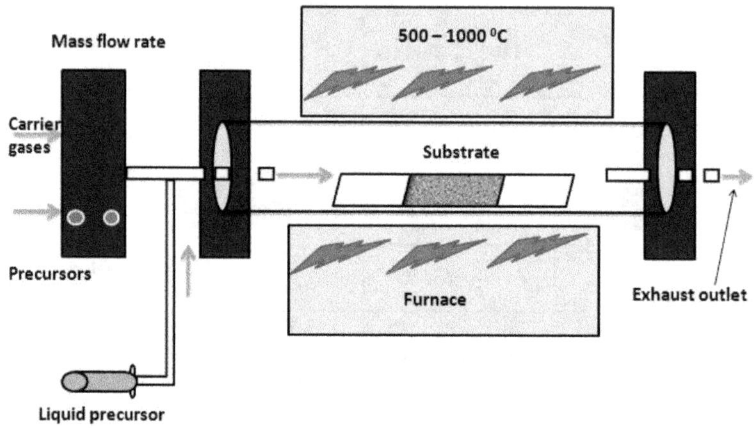

or in solid form for large scale synthesis if loaded on large surface area template as well as a well patterned growth with nanoparticles (Allen et al., 2009; Hart and Slocum, 2006).

Graphene nanocarbon is commonly prepared by four techniques (Anton et al., 2007; EMTSEV, 2009; Reina, 2009; Lomeda, 2008) listed below:

- Direct sonication of graphite or scoth tape approach (known as mechanical cleavage of graphite)
- Growth on SiC (epitaxial)
- Metal substrates CVD
- Graphite oxide reduction

Some other approaches being reported are; by an electrochemical process using graphite electrode and unwrapping of carbon nanotubes (Kosynkin et al., 2009; Liu et al., 2008). The most common approach for large scale preparation is the reduction of graphite oxide by, firstly, oxidize graphite with acids then convert the oxide formed to graphene by reduction. The reduction process can be carried out thermally, chemically or photocatalytically (McAllister et al., 2007; Lomeda et al., 2008; Williams et al).

Figure 7. Schematic representation of a graphene sheet with unit hexagonal lattice vectors (a_1 and a_2)

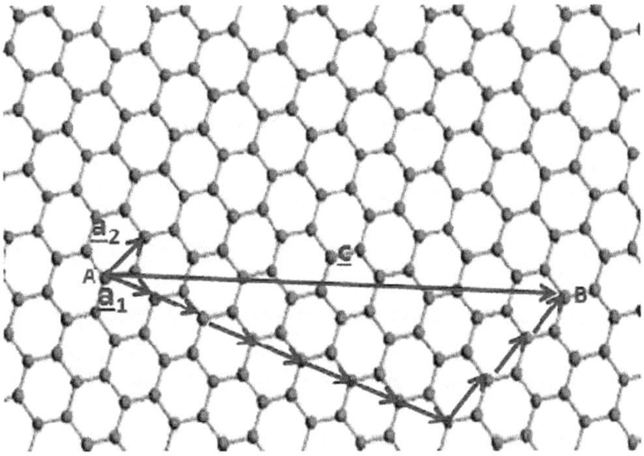

Properties

Carbon nanomaterial, graphene, for example, is a monoatomic layer of carbon atoms that is sp^2 hybridized (Geim and Novoselov, 2007). The good electrical properties in ballistic transport of electrons can be traced to its 2D pattern (Novoselov et al., 2004). Graphene also possesses large surface-volume ratio, fast electron transfer mechanism, good tensile strength and very interesting elastic character (Park and Ruoff, 2009; Rao et al., 2009; Yang et al., 2010). A very good stacking is possible with graphene, thus, layers can be varied from single to many of up to 10 layers and the resulting materials are equally interesting.

Multi-walled carbon nanotubes (MWNTs) comprise several single-walled nanotubes Single-walled carbon nanotubes (SWNTs). The one-dimensional SWNTs like graphene sheet (0.4 to 2.0 nm diameter) are rolled-up to form MWNTs (2 to 100 nm diameter) (Baughman et al., 2002; Yang et al., 2010). Largely, MWNTs can posses several surface morphologies such as hollow tubes, stacked cups, carbon filaments etc (Kiselev et al.,1998; Iijima, 1991; Delgado et al., 2008; Allen et al., 2008).

The classification of SWNTs can be done based on chirality vectors ($\mathbf{c}_h = n\mathbf{a}_1 + m\mathbf{a}_2 = (n, m)$) (Avouris, 2002) as shown in figure 7. The following are few classes:

- Zig-zag (m = 0)
- Arm-chair (n = m)
- Chiral (n ≠ m)

The particular class and vectors provide information about electronic properties; if (n – m) gives a multiple of 3, then the SWNTs of arm-chair type is metallic.

Depending on the diameter of SWNTs which have inverse proportionality with a band gap, SWNTs are good semiconductors and also have metallic character. Normally, MWNTs also have similar characteristics with SWNTS because the effect of a cylinder on its adjacent cylinder is very minute (Baughman et al.,2002; Charlier, 2002). However, the lack of definite control of chiralities for individual nano-cylinders results into MWNTs with SWNTs of different chiralities, thus a metallic property similar to graphite (turbostratic) will be observed (Cao, 2004).

Applications in the Desulfurization Process

Electrochemically, graphene and CNTs are very interesting (Pumera, 2009 and 2010). The very unique dimensions of CNTs (1-D) and graphene(2-D) are enabling features for their outstanding electrical conductivity, stability, and good surface chemistry for easy decoration with desired functionalities which in turn guaranteed their electrochemical applications. Such applications are; energy generation via fuel cells and photovoltaic (Brandon et al., 2003; Kauffman et al., 2010), energy storage (batteries, supercapacitors, hydrogen storage) (An et al., 2001) field-effect transistors (Katz, 2004; Kruger et al., 2001) sensors and biosensors (Ronkainen et al.,2010; Yang et al., 2010). However, there are challenges, like the non-uniformity of materials and their hybrids (Jacobs et al., 2010), lack a detailed understanding of structure as related to property and possible contaminants from synthetic precursors/catalysts(Dai, 2002; Kruusma,2007; Pumera and Miyahara, 2009). These are tasking challenges but an avenue for further detailed studies.

In Batch System

Adsorptive desulfurization studies of organosulfur compound removal from naphtha (550 ppmS) feedstocks can be carried out in a batch system. The use of adsorbents such as zeolites (5A and 13X) and activated carbon is effective. These two adsorbents (carbon and zeolites) demonstrated a unique performance in their adsorption capabilities; activated carbon has larger capacity but the percentage of sulfur removal (65% at 80 0C) is lower compared to the zeolitic adsorbent with relatively smaller capacity but a larger percentage of sulfur (100% at room temperature and 800g adsorbent/L feed) can be removed. In line with these observations and for

Figure 8. classification of desulfurization processes based on the nature of the process

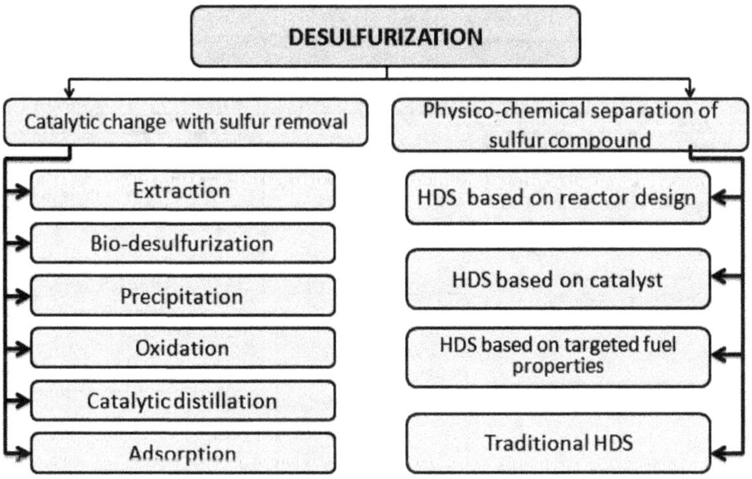

industrial application, a two-bed approach gives a better desulphurization process (Salem and Hamid, 1997).

In the Fixed Bed System

Adsorption studies in a fixed bed set-up can be achieved also with the use of different adsorbents. Alumina, carbon-supported nickel, as well as activated carbon, can to a large extent remove sulfur and nitrogen from diesel feed with aromatics, nitrogen and sulfur compounds. Activated carbon shows better selectivity in removing refractory sulfur and nitrogen compounds especially with alkylated dibenzothiophene (4,6-DMDBT) (Kim et al., 2006; Ko et al.,2007). A superior selectivity for nitrogen removal is observed with alumina as compare to the other two adsorbents. This selectivity with alumina is great when basic nitrogen compound is involved in hydrocarbon streams.

Types of Process

There is no generalized classification for desulfurization techniques. However, some classification can be made based on the involvement of hydrogen, what happened to sulfur compound during desulfurization and whether desulfurization occurs physically or chemically or both (Babich and Moulijn, 2003). Considering the change that happened to sulfur compounds; three situations are possible; either the sulfur compounds get decomposed and removed, removed from process stream without decomposing or decomposed only.

Figure 9. classes of desulfurization process according to the changes that occur with sulfur compounds

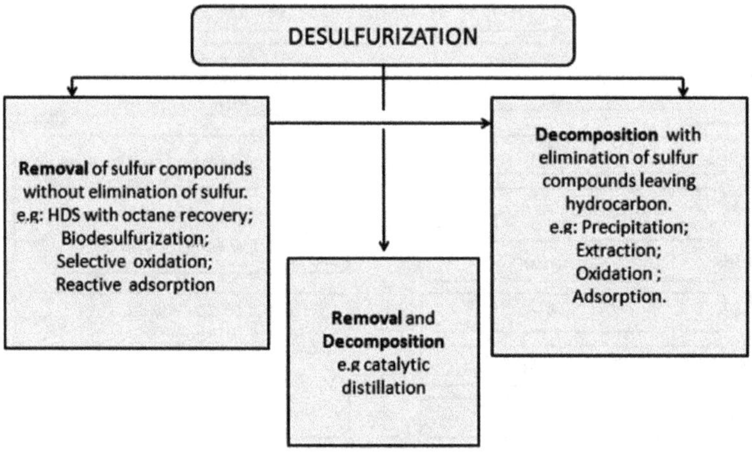

From the perspective of the hydrogen involvement; a broad classification desulfurization process can be made; hydrodesulfurization (HDS)-based and non HDS-based groups (Babich and Moulijn, 2003). Hydrogen is involved in the decomposition of sulfur compounds for all HDS-based processes while the non HDS-based processes do not involve hydrogen for desulfurization to occur. But, to eliminate sulfur, in most cases, hydrogen is involved with few exceptions as in the case of selective oxidation. Another way of classifying desulfurization processes is whether it occurs physically or chemically as shown in figure 8.

Catalytic conversion of sulfur compounds and subsequent removal of sulfur is a well-established desulfurization process. Such processes are hydroprocessing, distillation, extraction, adsorption, etc. Their different combinations have been identified as promising. Figure 9 depicts classes of desulfurization process according to the changes that occur with sulfur compounds.

During the process of desulfurization, sulfur compounds, for instance, benzothiophene or dibenzothiophene can be removed from the stream and then decomposed separately in a vessel. A very low level of sulfur in transportation fuels can be achieved through this process. This process provides insight for removing sulfur compounds in fuels through distillation with the help of catalysts. It's also possible that some sulfur products (solid or gas) are generated with just decomposition of organosulfur compounds with only the hydrocarbons (without sulfur) remaining in the process stream (an example is hydrodesulfurization). The last class is just a simple removal of organosulfur compounds from the process stream. This can be done by conversion of the organosulfur compounds to different compounds that can be removed easily if direct removal is difficult or impossible; however, disposal of the removed sulfur compounds becomes a great challenge for this type of class.

Extractive Desulfurization

Extractive desulfurization employs the use of solubility as the key factor for organosulfur removal. Organosulfur compounds tend to be more soluble in some specific solvents. The sulfur compounds, due to this reason, are transferred from oil feed into the solvent and therefore hydrocarbons are easily separated with the aid of a separator. The sulfur compound free hydrocarbons can now be further charged for the next process or discharged as a final product. The mixture of organosulfur compounds and the solvent used for its extraction will then be subjected to distillation thus solvent is recovered and can be re-used.

Depending on the number of extraction cycles, several solvents have been identified to be suitable for this purpose with sulfur compound removal of up to 90%. Acetone, glycols, nitrogenous-solvent (Horri, 1996), ethanol are good solvents for extractive desulfurization because they have their boiling points different from those of organosulfur compounds and relatively cheap (Forte, 1996; Funakoshi and Aida, 1998).

Desulfurization by Precipitation

Desulfurization can also be achieved via precipitation by forming an insoluble complex of organosulfur compounds and them eliminate them. When a pi acceptor (e.g. tetranitro fluorine) is missed with alkylated-DBT(e.g.4,6-DMDBT) in fuels, an insoluble complex is formed which can be conveniently removed by filtration and the tetranitro fluorine is recovered by adsorption process with the aid of suitable adsorbent(Meille et al., 1997; Milenkovic et al., 1999). This approach does not remove a significant amount of sulfur (< 25%) as some other hydrocarbons in the feed compete in complex formation (Hernandez-Maldonado et al. 2004). The detailed studies of the process are yet to be done.

Bio-Desulfurization (BDS)

This process involves the use of micro-organisms (e.g. bacteria) for removal of organosulfur compound. In this process, the sulfur is removed from its parent compound under mild condition and with the availability of oxygen and water without degradation of carbon structure present. One very vital feature of this process is that it does not require hydrogen thus a cheaper process than HDS.

Selective Oxidative Desulfurization

Desulfurization by selective oxidation involves oxidation of organosulfur compounds and then purification (Aida et al., 2000). Processes like sulphCo and photochemical desulfurization employ oxidation as a vital operation. Organosulfur compound can be oxidized selectively and processes like adsorption, distillation or decomposition by heating can subsequently be used to separate the oxidized oganosulfur compounds from the process stream.

The combination of distillation with selective oxidation is not reported in any literature till date. Considering the principle involved, the approach is promising since sulfones and sulfuroxides are the products of oxidation of organosulfur compounds and they increase their boiling points. This process can, therefore, be likened to the conventional distillation process provided that the only separation of organosulfur compounds is desired and subsequent treatment will be done separately. However, the formation of carbonmonoxide and carbon (iv) oxide as by-products and process safety are major issues.

Adsorptive Desulfurization

This is the use of adsorbents for removing sulfur compounds in the fuels. The efficiency of this process lies in the ability of the adsorbents to selectively remove the target sulfur compounds from oil processing stream, durability and regenerability of adsorbent and capacity for adsorption. Desulfurization via adsorption (ADS) can be classified into two based on the nature of the interaction of the sulfur compounds and the adsorbent: physical adsorption where organosulfur compounds only adhere to the surface of the solid adsorbent. Adsorbent regeneration is done by passing a suitable solvent on the adsorbent or heating to get rid of adhered sulfur compounds. The second is chemical (reactive) adsorption where organosulfur compounds are adsorbed to the surface of the adsorbent and react with it. Sulfur is thus eliminated in the form of elemental sulfur, SO_2, SO_X or H_2S. Several good materials have been used as adsorbents, namely; activated carbon, transition metals supported on zeolites and many metal oxides (Ma et al., 2002 and 2005; Velu et al., 2003; Kim et al., 2006).

Carbon-based adsorbents can selectively remove nitrogen compounds from hydrocarbon stream (Almarri et al., 2009). Very interesting adsorbent silica-supported nickel phosphides is also very efficient for adsorptive desulfurization of diesel fuel to a very low sulfur level (Landau et al., 2009). Enhanced performance can be achieved with combined adsorption and HDS (Landau et al., 2008) for denitrogenation and desulfurization as illustrated in figure 10.

Figure 10. An illustration of combined adsorption and hydrodesulfurization (two-step) process

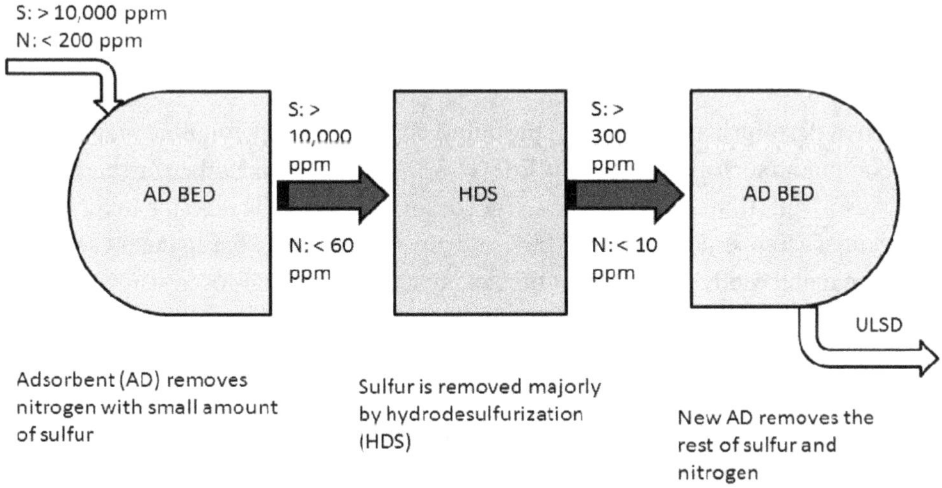

Both denitrogenation and desulfurization can be done simultaneously using activated carbon for the adsorption process. For every 1 gram of adsorbent, 0.04 g and 0.1 g of nitrogen and sulfur can be removed from fuel respectively at a temperature close to the room (Sano et al., 2004a; 2004b). Nitrogen and organosulfur compounds are removed in the first stage of the two-step process. The only very little amount of

Figure 11. Simplified adsorptive desulfurization process flow

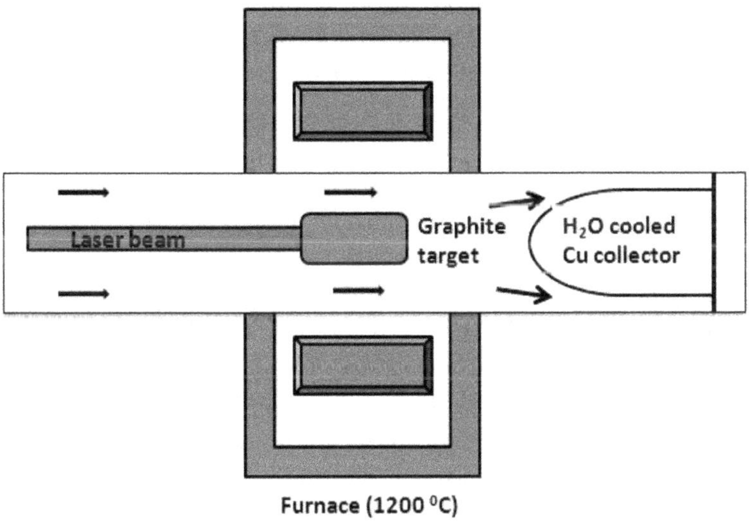

sulfur is removed at this stage but nitrogen is removed to a large extent. The HDS step removes most of the sulfur and the final adsorption step gets rid of remnant nitrogen and sulfur present.

Chemical (Reactive) Adsorption

In reactive desulfurization process, the sulfur in the fuel feed combine and react to form compound such as hydrogen sulfide (H_2S) which is then adsorbed on the surface of the adsorbent (figure 11). For diesel fuel desulfurization, the reaction to form H_2S can occur at a temperature of 500-700K and pressure of 19-35 bar in the presence of transition metal catalyst. In view of this, many hybrid materials of transition metals supported on carbon, zeolites and several other solid porous materials have been prepared (Slater et al.,2002; Park et al., 2008; Landau et al., 2008).

The concentration of organosulfur compounds is only decreased in by adsorptive desulfurization processes. A further downstream operation like hydroprocessing needs to be performed at high pressure in order to remove sulfur to achieve ultra-low diesel fuel. For commercial application, certain process parameters have to be set at optimum points. These parameters include but not limited to; fuel feed to adsorbent ratio, the particle size of the adsorbent, adsorption cycles, the temperature of reactivation etc.

Summary and Outlook

Recently, there has been a remarkable growth in research on achieving an ultra low sulfur level in transportation fuels. Several research groups from different academic institutions and industrial research laboratories have published a great number of rich scientific papers. This special interest in ultra-low sulfur diesel (ULSD) is informed by the need to comply with stricter government policy on the low sulfur content of transportation fuels; better knowledge of the different factors that concern deep desulfurization of diesel to achieve lower sulfur content; the cheaper way of producing ULSD etc. The ULSD is true, will be the global product that will dominate the market. Lower harmful exhaust emissions can easily be achieved with ULSD thereby enhancing air quality. The desulfurization of least active dibenzothiophene (DBT) derivative like 4, 6-DMDBT (sterically hindered) can now be easily achieved by optimizing certain factors such as; kinetics and thermodynamics, feed quality, inhibitory effects, catalytic sites, etc. In the future, the need to produce more volume ULSD from low-grade feedstocks like heavy oil and light cycle oil to meet up with the global ULSD demand is pertinent. Other needs like better calorific value, reduced aromatics content and density are also worthy of expectation. These, however, are challenging but with enormous economic gain. In order to achieve this great milestone, adsorptive method of desulfurization can play a very big role. ADS does not require hydrogen so it's cheap. It is capable of desulfurizing organ-sulfur compound (e.g. DBT) preferentially. The ADS can be used to achieve a sulfur level of less than 10ppm from initial sulfur content of 500ppm in diesel fuel.

ACKNOWLEDGMENT

The author would like to thank the editor, Dr. Saleh, for their assistance in the writing of this chapter.

REFERENCES

Agarwal, P., & Sharma, D. K. (2010). Comparative studies on the biodesulfurization of crude oil with other desulfurization techniques and deep desulfurization through integrated processes. *Energy & Fuels*, *24*(1), 518–524. doi:10.1021/ef900876j

Aida, T., Yamamoto, D., Iwata, M., & Sakata, K. (2000). Development of oxidative desulfurization process for diesel. *Review on Heteroatom Chemistry*, *22*, 241–256.

Allen, B. L., Kichambare, P. D., & Star, A. (2008). Synthesis, characterization, and manipulation of nitrogen-doped carbon nanotube cups. *ACS Nano*, *2*(9), 1914–1920. doi:10.1021/nn800355v PMID:19206432

Allen, B. L., Shade, C. M., Yingling, A. M., Petoud, S., & Star, A. (2009). Graphitic nanocapsules. *Advanced Materials*, *21*, 4692–4695.

Almarri, M., Ma, X., & Song, C. (2009). Role of surface oxygen-containing functional groups in liquid-phase adsorption of nitrogen compounds on carbon-based adsorbents. *Energy & Fuels*, *23*(8), 3940–3947. doi:10.1021/ef900051r

An, K. H., Kim, W. S., Park, Y. S., Moon, J.-M., Bae, D. J., Lim, S. C., ... Lee, Y. H. (2001). Electrochemical Properties of High-Power Supercapacitors Using Single-Walled Carbon Nanotube Electrodes. Journal of Advanced Functional. *Materials (Basel)*, *11*, 387–392.

Anton, N. S., Mehdi, M. Y., Romanch, J., Ouseph, P. J., Cohn, R. W., & Sumanasekera, G. U. (2007). Electrostatic deposition of graphene. *Nanotechnology*, *18*, 13530–13534. PMID:21730375

Avouris, P. (2002). Molecular electronics with carbon nanotubes. *Accounts of Chemical Research*, *35*(12), 1026–1034. doi:10.1021/ar010152e PMID:12484790

Babich, I. V., & Moulijn, J. A. (2003). Science and technology of novel processes for deep desulfurization of oil refinery streams: A review. *Fuel, 82*(6), 607–631. doi:10.1016/S0016-2361(02)00324-1

Bai, S., & Shen, X. (2012). Graphene-inorganic nanocomposites. *Royal Society of Chemistry Advances, 2*, 64–98.

Baker, R. T. K. (1995). A review of in-situ electron microscopy studies of metal/metal oxide graphite interactions. *Journal of Adhesion and Adhensives, 52*(1-4), 13–40.

Baker, R. T. K., & Chludzinski, J. J. Jr. (1981). Catalytic gasification of graphite by chromium and copper in oxygen, steam and hydrogen. *Carbon, 19*(2), 75–82. doi:10.1016/0008-6223(81)90110-X

Baughman, R. H., Zakhidov, A. A., & de Heer, W. A. (2002). Carbon nanotubes--the route towards applications. *Science, 297*(5582), 787–792. doi:10.1126cience.1060928 PMID:12161643

Bernier, P., & Ajayan, P. M. (1994). Application of Carbon Nanotubes as Supports in Heterogeneous Catalysis. *Journal of the American Chemical Society, 116*(17), 7935–7936. doi:10.1021/ja00096a076

Bhatia, S., & Sharma, D. K. (2006). Emerging role of bio-refining of heavier crude oils and integration of bio-refining with petroleum refineries in the future. *Petroleum Science and Technology, 24*(10), 1125–1159. doi:10.1080/10916460600817447

Blanco-Lopez, M. C., Gutierrez-Fernandez, C., Lobo-Casta, Ç. M. J., Miranda-Ordieres, A. J., Tucon-Blanco, P. (2004). Electrochemical sensing with electrodes modified with molecularly imprinted polymer films. *Analytical and Bioanalytical Chemistry, 378*(8), 1922 –1928.

Brandon, N. P., Skinner, S., & Steele, B. C. (2003). Recent advances in materials for fuel cells. *Annual Review of Materials Research, 33*(1), 183–213. doi:10.1146/annurev.matsci.33.022802.094122

Cao, G. (2004). *Nanostructures and Nanomaterials: Synthesis, properties and applications*. Imperial College. doi:10.1142/p305

Charlier, J. C. (2002). Defects in carbon nanotubes. *Accounts of Chemical Research, 35*(12), 1063–1069. doi:10.1021/ar010166k PMID:12484794

Cheng, G., & Guo, T. (2002). Surface segregation in Ni/Co bimetallic nanoparticles produced in single-walled carbon nanotube synthesis. *The Journal of Physical Chemistry B, 106*(23), 5833–5839. doi:10.1021/jp025530r

Cui, R. L., Zhang, Y., Wang, J. Y., Zhou, W. W., & Li, Y. (2010). Comparison between copper and iron as catalyst for chemical vapor deposition of horizontally aligned ultralong single-walled carbon nanotubes on silicon substrates. *Journal of Physical Chemistry*, *114*(37), 15547–15552.

Dai, H. (2002). Carbon nanotubes: Opportunities and challenges. *Surface Science*, *500*(1-3), 218–241. doi:10.1016/S0039-6028(01)01558-8

Daimler–Chrysler, A.G. (2000). *Method for desulfurization engine fuel on board a motor vehicle*. W.O 2000/20531.

Deck, C. P., & Vecchio, K. (2006). Prediction of carbon nanotube growth success by the analysis of carbon-catalyst binary phase diagrams. *Carbon*, *44*(2), 267–275. doi:10.1016/j.carbon.2005.07.023

Delgado, J. L., Herranz, M. A., & Martin, N. (2008). The nano-forms of carbon. *Journal of Materials Chemistry*, *18*(13), 1417–1426. doi:10.1039/b717218d

DeVolder, M. F. L., Tawfick, S. H., Baughman, R. H., & Hart, A. J. (2013). Carbon nanotubes: Present and future commercial applications. *Science*, *339*(6119), 535–539. doi:10.1126cience.1222453 PMID:23372006

Ding, F., Larsson, P., Larsson, J. A., Ahuja, R., Duan, H. M., Rosen, A., & Bolton, K. (2008). The importance of strong carbon-metal adhesion for catalytic nucleation of single-walled carbon nanotubes. *Nano Letters*, *8*(2), 63–68. doi:10.1021/nl072431m PMID:18162001

Ding, M., Tang, Y., & Star, A. (2012). Understanding Interfaces in Metal–Graphitic Hybrid Nanostructures. *The Journal of Physical Chemistry Letters*, *4*(1), 147–160. doi:10.1021/jz301711a PMID:26291227

Dribinskii, A. V., Tarasevich, M. R., & Kazarinov, V. E. (1989). Electrocatalysis on carbon materials. *Materials Chemistry and Physics*, *22*(3-4), 377–400. doi:10.1016/0254-0584(89)90006-0

Ebbesen, T. W., & Ajayan, P. M. (1992). Large scale synthesis of carbon nanotubes. *Nature*, *358*(6383), 220–222. doi:10.1038/358220a0

Ebbesen, T. W., Hiura, H., Bisher, M. E., Treacy, M. M. J., Shreeve-Keyer, J. L., & Haushalter, R. C. (1996). Decoration of carbon nanotubes. *Advanced Materials*, *8*(2), 155–157. doi:10.1002/adma.19960080212

Eder, D. (2010). Carbon Nanotube–Inorganic Hybrids. *Chemical Reviews*, *110*(3), 1348–1385. doi:10.1021/cr800433k PMID:20108978

Edwards, R. S., & Coleman, K. S. (2013). Graphene film growth on polycrystalline metals. *Accounts of Chemical Research, 46*(1), 23–30. doi:10.1021/ar3001266 PMID:22891883

Emtsev, K. V., Bostwick, A., Horn, K., Jobst, J., Kellogg, G. L., Ley, L., ... Seyller, T. (2009). Towards wafersize graphene layers by atmospheric pressure graphitization ofsilicon carbide. *Nature Materials, 8*(3), 203–207. doi:10.1038/nmat2382 PMID:19202545

Forte, P. (1996). *Process for the Removal of Sulfur from Petroleum Fractions*. US Patent, 5582714.

Funakoshi, I., & Aida, T. (1998). *Process for recovering organic sulfur compounds from fuel oil*. US Patent, 5753102.

Geim, A. K. (2009). Graphene: Status and prospects. *Science, 324*(5934), 1530–1534. doi:10.1126cience.1158877 PMID:19541989

Geim, A. K., & Novoselov, K. S. (2007). The rise of graphene. *Nature Materials, 6*(3), 183–191. doi:10.1038/nmat1849 PMID:17330084

Guo, T., Nikolaev, P., Thess, A., Colbert, D. T., & Smalley, R. E. (1995). Catalytic growth of single-walled nanotubes by laser vaporization. *Chemical Physics Letters, 243*(1-2), 49–54. doi:10.1016/0009-2614(95)00825-O

Hart, A. J., & Slocum, A. H. (2006). Rapid growth and flow-mediated nucleation of millimeter-scale aligned carbon nanotube structures from a thin-film catalyst. *The Journal of Physical Chemistry B, 110*(16), 8250–8257. doi:10.1021/jp055498b PMID:16623503

Hernandez-Maldonado, A. J., & Yang, R. T. (2004b). Desulfurization of diesel fuels via pi-complexation with nickel(JI)-exchanged X- and Y-zeolites. *Industrial & Engineering Chemistry Research, 43*(4), 1081–1089. doi:10.1021/ie034206v

Hernandez-Maldonado, A. J., & Yang, R. T. (2004c). New sorbents for desulfurization of diesel fuels via pi-complexation. *The American Institute of Chemical Engineers Journal, 50*(4), 791–801. doi:10.1002/aic.10074

Hsu, W. K., Terrones, M., Hare, J. P., Terrones, H., Kroto, H. W., & Walton, D. R. M. (1996). Electrochemical formation of carbon nanostructures. *Chemical Physics Letters, 262*(1-2), 161–166. doi:10.1016/0009-2614(96)01041-X

Huang, X., Qi, X., Boey, F., & Zhang, H. (2012). Graphene-based composites. *Chemical Society Reviews, 41*(2), 666–686. doi:10.1039/C1CS15078B PMID:21796314

Iijima, S. (1991). Helical microtubules of graphitic carbon. *Nature*, *354*(6348), 56–58. doi:10.1038/354056a0

Iwamoto, R., & Grimblot, J. (1997). Acidity and hydrogenation properties of Mo-P-alumina catalysts prepared by a sol-gel method. *Journal of Catalysis*, *172*(1), 252–255. doi:10.1006/jcat.1997.1855

Jacobs, C. B., Peairs, M. J., & Venton, B. J. (2010). Carbon nanotubes electrochemical sensors for biomolecules. *Analytica Chimica Acta*, *662*(2), 105–127. doi:10.1016/j.aca.2010.01.009 PMID:20171310

Jourdain, V., & Bichara, C. (2013). Current understanding of the growth of carbon nanotubes in catalytic chemical vapour deposition. *Carbon*, *58*, 2–39. doi:10.1016/j.carbon.2013.02.046

Journet, C., Maser, W. K., Bernier, P., Loiseau, A., de la Chapelle, M. L., Lefrant, S., ... Fischer, J. E. (1997). Large-scale production of single-walled carbon nanotubes by the electric-arc technique. *Nature*, *388*(6644), 756–758. doi:10.1038/41972

Katz, H. E. (2004). Chemically Sensitive Field-Effect Transistors and Chemiresistors: New Materials and Device. Structures. *Journal of Electroanalytical Chemistry*, *16*(22), 1837–1842. doi:10.1002/elan.200403071

Kauffman, D. R., Tang, Y., Kichambare, P. D., Jackovitz, J. F., & Star, A. (2010). Long-Term Performance of Pt-Decorated Carbon Nanotube Cathodes in Phosphoric Acid Fuel Cells. *Energy & Fuels*, *24*(3), 1877–1881. doi:10.1021/cf100013v

Kim, J. H., Ma, X., Zhou, A., & Song, C. (2006). Ultra clean desulfurization and denitrogenation of diesel fuel by selective adsorption over three different adsorbents: A study on adsorptive selectivity and mechanism. *Catalysis Today*, *111*(1-2), 74–83. doi:10.1016/j.cattod.2005.10.017

Kinoshita, K., & Bett, J. A. S. (1973). Potentiodynamic analysis of surface oxides on carbon blacks. *Carbon*, *11*(4), 403–411. doi:10.1016/0008-6223(73)90080-8

Kiselev, N. A., Sloan, J., Zakharov, D. N., Kukovitskii, E. F., Hutchison, J. L., Hammer, J., & Kotosonov, A. S. (1998). Carbon nanotubes from polyethylene precursors: Structure and structural changes as revealed by HREM. *Carbon*, *36*, 1149–1157. doi:10.1016/S0008-6223(98)00092-X

Kosynkin, D. V., Higginbotham, A. L., Sinitskii, A., Lomeda, J. R., Dimiev, A., Price, B. K., & Tour, J. M. (2009). Mechanically Assisted Exfoliation and Functionalization of Thermally Converted Graphene Sheets. *Nature*, *458*, 872–876. doi:10.1038/nature07872 PMID:19370030

Kruger, M., Buitelaar, M. R., Nussbaumer, T., Schonenberger, C., & Forro, L. (2001). Electrochemical carbon nanotube field-effect transistor. *Applied Physics Letters*, *78*(9), 1291–1293. doi:10.1063/1.1350427

Kruusma, J., Mould, N., Jurkschat, K., Crossley, A., & Banks, C. E. (2007). Single walled carbon nanotubes contain residual iron oxide impurities which can dominate their electrochemical activity. *Electrochemistry Communications*, *9*(9), 2330–2333. doi:10.1016/j.elecom.2007.06.024

Kumar, S., Srivastava, V. C., & Badoni, R. P. (2012). Oxidative desulfurization by chromium promoted sulfated zirconia. *Fuel Processing Technology*, *93*(1), 18–25. doi:10.1016/j.fuproc.2011.08.017

Lam, V., Li, G., Song, C., Chen, J., Fairbridge, C., Hui, R., & Zhang, J. (2012). A review of electrochemicaldesulfurization technologies for fossil fuels. *Fuel Processing Technology*, *98*, 30–38. doi:10.1016/j.fuproc.2012.01.022

Laplaze, D., Bernier, P., Maser, W. K., Flamant, G., Guillard, T., & Loiseau, A. (1998). carbon nanotubes: The solar approach. *Carbon*, *36*(5-6), 685–688. doi:10.1016/S0008-6223(98)00025-6

Li, X., Cai, W., An, J., Kim, S., Nah, J., Yang, D., ... Ruoff, R. S. (2009). Large-area synthesis of high-quality anduniform graphene films on copper foils. *Science*, *324*(5932), 1312–1314. doi:10.1126cience.1171245 PMID:19423775

Li, X., Qin, Y., Picraux, S. T., & Guo, Z.-X. (2011). Noncovalent assembly of carbon nanotube-inorganic hybrids. *Journal of Materials Chemistry*, *21*(21), 7527–7547. doi:10.1039/c1jm10516g

Li, Y., Cui, R. L., Ding, L., Liu, Y., Zhou, W. W., Zhang, Y., ... Liu, J. (2010). How catalysts affect the growth of single-walled carbon nanotubes on substrates. *Advanced Materials*, *22*(13), 1508–1515. doi:10.1002/adma.200904366 PMID:20437500

Liu, N., Luo, F., Wu, H., Liu, Y., Zhang, C., & Chen, J. (2008). One-step ionic-liquid-assisted electrochemical synthesis of ionic-liquid-functionalized graphene sheets directly from graphite. *Advanced Functional Materials*, *18*(10), 1518–1525. doi:10.1002/adfm.200700797

Lomeda, J. R., Doyle, C. D., Kosynkin, D. V., Hwang, W. F., & Tour, J. M. (2008). Diazonium functionalization of surfactant-wrapped chemically converted graphene sheets. *Journal of the American Chemical Society*, *130*(48), 16201–16206. doi:10.1021/ja806499w PMID:18998637

Loo, A. H., Bonanni, A., & Pumera, M. (2013). Biorecognition on graphene: Physical, covalent, and affinity immobilization methods exhibiting dramatic differences. *Asian Journal of Chemistry*, *8*(1), 198–203. doi:10.1002/asia.201200756 PMID:23090869

Ma, X., & Schobert, H. H. (2000). Molecular simulation on hydrodesulfurization of Thiophenic compounds over MoS_2 using ZINDO. *Journal of Molecular Catalysis A Chemical*, *160*(2), 409–427. doi:10.1016/S1381-1169(00)00278-8

Ma, X., Sun, L., & Song, C. (2002). A new approach to deep desulfurization of gasoline, diesel fuel and jet fuel by selective adsorption for ultra-clean fuels and for fuel cell applications. *Catalysis Today*, *77*(1-2), 107–116. doi:10.1016/S0920-5861(02)00237-7

Mattevi, C., Kim, H., & Chhowalla, M. (2011). A review of chemical vapour deposition of grapheme on copper. *Journal of Materials Chemistry*, *21*(10), 3324–3334. doi:10.1039/C0JM02126A

Milenkovic, A., Shulz, E., Meille, V., Lofferda, D., Forissier, M., Vrinat, M., ... Echegoyen, L. (2013). Preparation and characterization of soluble carbon nano-onions by covalent functionalization, employing a Na-K alloy. *Chemical Communications (Cambridge)*, *49*(24), 2406–2408. doi:10.1039/c3cc39077b PMID:23411670

Nagai, M., Goto, Y., Ishii, H., & Omi, S. (2000). XPS and TPSR study of nitrided molybdena-alumina catalyst for the hydrodesulfurization of dibenzothiophene. *Applied Catalysis A, General*, *192*(2), 189–199. doi:10.1016/S0926-860X(99)00341-5

Naidich, J. V. (1981). The wettability of solids by liquid metals. *Progress in Surface Membrane Science*, *14*, 353–484. doi:10.1016/B978-0-12-571814-1.50011-7

Park, J. G., Ko, C. H., Yi, K. B., Park, J. H., Han, S. S., Cho, S. H., & Kim, J. N. (2008). Reactive adsorption of sulfur compounds in diesel on nickel supported on mesoporous silica. *Applied Catalysis B: Environmental*, *81*(3-4), 244–250. doi:10.1016/j.apcatb.2007.12.014

Park, S., & Ruoff, R. S. (2009). Chemical methods for the production of graphenes. *Nature Nanotechnology*, *4*(4), 217–224. doi:10.1038/nnano.2009.58 PMID:19350030

Perathoner, S., Gangeri, M., Lanzafamc, P., & Centi, G. (2007). Nanostructured electrocatalytic Pt-carbon materials for fuel cells and CO2conversion. *Kinetics and Catalysis*, *48*(6), 877–883. doi:10.1134/S0023158407060171

Popov, V. N. (2004). Carbon nanotubes: Properties and applications. *Materials Science and Engineering R Reports*, *43*(3), 61–102. doi:10.1016/j.mser.2003.10.001

Pumera, M. (2009). The electrochemistry of carbon nanotubes: Fundamentals and applications. *Chemistry (Weinheim an der Bergstrasse, Germany)*, *15*(20), 4970–4978. doi:10.1002/chem.200900421 PMID:19360829

Pumera, M. (2010). Graphene-based nanomaterials and their electrochemistry. *Chemical Society Reviews*, *39*(11), 4146–4157. doi:10.1039/c002690p PMID:20623061

Rana, M. S., Sa'mano, V., Ancheyta, J., & Diaz, J. A. I. (2007). A review of recent advances on processing technologies for upgrading of heavy oils and residual. *Fuel*, *86*(9), 1216–1231. doi:10.1016/j.fuel.2006.08.004

Rao, C. N. R., Sood, A. K., Subrahmanyam, K. S., & Govindaraj, A. (2009). Graphene: The new two-dimensional nanomaterial. *Angewandte Chemie International Edition*, *48*(42), 7752–7777. doi:10.1002/anie.200901678 PMID:19784976

Reina, A., Jia, X., Ho, J., Nezich, D., Son, H., Bulovic, V., ... Kong, J. (2009). Large area, few-layer graphene films on arbitrary substrates by chemical vapor deposition. *Nano Letters*, *9*(1), 30–35. doi:10.1021/nl801827v PMID:19046078

Ronkainen, N. J., Halsall, H. B., & Heineman, W. R. (2010). Electrochemical biosensors. *Chemical Society Reviews*, *39*(5), 1747–1763. doi:10.1039/b714449k PMID:20419217

Salem, A. B. S. H. (1994). Naphtha desulfurization by adsorption. *Industrial & Engineering Chemistry Research*, *33*(2), 336–340. doi:10.1021/ie00026a025

Salem, A. B. S. H., & Hamid, H. S. (1997). Removal of sulphur compounds from naphtha solutions using solid adsorbents. *Chemical Engineering & Technology*, *20*(5), 342–347. doi:10.1002/ceat.270200511

Sano, Y. (2004a). Adsorptive removal of sulfur and nitrogen species from a straight run gas oil over activated carbons for its deep hydrodesulfurization. *Applied Catalysis B: Environmental*, *49*(4), 219–225. doi:10.1016/j.apcatb.2003.12.007

Sano, Y., Choi, K. K., Korai, Y., & Mochida, I. (2004b). Selection further activation of activated carbons for removal of nitrogen species in gas oil as a pretreatment for its deep hydrodesulfurization. *Energy & Fuels*, *18*(3), 644–651. doi:10.1021/ef034080e

Seah, C. M., Chai, S. P., & Mohamed, A. R. (2014). Mechanisms of graphene growth by chemical vapour deposition on transition metals. *Carbon*, *70*, 1–21. doi:10.1016/j.carbon.2013.12.073

Seredych, M., Lison, J., Jans, U., & Bandosz, T. J. (2009). Textural and chemical factors affecting adsorption capacity of activated carbon in highly efficient desulfurization of diesel fuel. *Carbon*, *47*(10), 2491–2500. doi:10.1016/j.carbon.2009.05.001

Song, C. (2003). An overview of new approaches to deep desulfurization for ultra-clean gasoline, diesel fuel and jet fuel. *Catalysis Today*, *86*(1-4), 211–263. doi:10.1016/S0920 5861(03)00412-7

Stephanie, N. S., Goubert, R., Xiuling, Z., & Andrzej, W. (2010). Synthesis and characterization of carbon-supported transition metal oxide nanoparticles-Cobalt porphyrin as catalysts for electroreduction of oxygen in acids. *Electrochemistry Communications*, *12*(11), 1457–1461. doi:10.1016/j.elecom.2010.06.004

Sundaraman, R., Peduzzi, E., & Song, C. (2009). Catalytic cracking of oxidized sulfur compounds in gas oil stream. *Prepr Pap Am Chem Soc Div Fuel Chem*, *54*(2), 954–955.

Svoboda, K., Lin, W., Hannes, J., Korbee, R., & van den Beek, C. M. (1994). Low temperature flue gas desulfurization by alumina-CaO regenerable sorbents. *Fuel*, *73*(7), 1144–1150. doi:10.1016/0016-2361(94)90251-8

Terrones, M. (2003). Science and technology of twenty-five century: Synthesis, properties and applications. *Annual Review of Materials Research*, *33*, 419. doi:10.1146/annurev.matsci.33.012802.100255

Thess, A., Smalley, R. E., Roland, L., Nikolaev, P., Dai, H., Petit, P., ... Fischer, J. E. (1996). Crystalline ropes of metallic carbon nanotubes. *Science*, *273*(5274), 483–487. doi:10.1126cience.273.5274.483 PMID:8662534

Velu, S., Ma, X., & Song, C. (2003). Selective adsorption for removing sulfur from jet fuel over zeolite based adsorbents. *Industrial & Engineering Chemistry Research*, *42*(21), 5293–5304. doi:10.1021/ie020995p

Wang, Y. H., Yang, F. H., Yang, R. T., Heinzel, J. M., & Nickens, A. D. (2006). Desulfurization of high sulfur jet fuel by π-Complexation with copper and palladium halide sorbents. *Industrial & Engineering Chemistry Research*, *45*(22), 7649–7655. doi:10.1021/ie060922w

Wang, Y. H., Yang, R. T., & Heinzel, J. M. (2008). Desulfurization of jet fuel by π-complexation with metal halides supported on MCM-41 and SBA-15 mesoporous materials. *Chemical Engineering Science*, *63*, 356–365. doi:10.1016/j.ces.2007.09.002

Wildgoose, G. G., Banks, C. E., & Compton, R. G. (1996). Metal Nanoparticles and Related Materials Supported on Carbon Nanotubes: Methods and Applications. *Small*, *2*(2), 182–193. doi:10.1002mll.200500324 PMID:17193018

Williams, G., Seger, B., & Kamat, P. V. (2008). TiO_2-graphene nanocomposites. UV-assisted photocatalytic reduction of graphene oxide. *ACS Nano*, *2*(7), 1487–1491. doi:10.1021/nn800251f PMID:19206319

Wondrous World of Carbon Nanotubes. (n.d.). *A Review of Current Carbon Nanotube Technologies*. Eindhoven University of Technology.

Wu, B., Kuang, Y., Zhang, X., & Chen, J. (2011). Noble metal nanoparticles /carbon nanotubes nanohybrids: Synthesis and applications. *Nano Today*, *6*(1), 75–90. doi:10.1016/j.nantod.2010.12.008

Xie, X., Ju, L., Feng, X., Sun, Y., Zhou, R., Liu, K., ... Jiang, K. (2009). Controlled fabrication of high-quality carbon nanoscrolls from monolayer graphene. *Nano Letters*, *9*(7), 2565–2570. doi:10.1021/nl900677y PMID:19499895

Yang, F. H., Hernandez-Maldonado, A. J., & Yang, R. T. (2004). Selective adsorptipion of organosulfur compounds from transportation fuels by π-complexation. *Separation Science and Technology*, *39*(8), 1717–1731. doi:10.1081/SS-120035937

Yang, W., Ratinac, K. R., Ringer, S. P., Thordarson, P., Gooding, J. J., & Braet, F. (2010). Carbon nanomaterials in biosensors: Should you use nanotubes or graphene? *Angewandte Chemie International Edition*, *49*(12), 2114–2138. doi:10.1002/anie.200903463 PMID:20187048

Yang, X., Cao, C., Klabunde, K. J., Hohn, K. L., & Erickson, L. E. (2007). Adsorptive desulfurization with xerogel-derived zinc-based nanocrystalline aluminum oxide. *Industrial & Engineering Chemistry Research*, *46*(14), 4819–4823. doi:10.1021/ie0701928

Yoshida, H., Shimizu, T., Uchiyama, T., Kohno, H., Homma, Y., & Takeda, S. (2009). Atomic-scaleanalysis on the role of molybdenum in iron-catalyzed carbon nanotube growth. *Nano Letters*, *9*(11), 3810–3815. doi:10.1021/nl9019903 PMID:19731921

Yuan, L., Saito, K., Hu, W., & Chen, Z. (2001). Ethylene flame synthesis of well-aligned multi-walled carbon nanotubes. *Chemical Physics Letters*, *346*(1-2), 23–28. doi:10.1016/S0009-2614(01)00959-9

Yudasaka, M., Komatsu, T., Ichihashi, T., & Iijima, S. (1997). Single wall carbon nanotube formation by laser ablation using double-targets of carbon and metal. *Chemical Physics Letters*, *278*(1-3), 102–106. doi:10.1016/S0009-2614(97)00952-4

Zhang, Q., Huang, J. Q., Qian, W. Z., Zhang, Y. Y., & Wei, F. (2013). The Road for Nanomaterials Industry: A Review of Carbon Nanotube Production, Post-Treatment, and Bulk Applications for Composites and Energy Storage. *Small*, *9*(8), 1237–1265. doi:10.1002mll.201203252 PMID:23580370

Zhang, Q., Yu, Z., Du, P., & Su, C. (2010). Carbon nanomaterials used as conductive additives in lithium ion batteries. *Recent Patents on Nanotechnology*, *4*(2), 100–110. doi:10.2174/187221010791208803 PMID:20415660

Zhou, H., Li, G., Wang, X., Jin, C., & Chen, Y. (2009). Preparation of a kind of mesoporous carbon and its performance in adsorptive desulfurization. *Journal of Natural Gas Chemistry*, *18*(3), 365–368. doi:10.1016/S1003-9953(08)60117-6

Zhou, W. W., Han, Z. Y., Wang, J. Y., Zhang, Y., Jin, Z., Sun, X., ... Li, Y. (2006). Copper catalyzing growth of single-walled carbon nanotubes on substrates. *Nano Letters*, *6*(12), 2987–2990. doi:10.1021/nl061871v PMID:17163745

Chapter 4
Polyoxometalates-Based Nanocatalysts and Their Efficiency for Production of Sulfur-Free Diesel

Diana Julião
LAQV/REQUIMTE, Departamento de Química e Bioquímica, Faculdade de Ciências, Universidade do Porto, Portugal

Susana Ribeiro
LAQV/REQUIMTE, Departamento de Química e Bioquímica, Faculdade de Ciências, Universidade do Porto, Portugal

Baltazar de Castro
LAQV/REQUIMTE, Departamento de Química e Bioquímica, Faculdade de Ciências, Universidade do Porto, Portugal

Luís Cunha-Silva
LAQV/REQUIMTE, Departamento de Química e Bioquímica, Faculdade de Ciências, Universidade do Porto, Portugal

Salete S. Balula
LAQV/REQUIMTE, Departamento de Química e Bioquímica, Faculdade de Ciências, Universidade do Porto, Portugal

ABSTRACT

Polyoxometalates have been demonstrated to be efficient catalysts for the activation of oxidants in desulfurization processes. Successful results on desulfurization using polyoxometalates and hydrogen peroxide to desulfurize model oils and liquid fuels were reported and can be found in the literature. The desulfurization is an actual subject with notable interest for refineries and fuel cost, and consequently, it is

DOI: 10.4018/978-1-7998-2146-5.ch004

important to focus the scientific community to work in desulfurization technology in order to develop catalytic systems based on polyoxometalates capable to be reused, stable, efficient, and selective. Therefore, the main goal is the design of heterogeneous polyoxometalate based catalysts. This chapter pretends to inform the research society about the scientific directions that have been taken using heterogeneous polyoxometalate catalysts in oxidative desulfurization of simulated and real liquid fuels. In addition, future perspectives are proposed to cover the actual needs of this area.

Figure 1. Schematic representation of some sulfur-compounds present in crude oil

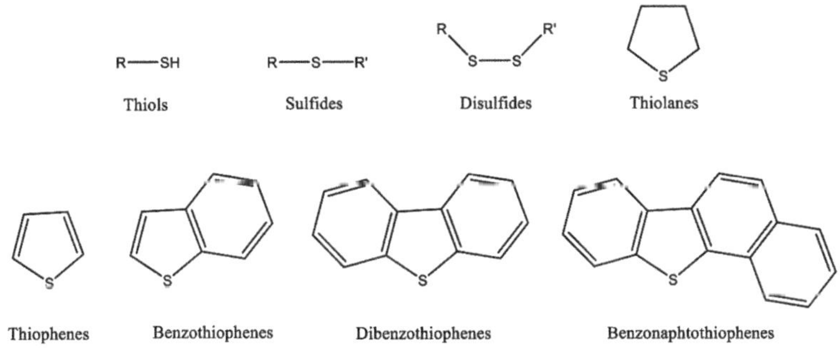

INTRODUCTION

The specifications that govern transportation fuels have become increasingly stringent with respect to sulfur content, over the years. Consequently, the removal of sulfur from oil is consequently one of the central requirements in most refineries and the price of a crude oil is influenced by its sulfur content. The refining cost (financial and environmental) increases as heavier and more sulfur-rich crude oils are being processed. Sulfur is the most abundant element in petroleum after carbon and hydrogen. The average sulfur content varies from 0.03 to 7.89% (in mass) in crude oil.(Soleimani, Bassi, & Margaritis, 2007) Organic sulfur molecules such as thiols, sulfides, and thiophenic compounds represent the main source of sulfur found in crude oil (Figure 1). Desulfurization of compounds that contain aliphatic sulfur, i.e. thiols and sulfides, is easier than desulfurization of compounds that contain aromatic sulfur, i.e. thiophenics.

The actual desulfurization process applied in petroleum refining for treating sulfur content in transportation fuel, is the hydrodesulfurization (HDS), which is a simple and proven process for the refiners. The HDS technology has been adjusted to meet the tight specifications of the current limit imposed by government directives for liquid fuel transportation of 10 ppm.(Lissner, de Souza, Ferrera, & Dupont, 2009; Y. Liu et al., 2013) However this requirement has forced the HDS units to operate at extreme severe conditions (high temperature, pressure and consume of large amounts of hydrogen) and/or with higher catalyst volumes, do not only affect the economic viability of the process but most importantly the fuel specifications are affected.

The strict sulfur directives associated to the high cost of the actual HDS process requires the development of alternative or complementary desulfurization technologies, mainly for the removal of the aromatic sulfur-compounds which are the most difficult to eliminate from diesel fuels.

The oxidative desulfurization (ODS) is considered one of the most promising alternative or supplementary process, offering several advantages over HDS, such as mild reaction conditions (operating under atmospheric pressure), high selectivity and economic viability. In the ODS process the sulfur-containing compounds are oxidized to the corresponding sulfoxides and/or sulfones which are more easily removed from fuels by extraction, adsorption or other appropriate techniques.(Lü, Ren, Wang, et al., 2013; Nie, Dong, Bai, Dong, & Zhang, 2013)

Early-transition metal oxygen anion clusters, also known as polyoxometalates (POMs), are a class of compounds with high interest for application in different areas such as, medicine,(Rhule, Hill, Judd, & Schinazi, 1998) materials science,(Coronado & Gómez-García, 1998) photochemistry,(Katsoulis, 1998; Yamase, 1998) biology(Yamase, 2005) and catalysis(Mizuno & Misono, 1998; Proust, Thouvenot, & Gouzerh, 2008). In particular, POMs have been extensively used as catalysts due to their unusual versatility and compatibility with environmental friendly conditions and operations, being designated by "green catalysts".(Hill & Prosser-McCartha, 1995; R. Wang, Zhang, & Zhao, 2010) The first POMs applied as catalysts in ODS systems were the tetrabutylammonium salts of $[W_6O_{19}]^{2-}$, $[V(VW_{11})O_{40}]^{4-}$, $[PVW_{11}O_{40}]^{4-}$, and $[PV_2Mo_{10}O_{40}]^{4-}$.(Komintarachat & Trakarnpruk, 2006) Since 2011 the use of POMs in ODS has received particular attention due to their effectiveness and efficiency as oxidative catalysts. Despite their good activity in ODS systems, their solubility in polar media became a disadvantage, precluding their application at industrial scale where efficiency, robustness and recycling are required. To overcome this drawback, considerable efforts to promote their recovering and recycling have been made through the immobilization of POMs using various solids supports.

Various POMs based heterogeneous catalysts have been prepared using a variety of support materials, such as: alumina,(Thanasilp, Schwank, Meeyoo, Pengpanich, & Hunsom, 2013) titania,(K. Li et al., 2010; Pandiyarajan, Pandikumar, & Ramaraj, 2013; Xie, 2006) zirconia,(G. S. Armatas, Bilis, & Louloudi, 2011; Salavati, Tavakkoli, & Hosseinpoor, 2012) zeolites, silica(Farsani & Yadollahi, 2014; Okada, Miyamoto, Sakai, & Mishima, 2014) and porous Metal-Organic Frameworks (MOFs). (Gerasimos S. Armatas, 2013; Du, Qin, Li, Su, & Lan, 2014; Lefebvre, 2013; Pires et al., 2014; F. Yu et al., 2010) The heterogeneous catalysts based POMs most used in ODS processes have been prepared with MOFs and silica as supports,(Nogueira et al., 2014; Ribeiro, Granadeiro, et al., 2013; X. M. Yan et al., 2013; X. M. Yan, Mei, Mei, & Yang, 2014) and incorporate different structures of POMs, mainly Keggin- ($[XM_{12}O_{40}]^{n-}$) and Lindqvist-type $[M_5O_{18}]^{p-}$. The catalyst performance seems to be influenced by various fundamental parameters, such as the support, immobilization methodology and the presence of functional groups.

MAIN FOCUS OF THE CHAPTER

The present chapter aims to identify and also to compare the different solid nanocatalysts (nanoporous MOFs, silica nanoparticles, carbon nanotubes, nano-sized hybrid POMs, etc) based on POMs as heterogeneous catalysts for the oxidative desulfurization of a model oil (containing one or a mixture of various sulfur compounds in a non-polar solvent to simulate a liquid fuel) already reported. The desulfurization efficiency and recycling capability in various desulfurization cycles will be discussed and evaluated. Furthermore, the few examples described in the literature using ODS systems involving real fuel samples, catalyzed by heterogeneous POMs catalysts, will be described and compared. Finally, future research guides and perspectives will be proposed.

CATALYTIC OXIDATIVE DESULFURIZATION STUDIES

Desulfurization studies using Model Oils

POMs@MOFs Composites

Various POMs, mainly based in Keggin-type derivatives have been immobilized in the cavities of porous coordination polymers, also known as metal-organic frameworks (MOFs), forming the composite materials POMs@MOFs. In fact, porous MOFs are crystalline materials constructed from metal ions and organic ligands linked through coordination bonds, with regular, large and accessible nanocages and tunnels that allow the accommodation of catalytically active molecules, making them attractive candidates as solid supports to design novel heterogeneous catalysts.(Granadeiro et al., 2013; Jana Juan-Alcaniz, Gascon, & Kapteijn, 2012; Julião et al., 2015; Kholdeeva, Maksimchuk, & Maksimov, 2010; Ribeiro, Barbosa, Gomes, Pillinger, Gonçalves, et al., 2013; Ribeiro, Granadeiro, et al., 2013; Q.-L. Zhu & Xu, 2014)

Figure 2. Schematic comparison of Keggin POM structure with MIL-101(Cr) structural features: (Left) Hexagonal windows; (center) zeotype structure; and (right) pentagonal windows. Atom color scheme: red: oxygen; green: chromium; and white: carbon.
(J. Juan-Alcaniz et al., 2010)

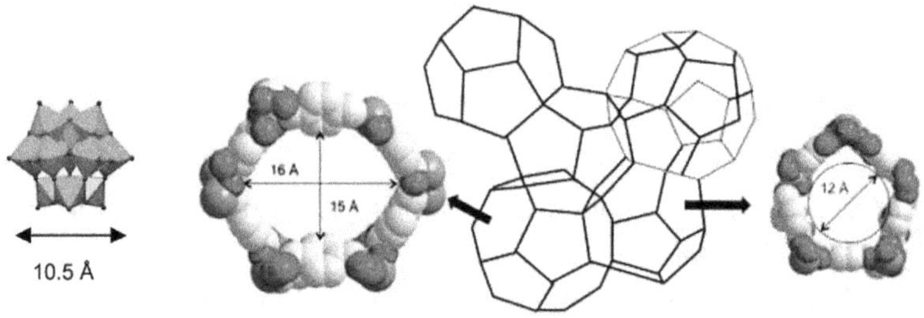

Different strategies to incorporate POMs in MOFs cavities have been described in the literature.(Férey et al., 2005; Jana Juan-Alcaniz et al., 2012; J. Juan-Alcaniz, Ramos-Fernandez, Lafont, Gascon, & Kapteijn, 2010) The most used methodologies to prepare POM@MOFs composites are the impregnation and one-pot encapsulation methods. The impregnation method consists in the addition of MOFs on a POM solution. This method favors the occupancy of the large cavities of MOFs with the POMs. The one-pot encapsulation is based on the synthesis of MOFs in the presence of the POM in solution. In contrast to the first method, the immobilized POMs can also occupy the smaller cavities of MOFs and the loss of POMs from the composite can be avoided. Férey *et al.* reported the first example of the impregnation method using the large porous structure MIL-101. In this original work, the mono-lacunar POM $K_7PW_{11}O_{39}$ was incorporated in the MIL-101 cavities.(Férey et al., 2005) Later on, several other Keggin-type POMs were successfully incorporated into MIL-101 frameworks using the impregnation method and these composites were used as heterogeneous catalysts in various oxidative reactions.(Balula, Granadeiro, Barbosa, Santos, & Cunha-Silva, 2013; Granadeiro et al., 2013; Kholdeeva et al., 2010; Wee, Bonino, Lamberti, Bordiga, & Martens, 2014) Taking into account that medium-sized cavities present in the MIL-101 structure are in greater number than the large cavities, and that the window openings of these cavities are with equivalent dimension of Keggin POM structures (10-14 Å diameters) (Figure 2), Alcañiz and coworkers have developed the second methodology by adding the Keggin structure $H_3PW_{12}O_{40}$ to the synthetic reaction medium of MIL-101.(J. Juan-Alcaniz et al., 2010)

The application of POMs@MOFs in desulfurization area is relatively limited, with their use only being reported in adsorptive and oxidative desulfurization (ODS). (Granadeiro et al., 2014; Hu, Lu, Dai, Liu, & Liu, 2013; Jia et al., 2013; Julião et al., 2015; Y. Liu et al., 2013; Ribeiro, Barbosa, Gomes, Pillinger, Goncalves, et al., 2013;

Ribciro, Granadciro, ct al., 2013; X.-S. Wang, Huang, Lin, & Cao, 2014) The first work using POM@MOFs in ODS was published in 2013 by *Hu* and collaborators, with the composite material being prepared by one-pot encapsulation of the the Keggin POM $H_3PW_{12}O_{40}$ into MIL-101(Cr) with different loadings.(Hu et al., 2013) The composite material with large amount of POM (approximately 17 - 50 wt.%) showed the best catalytic activity (after 4 h of reaction at 50 ^{0}C, using H_2O_2 as oxidant) for the oxidation of different model oils, each one containing different refractory sulfur compounds: 1-benzothiophene (BT) or dibenzothiophene (DBT) or 4,6-dimethyldibenzothiophene (4,6-DMDBT). The recyclability and reusability of the best catalyst was investigated in the oxidation of DBT and it was found that the DBT conversion dropped from 91% to 87% after four cycles, possibly due to the leaching of small amounts of POM. With this work it was possible to verify that the sulfur compound that is more easily desulfurized from oil during the catalytic stage is the DBT, followed by the 4,6-DMDBT and for last the BT. This order of desulfurization is mainly attributed to the significant lower electron density of BT compared with the two other sulfur compounds. The electron density on the sulfur atom is similar in the DBT and 4,6-DMDBT, consequently the difference of reactivity observed was essentially due to the steric hindrance of the methyl groups.

In 2013, another investigation using the same PW_{12}@MIL-101(Cr) composite but prepared by the impregnation method was reported.(Ribeiro, Barbosa, Gomes, Pillinger, Goncalves, et al., 2013) This work showed better results than the previous, reaching a total removal of DBT, BT and 4,6-DMDBT from a model oil (prepared with 500 ppm of each sulfur compound) after 4 h of reaction at 50 ^{0}C, using an ionic liquid (IL; 1-butyl-3-methylimidazolium hexafluorophosphate) as extraction solvent and H_2O_2 as oxidant. The recyclability of the composite was evaluated through the execution of three consecutive ODS cycles, and it was found a slight decrease of sulfur removal, probably due to the leaching of the active center (POM) from the composite to the extraction solvent. Afterwards, other Keggin-type derivatives were incorporated into MIL-101(Cr) by the impregnation method and successfully used as proper heterogeneous catalysts for desulfurization of multicomponent model oils containing various refractory sulfur compounds present in fuels. Most of these studies, reported by Balula´s research group, were performed in the presence of an extraction solvent, and conciliating extraction and oxidation steps in the ODS process. Therefore, extraction and oxidative catalytic steps were evaluated and optimized in each ODS system to achieve complete desulfurization after a few hours.(Granadeiro et al., 2014; Julião et al., 2015; Ribeiro, Barbosa, Gomes, Pillinger, Goncalves, et al., 2013; Ribeiro, Granadeiro, et al., 2013) Preceding the catalytic stage, an initial extraction was performed, where the sulfur compounds were transferred from the model oil to the extraction solvent. Regardless the solvent used, the extraction from oil to solvent followed the order BT > DBT > 4,6-DMDBT, according to molecular

Figure 3. Oxidative desulfurization system exemplified for PW_{12}@MIL-101 composite, using an ionic liquid as extraction solvent and reaction medium.

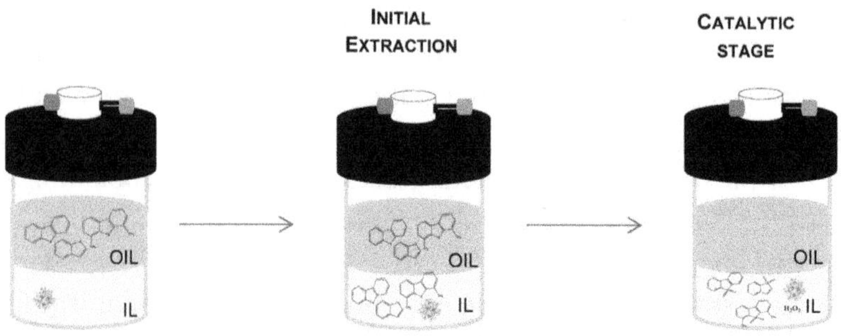

size, geometry and polarity of these sulfur molecules. The BT possesses the smaller size and the higher polarity which facilitate its transfer to the polar extraction phase. The oxidative catalytic stage was always initiated after the addition of the oxidant, and this step also occurred in the presence of a polar extraction solvent. At the end of the ODS process the oxidized products (sulfones or sulfoxides) were not detected in the oil phase (Figure 3).

One of the studies performed by Balula and co-workers was executed using a terbium-POM incorporated into the solid support MIL-101(Cr), $Tb(PW_{11})_2$@MIL-101(Cr).(Ribeiro, Granadeiro, et al., 2013) The ODS system was carried out using a model oil with DBT, BT and 4,6-DMDBT (500 ppm of each in *n*-octane). The best result was achieved with acetonitrile as extraction solvent yielding sulfur-free model oil after 5 h of reaction at 50 ^{0}C in the presence of H_2O_2, and emphasizing that after 3 h only 17 ppm of sulfur remained in the model oil. The recycling capability of $Tb(PW_{11})_2$@MIL-101(Cr) was verified for three consecutive ODS cycles and the loss of activity found was not considerable. The influence of the extraction solvent, using three different polar solvents (acetonitrile, dimethylformamide and isopropanol), was also carefully analyzed and it was possible to observe that during the initial extraction the best solvent was dimethylformamide; however, during the catalytic stage, the performance of the catalyst was higher in the presence of the two other solvents. This distinct behavior can be related with the different affinity of the solvents for the catalyst. Another polyoxometalate that was incorporated into MIL-101(Cr) was the trivacant polyoxotungstate, $[A\text{-}PW_9O_{34}]^{9-}$.(Granadeiro et al., 2014) As in the previous work, the ODS process was performed using a model oil containing three different sulfur compounds (DBT, BT and 4,6-DMDBT) in *n*-octane with a concentration of 500 ppm for each molecule. The composite PW_9@MIL-101(Cr) showed to be efficient for the oxidation of the three sulfur compounds in the presence of acetonitrile as extraction solvent, at 50 ^{0}C using H_2O_2 as oxidant. In

Figure 4. Schematic representation of the electrostatic interaction between NH_3^+ group and $[PW_{12}O_{40}]^{3-}$ anions.
(X.-S. Wang et al., 2014)

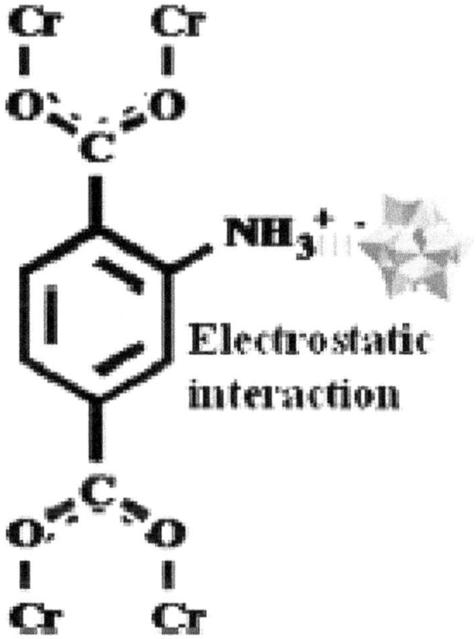

fact, after 1 h of reaction only 5 ppm of sulfur compounds remained in the model oil. The composite was also recycled without a significant loss of activity.

Besides MIL-101(Cr) framework, also NENU-9 ($[(CH_3)_4N]_2\{[Cu_2(BTC)_{4/3}(H_2O)_2]_6]$) has been used as solid support to accommodate the polyoxometalate PV_2Mo_{10} ($H_5PV_2Mo_{10}O_{40}$).(Y. Liu et al., 2013) This last POM has been extensively used as homogenous catalyst for oxidative desulfurization of model oils. The catalytic activity of the composite PV_2Mo_{10}@NENU-9 was investigated for the oxidation of model oil with 500 ppm of DBT in decalin, using molecular oxygen as oxidant in the presence of isobutyraldehyde as sacrificial agent, at 80 ^{0}C. After 1.5 h of reaction a complete conversion of DBT to its corresponding sulfone was obtained.

Recently, Wang *et al.* reported the functionalization of MIL-101(Cr) cavities with amine groups (MIL-101(Cr)-NH_2, Figure 4) to increase the stability of POMs inside the porous structure and to avoid their leaching.(X.-S. Wang et al., 2014) Using $H_3PW_{12}O_{40}$ incorporated into this functional support as heterogeneous catalyst in model oil with 950 ppm of DBT, the complete conversion was obtained after 1 h of reaction using acetonitrile as extraction solvent at 50 ^{0}C. In the same reaction time a conversion of approximately 75% for BT and 90% for 4,6-DMDBT was also obtained. The recycling potential of this heterogeneous catalyst was evaluated for various ODS cycles, and was not observed leaching or loss of activity. This fact, can be related with of the electrostatic interaction between the POM and the amine groups of the solid support (Figure 4).

One of the most recent publications using POM@MOFs composites as heterogeneous catalysts for ODS systems was reported by Balula *et al.*, using a zinc-substituted polyoxotungstate anion $[PW_{11}Zn(H_2O)O_{39}]^{5-}$, immobilized in the nanocages of MIL-101(Cr) as heterogeneous catalyst for oxidative desulfurization of a model oil. In this study the extraction performance of an IL (1-butyl-3-methylimidazolium phosphate, $BMIPF_6$) and acetonitrile was studied and compared. Using model oil with three different sulfur compounds (DBT, BT and 4,6-DMDBT) with 500 ppm of each in *n*-octane, a sulfur-free model oil was achieved after 2 h of reaction, using the IL as extraction solvent and H_2O_2 as oxidant at 50 ^{0}C. However, in the presence of acetonitrile as extraction solvent, slight amounts of sulfur compounds were still present in the model oil after 6 h of reaction. The reutilization of the composite material in three consecutive ODS processes for both oil/$BMIPF_6$ and oil/acetonitrile was investigated. A small decrease in desulfurization efficiency was noticed for the multicomponent oil/$BMIPF_6$, probably due to some leaching of the POM to the extraction solvent. Interestingly, in the presence of the acetonitrile there was no loss of activity. In fact, the system $BMIPF_6$/PW_{11}Zn@MIL-101(Cr) revealed to be more active but less stable (active POM leaching occurred) than the equivalent system using acetonitrile as extraction solvent (no leaching occured).(Julião et al., 2015)

Silica-Based Catalysts: POMs@SiO_2

A variety of mesoporous silica (M41S, HMS-n, SBA-n, among others) have been successfully used as support materials to immobilize various POMs, mainly the Keggin-type derivatives ($[PM_{12}O_{40}]^{n-}$, M = W, Mo, V, Ti). The silica supports with adequate porosity have the advantage of easy surface modification by the simple reaction with organosilanes presenting the appropriate functional groups. The variety of silica structures, associated with the various possible surface modifications by the incorporation of several functional groups create a large diversity of heterogeneous catalyst based on POMs@SiO_2. The structural robustness of these catalysts depends mainly on the adopted immobilization methodology. Many strategies to prepare POMs@SiO_2, such as impregnation, sol-gel techniques, electrostatic interactions, ion-exchange and covalent bonding, can be found in the literature.(Zhou, Chen, Long, & Wang, 2014) The application of POMs@SiO_2 in ODS processes is recent and the first publication dates back to 2009, reporting the immobilization of the mono-lacunar POM $(C_{19}H_{42}N)_4H_3[PW_{11}O_{39}]$ by its dispersion in the silica matrix via a co-condensation sol-gel method.(Abdalla, Li, & Tufail, 2009) This heterogeneous catalyst was used to desulfurize a model oil containing DBT and applying H_2O_2 as oxidant (oxidant / S = 4). The reduction of the sulfur content of this oil from 500 ppm to 2 ppm was attained in 2 h of reaction at 60ºC. The authors referred that the presence of quaternary ammonium salt facilitates a higher distribution of the catalyst which accelerates the reaction rate. However, the recyclability studies of this heterogeneous catalyst resulted in some loss of activity due to the loss of some POM from the surface of silica support.

In the same year (2009), another research group immobilized $H_3PW_{12}O_{40}$ on the surface of mesoporous silica modified with Ag^+ for oxidative desulfurization of a model oil (containing only DBT) and diesel oil. The authors reported that this composite functionalized with Ag^+ combines the absorptive capacity of sulfur compounds by the free Ag^+ centers and the catalytic oxidative properties represented by the POM, increasing the desulfurization efficiency when compared with the same composite without Ag^+ centers. The recyclability of this catalyst shows negligible activity loss after three consecutive cycles. In this study the complete desulfurization of the model oil was not achieved even after 4 h. Furthermore, the efficiency of the catalyst decreased when tested in diesel oil, due to the presence of alkenes and aromatic hydrocarbons in which tend to adsorb on the heterogeneous catalytic sites. (X.-m. Yan, Su, & Xiong, 2009)

In 2012, Hu and co-workers reported the original application of magnetic heterogeneous catalysts based on silica nanoparticles functionalized with a long cationic carbon chain and the Keggin POM ($[PW_{12}O_{40}]^{3-}$) immobilized by ionic interaction. This heterogeneous catalyst was applied to the oxidative desulfurization of DBT model oil using H_2O_2 as oxidant. The main advantage of this catalyst was the easy recovery of the solid material from reaction medium by an external magnetic field that facilitates its recyclability. However, 10 h of reaction were necessary to obtain the approximately complete desulfurization of model oil. This work will be discussed in more detail in this chapter in section 1.4.(Cui, Yao, Li, Yang, & Hu, 2012) In the same year, Li and Abdalla published another example of the impregnation of the mono-lacunar $[PW_{11}O_{39}]^{7-}$ polyanion in a highly ordered mesoporous material MCM-41.(Abdalla & Li, 2012) Model oils of DBT and thiophene were used to test the efficiency of the catalyst in the presence of H_2O_2. The extraction of oxidized products from oil was only performed after the catalytic step using methanol. The POM loading and the use of reaction temperature higher than 50ºC and bellow 70ºC demonstrated to have a remarkable influence on the desulfurization results. The complete desulfurization of the model oils was achieved after 1 h at 70ºC in the presence of $H_2O_2/S = 4$.

More recently, in 2013, three distinct silica materials containing POMs were reported. Yan *et al.* used a mesoporous support containing a mixture of titania and silica, and the Keggin polyanion $[PW_{12}O_{40}]^{3-}$ was incorporated by evaporation-induced self-assembly method.(X. M. Yan et al., 2013) This heterogeneous catalyst

Figure 5. Schematic representation of LaW_{10} immobilized on the surface of functionalized mesoporous silica.
(Y. Chen et al., 2013)

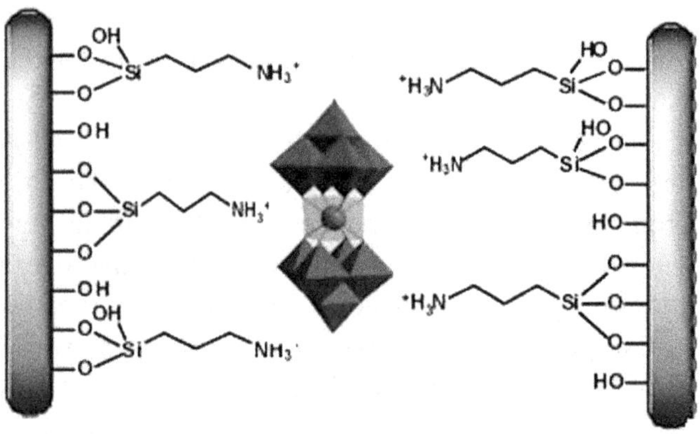

was used to desulfurize a model oil containing only DBT, using H_2O_2 as oxidant, in the presence of acetonitrile as extraction solvent, at 60ºC. Complete desulfurization was attained after 2 h using high excess of oxidant (O/S = 12). The recyclability was performed for three consecutive cycles without loss of significant activity suggesting the stability of the catalyst. A surfactant peroxopolyoxotungstate ($[C_{16}H_{3}{}_{3}(CH_3)_2NOH]_3\{PO_4(WO(O_2)_2]_4\}$ was encapsulated in the channels of silica used by Jiang *et al.*.(Zheng et al., 2013) This catalyst was tested for oxidative desulfurization of DBT model oil, at 70ºC, with a H_2O_2/S molar ratio 4:1 in the presence of *tert*-butyl alcohol as additive. The desulfurization was only completed after 3 h, and the catalyst could be recycled for six ODS cycles with only a small loss of activity after the first cycle. Song and co-workers reported the first work using silica based POMs heterogeneous catalyst for oxidative desulfurization in the presence of an IL (1-butyl-3-methylimidazolium) as extraction solvent and H_2O_2 as oxidant.(Y. Chen, Zhao, & Song, 2013) The lanthanum tungstate ($[LaW_{10}O_{36}]^{9-}$, LaW_{10}) was chemically anchored to an amino-modified mesoporous silica (Figure 5). The ODS system was formed by two immiscible phases (oil and IL) and a solid catalyst. To date, this system showed the best oxidative desulfurization performance at least for a BT model oil, revealing total desulfurization after only 35 min, at 70ºC, using 5:1 oil/IL and H_2O_2/DBT/POM 150:30:1. Another notable advantage in this work was the ability to recycle the solid catalyst as well as the IL used in the ODS system. The solid catalyst was recycled for ten times without decreasing the catalytic efficiency.

Figure 6. Schematic representation of the ODS process using a DBT model oil catalyzed by LaW_{10}@IL-functionalized silica. Step 1 is the extraction of the substrate by the IL on the surface of material and step 2 is the catalytic oxidation having as active center the LaW_{10} and the formation of the active peroxotungstate intermediate ($R=C_8H_{17}$).
(Y. Chen & Song, 2014)

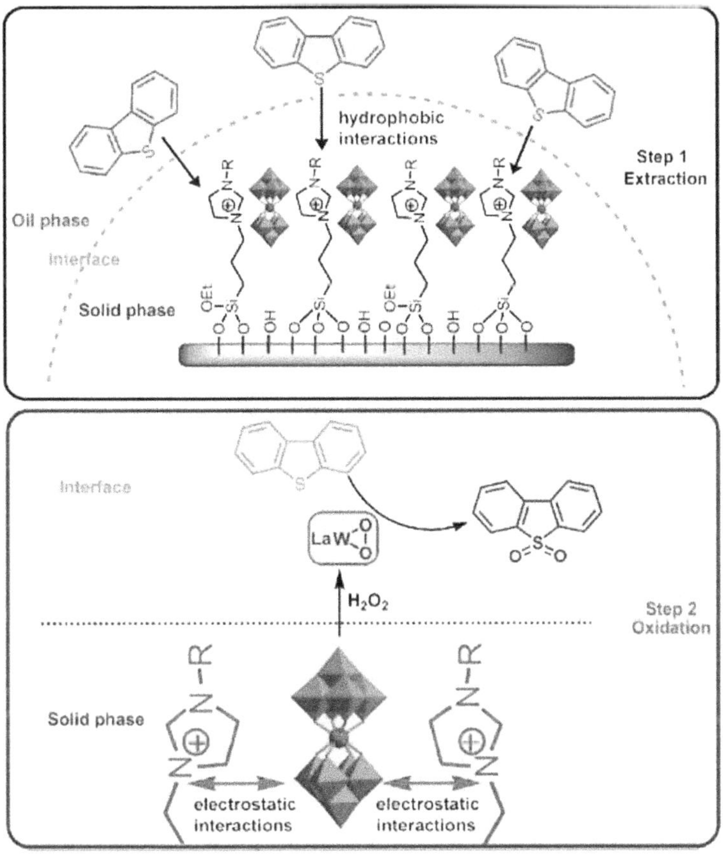

After this notable result Song *et al.*(Y. Chen et al., 2013) explored the same lanthanum tungstate LaW_{10}, but this time immobilize the POM on the surface of a mesoporous silica functionalized by an IL (dihydroimidazolium-based IL).(Y. Chen & Song, 2014) In fact, the IL has shown to have a remarkable influence in the oxidative desulfurization process.(Y. Chen et al., 2013) Notably, when the IL was present on the surface of the mesoporous silica the efficiency of desulfurization of a model oil containing various sulfur compounds (DBT, BT, and 4,6-DMDBT) was even higher than the previous case where the IL was used as extraction solvent under the same reaction conditions (70ºC and H_2O_2/DBT/POM = 150:30:1). In the

Figure 7. Schematic representation of encapsulation of $PW_{11}Zn$ into silica nanoparticles using a cross-linked organic–inorganic core formed by POM and (3-aminopropyl)triethoxysilane, APTES.
(Nogueira et al., 2014)

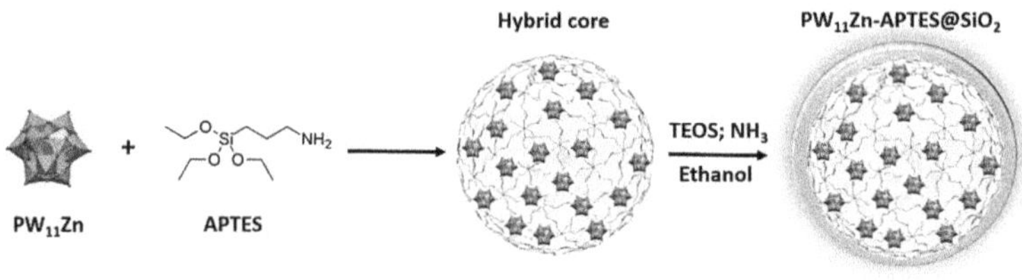

catalytic step an intermediate peroxotungstate was formed by the interaction of the LaW_{10} and the oxidant H_2O_2 (Figure 6). The solid catalyst LaW_{10}@IL-functionalized silica could be recycled ten times without losing activity.(Y. Chen & Song, 2014)

In 2014, other three POMs@SiO_2 ODS systems mainly based on Keggin derivatives POMs were published. Mahjoub *et al.* reported the use tert-butyl hydroperoxide as oxidant in the presence of POMs@SiO_2 catalyst for the first time. The desulfurization of a DBT model oil at 60ºC, catalyzed by a mixed addenda POMs $[PMo_{12-y}W_yO_{40}]^{3-}$ (y = 2-10) with different ratio Mo/W, supported on platelet SBA-15 without the presence of an extraction solvent, was carried out. (Chamack, Mahjoub, & Aghayan, 2014) The highest performance was achieved using $[PMo_8W_4O_{40}]^{3-}$, since the complete desulfurization was attained after 1.3 h, however it was needed high excess of oxidant (O/S = 5). The catalyst was recycled for four ODS cycles with just a slight decrease of activity.(Chamack et al., 2014) The commercial phosphotungstic acid, $H_3[PW_{12}O_{40}]$, was used by Yan and co-workers and immobilized on a porous silica-alumina SiO_2-Al_2O_3 to desulfurize a BT model oil using H_2O_2 as oxidant (O/S molar 2:1) and acetonitrile as extraction solvent, at 60ºC.(X. M. Yan et al., 2014) The alumina centers of material acted as Lewis acid sites facilitating the adsorption of BT on catalyst surface and accelerating the desulfurization. The desulfurization of the model oil was almost total after 2 h of reaction and the heterogeneous catalyst was recycled by three ODS cycles without losing activity. Balula and co-workers reported an unprecedented methodology to encapsulated POMs into silica nanoparticles using a cross-linked organic–inorganic core (Keggin derivative POM and (3-aminopropyl)triethoxysilane, APTES, Figure 7).(Nogueira et al., 2014) In this investigation, a zinc-substituted Keggin POM $[PW_{11}Zn(H_2O)O_{39}]^{5-}$ ($PW_{11}Zn$) was encapsulated into silica nanoparticles (PW_{11}Zn-APTES@SiO_2). The ODS studies were performed using a model oil containing DBT and 4,6-DMDBT, in the presence of H_2O_2 and acetonitrile as extraction solvent. The DBT desulfurization were less efficient (DBT was completely desulfurized after 3 h for the first ODS cycle and after 2 h for the second cycle) than the previous silica materials published in the same year, as well as when the same zinc-substituted Keggin POM was incorporated into MIL-101(Cr).(Julião et al., 2015) However, the nanocomposite revealed to be structural stable for several ODS cycles and the leaching of POM was not observed. The higher catalytic performance verified after the first ODS cycle must be related with the need of active catalytic intermediate formation during the first cycle by the interaction of POM with H_2O_2 (as referred before for LaW_{10}). This active intermediate is a peroxotungstate compound which it is capable to be regenerated without affecting the structure of the POM.

In general, it is possible to verify and affirm that the efficiency of POMs@SiO_2 systems have considerably increased since the first applications in 2009. However, most the examples reported to date are based on DBT model oil oxidative desulfurization using H_2O_2 oxidant. The POMs@SiO_2 system more efficient was proposed by Song *et al.* in 2014 using the heterogeneous catalyst LaW_{10}@IL-functionalized porous silica to desulfurize a multicomponent model oil containing DBT, BT, and 4,6-DMDBT at 70°C, H_2O_2/DBT/POM = 150:30:1, without the presence of extraction solvent. With this ODS system deep desulfurization was achieved for DBT, BT and 4,6-DMDBT in 25, 30 and 25 min, respectively.(Y. Chen & Song, 2014)

Heterogenization of POMs With Activated Carbon

The immobilization of POMs in activated carbon or carbon nanotubes is practically unexplored, and only three examples of their application for oxidative desulfurization can be found in the literature. These investigations used the commercial phosphotungstic acid as catalytic centers ($H_3[PW_{12}O_{40}]$). The first report was presented by Wang *et al.* in 2010, where the POM $Cs_{2.5}H_{0.5}PW_{12}O_{40}$ was immobilized onto multi-walled carbon nanotubes (MWCN).(R. Wang, Yu, Zhang, & Zhao, 2010) This heterogeneous catalyst was tested using a DBT model oil (500 ppm) in a biphasic system with acetonitrile as extraction solvent and H_2O_2 as oxidant (O/S molar ratio 20), at 60°C. The authors also compared the catalytic performance of MWCN and activated carbon as support materials for the same POM: $Cs_{2.5}H_{0.5}PW_{12}O_{40}$@MWCN revealed higher performance than $Cs_{2.5}H_{0.5}PW_{12}O_{40}$@activated carbon, the support MWCN and the non-immobilized POM $Cs_{2.5}H_{0.5}PW_{12}O_{40}$. Using the $Cs_{2.5}H_{0.5}PW_{12}O_{40}$@MWCN catalyst the total desulfurization of DBT model oil was achieved after 2.7 h of reaction. Furthermore, this heterogeneous catalyst was recycled for three ODS cycles without significant loss of activity.

The remaining two works were published in 2014 and used the $H_3[PW_{12}O_{40}]$ immobilized on activated carbon as heterogeneous catalyst for oxidative desulfurization. Tang *et al.* tested this heterogeneous catalyst in a DBT (2000 ppm) model oil using the assist of ultrasound (100 W) at 70°C and H_2O_2:model oil volume = 1:10.(L. Liu, Zhang, & Tan, 2014) Interestingly, the complete desulfurization of DBT was achieved after only 9 min. Furthermore, the authors claimed that the present heterogeneous ODS system using ultrasound irradiation avoids the use of costly phase transfer agent. The catalyst was easily separated for recycling however, the activity decreased considerably with the number of recycling cycles. The authors suggested that the violent movement of solid particles under ultrasound can greatly boost the mass transfer in the reaction. Xiao *et al.* used the same catalyst $H_3[PW_{12}O_{40}]$@activated carbon to study the oxidative desulfurization of a model oil containing one of the sulfur compounds most difficult to oxidize, i.e. thiophene (489 ppm).(Xiao et al., 2014) The optimization of the experimental conditions lead to 90% desulfurization of this model oil attained after 2 h at 90ºC. In this publication, the authors were more interested in explore the performance of the studied heterogeneous catalyst using a real gasoline, and in the comparison of the ODS results using real gasoline and thiophene model oil (this evaluation is carefully detailed in section 2 of this chapter).

Magnetics Heterogeneous Catalysts Based POMs

The immobilization of catalytically active species, as POMs, on the surface of magnetic nanoparticles is emerging as an alternative method for the heterogenization of homogeneous catalysts. This method allows the separation and recovery of the catalyst from the reaction media by application of a simple magnetic field. The application of POM based magnetic heterogeneous catalysts in oxidative desulfurization is quite scarce and relatively new. There are only a few reports of phosphotungstic acid anions immobilized on the surface of functionalized magnetic nanoparticles as magnetic heterogeneous catalysts for ODS processes, using H_2O_2 as oxidant. One of these rare investigations was carried out by Wang and co-workers (in 2011) and reports the functionalization of the POM nanocone assemblies with magnetite (Fe_3O_4) nanocrystals, and its application as catalyst on the oxidation of sulfides (Figure 8) (Nisar, Lu, Zhuang, & Wang, 2011). The POM nanocones assemblies were prepared by ion exchange reaction between dimethyloctadecylammonium bromide and phosphotungstic acid, and their catalytic properties were evaluated for the oxidation of DBT, diphenyl sulfide (DPS) and dimethyl sulfide (DMS) into their correspondent sulfones. The conversion of DBT reached almost 100% after 38 min, and the reaction was also complete after 20 min for DPS and DMS. Besides, the catalyst could be easily recovered after reaction by applying an external magnetic field, the recycling of the catalyst was not performed.

Figure 8. a) Schematic representation of POM nanocone nanoreactor; b) oxidation of sulfides to sulfones in the presence of cones as reactors.
(Nisar, Lu, et al., 2011)

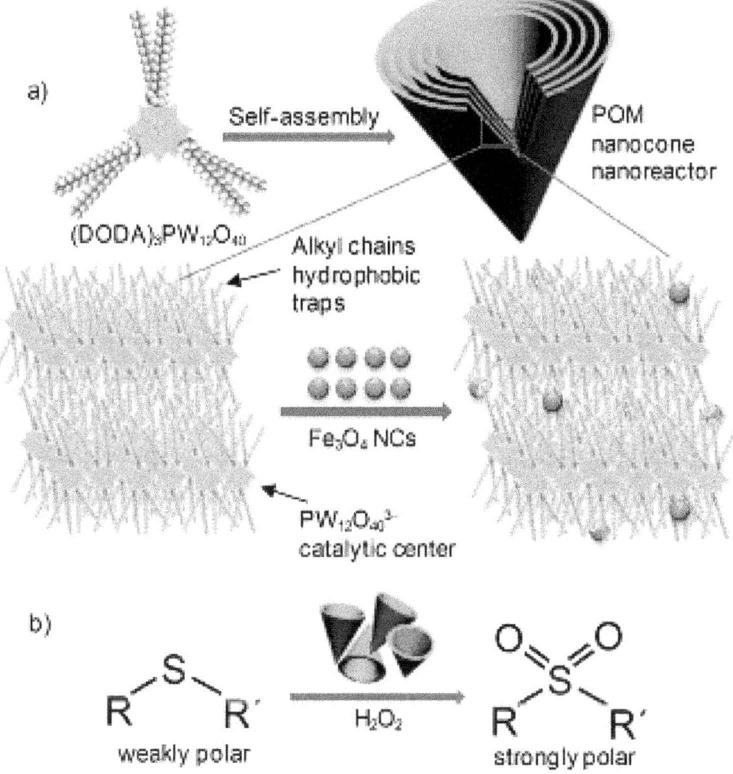

The immobilization of alkyl quaternary ammonium surfactant phosphotungstate ($[PW_{12}O_{40}]^{3-}$) compounds on nanomagnetic silica supports was the strategy used by Hu *et al.* to prepare a recyclable amphiphilic catalyst which was applied on the oxidative desulfurization of a DBT model oil, in 2012.(Cui et al., 2012) This catalyst was synthesized by the attachment of organosiloxane 3-(trimethoxysilyl)-propyldimethyloctadecyl ammonium chloride (AEM) on the surface of magnetic silica nanospheres (MSN), followed by the immobilization of the $[PW_{12}O_{40}]^{3-}$ polyanions via ion exchange (Figure 9). The composite nanospheres revealed high catalytic activity on the selective oxidation of DBT, and with the optimized conditions the sulfur level could be reduced from 487 ppm to less than 0.8 ppm. The advantageous easy recovery of this nanocatalyst by an external magnetic field was associated to the possible recyclability for various ODS cycles, and simultaneously with the separation of oxidized products from the oil.

Figure 9. Schematic representation of the oxidation of DBT catalyzed by MSN/AEM-[$PW_{12}O_{40}$] nanoparticles.
(Cui et al., 2012)

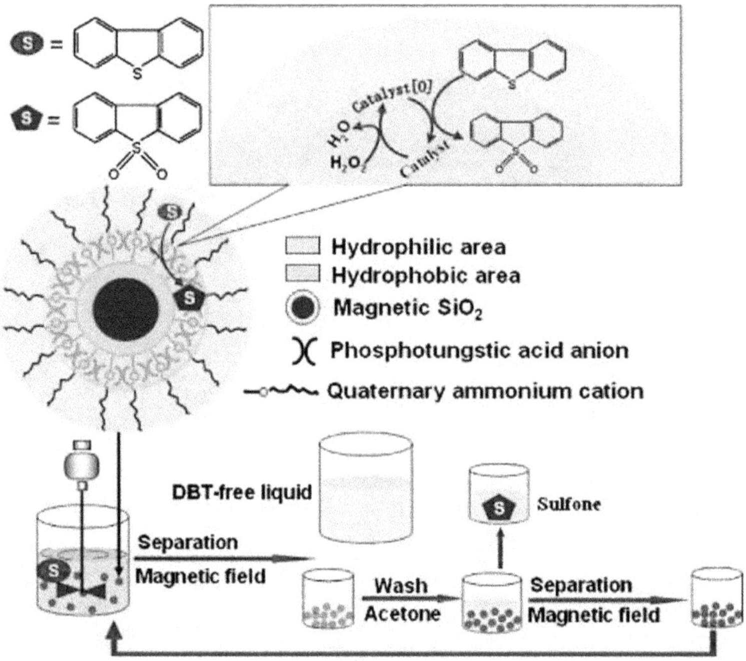

Heterogenization of POMs with Layered Double Hydroxides (LDHs)

Layered double hydroxides (LDHs) are a family of pillared solid matrices represented by the empirical formula $[M(II)_{1-x}M(III)_{x}.(OH)_2]^{x+}(A^{n-}).mH_2O$ that comprise an unusual class of layered materials with positively charged nanosheets / layers counterbalanced by exchangeable interlayer anions, such as POMs. Since the first scientific publication reported in 1988 by Pinnavaia,(Kwon, Tsigdinos, & Pinnavaia, 1988) POM@LDHs nanocomposites have attracted wide interests due to its superior application when compared to POMs or LDHs themselves. The immobilization of POMs in LDH layers can be made by different approaches, however the ion exchange, the reconstitution and co-precipitation are the most used.(Yang Chen, Yan, & Song, 2014; Omwoma, Chen, Tsunashima, & Song, 2014)

A very limited number of reports involving the use of POM@LDHs in ODS systems are known presently. Maciuca and co-workers prepared LDHs based $[M_xO_y]^{p-}$ catalysts by direct ion exchange.(Maciuca, Ciocan, Dumitriu, Fajula, & Hulea, 2008) The resulting materials were used in oxidative desulfurization of a model fuel containing methyl phenyl sulfide (MPS) and DBT dissolved in *n*-decane. The

Figure 10. Schematic representation of the directional self-assembly of LEuH nanosheets and $Na_9EuW_{10}O_{36}$.
(X. Wang et al., 2014)

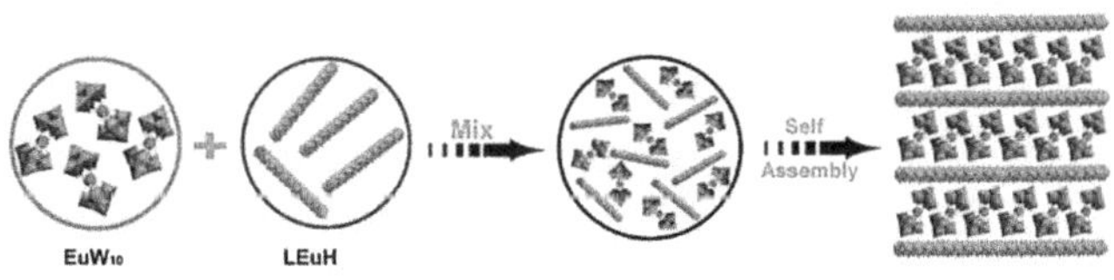

$[W_xO_y]^{p-}$ supported in LDH (W-LDH) showed to be a heterogeneous catalyst with high efficiency for the oxidation of MPS and DBT: after 3 h and 45 min, were removed from the oil phase 85% of DBT and 100% of MPS, respectively, using H_2O_2 as oxidant.

A recent publication reported new hybrid materials successfully prepared through the interconnection of the lanthano-POM $Na_9EuW_{10}O_{36}$ with the layered rare-earth hydroxo cation nanosheets, designed as LEuH-EuW_{10} (Figure 10).(X. Wang, Chen, & Song, 2014) This material was explored as heterogeneous catalyst in extractive catalytic oxidative desulfurization of model oil with a sulfur content of 1000 ppm for DBT, BT and 4,6-DMDBT, using an IL (1-octil-3-methylimidazolium hexafluorophosphate, [omim]PF_6) as extraction solvent and H_2O_2 as oxidant. This heterogeneous catalyst showed an efficient extractive catalytic desulfurization, capable to produce a model oil with only 6% of 4,6-DMDBT and 36% of BT, after the first 15 min. The influence of the temperature was also investigated, verifying that the extraction of sulfur compounds increases with the temperature, enabling shorter reaction times. In addition, the catalyst and the IL could be recycled and reused for ten consecutive cycles in oxidative desulfurization of a DBT model oil without decrease of the catalytic activity.(X. Wang et al., 2014)

Chen and co-workers reported the use of modified LDHs to facilitate the $Na_3PW_{12}O_{40}$ intercalation in the LDHs.(Yang Chen et al., 2014) The prepared material, Tris-LDH-PW_{12} (Tris = tris(hydroxymethyl)aminomethane), was investigated as heterogeneous catalyst for desulfurization of a model oil containing 1000 ppm of DBT, BT and 4,6-DMDBT in the presence of H_2O_2. Interestingly, a sulfur-free model oil was obtained after 1 h of reaction, and the catalyst showed to be capable of being recycled and reused for ten consecutive cycles on the oxidative desulfurization of DBT, practically without loss of catalytic activity.(Yang Chen et al., 2014)

Heterogenization of POMs With Other Supports

The families of heterogeneous catalysts used in oxidative desulfurization of model oils presented previously (sections 1.1 to 1.5) are the most significant; however other few examples were still reported and deserve their mention in this chapter.

To our knowledge, there is only one report of alumina-supported POM composite utilized in ODS technology. Alumina is an amphoteric material with both acidic and basic properties. It has been used as support for different POMs due to its remarkable stability resultant from the strong columbic and hydrogen bonding interactions between the POM anions and the coordinative unsaturated Al(δ<3)+ Lewis acids and Al-O- BrØnsted base sites.(Gerasimos S. Armatas, 2013) In this work the phosphomolybdic acid ($H_3PMo_{12}O_{40}$) was impregnated in γ-Al_2O_3 forming the Mo/Al_2O_3 composite, which was used as heterogeneous catalyst for desulfurization of a model oil and diesel fuel. The amount of sulfur compounds extracted from a multi-component model oil with thiophene, 2,5-dimethylthiophene (2,5-DMT), BT, DBT and 4,6-DMDBT (sulfur content of 320 ppm for each) after 1 h of reaction was 0.2, 15.3, 64.2, 98.7 and 99.7%, respectively. Using the same reaction conditions, the Mo/Al_2O_3 composite was further utilized in a diesel fuel containing 320 ppm of sulfur content, yielding an ultra-low sulfur diesel (7 ppm) after 75 min of reaction. (García-Gutiérrez et al., 2006)

Xu *et al.* reported recently the use of an interesting hybrid catalyst designed as peroxo-POM hydrogel catalyst, prepared from the peroxo-POM $[C_{16}H_{33}N(CH_3)_3]_3[PO_4\{MoO(O_2)_2\}_4]$ ($C_{16}PMo(O_2)_2$) dispersed around agarose matrix.(Jian Xu et al., 2014) The choice of agarose results from its peculiar properties such as flexibility, low cost, non-toxicity, freely solubility in hot water and the formation of a gel network. The use of agarose as support material in ODS process has particular advantages, such as the capability to improve the efficiency of the oxidant H_2O_2, through higher adsorption of H_2O_2 molecules in the proximity of the active catalytic centers. Furthermore, the ability of agarose to absorb the oxidative products from the reaction system simplifies the ODS process. The catalyst $C_{16}PMo(O_2)_2$@agarose was used in oxidative desulfurization of a model oil containing DBT with a sulfur concentration of 500 ppm. After 4.5 h of reaction at 70^{0}C, the DBT was completely oxidized to sulfone. These products were absorbed by agarose support during the reaction and consequently removed from ODS system, followed by the separation of catalyst to obtain clean model oil. Additionally, the heterogeneous catalyst could be reused for ten consecutive ODS cycles without any loss of catalytic activity.

The anatase titania crushed nano leaf is another interesting support material used to immobilize POMs. A particular advantage of the incorporation of POMs into the titania matrix is the increase of the catalyst density, making its separation easier.(Lefebvre, 2013) Three different publications describe the utilization of distinct POMs incorporated into anatase titania crushed nano leaf as heterogeneous catalysts for oxidative desulfurization of a model oil. The POMs used were $[P_2W_{18}Cd_4(Br)_2O_{68}]^{10-}$, $[SiV_3W_9O_{40}]^{7-}$ and $[PMo_{10}V_2O_{40}]^{5-}$. The performance of the catalysts was investigated using a model oil containing DBT, 4-MDBT, 4,6-DMDBT and BT with a sulfur content of 500 ppm. Using $(Bu_4N)_7H_3[P_2W_{18}Cd_4(Br)_2O_{68}]$-$TiO_2$ catalyst (Bu_4N = tetrabutyl ammonium) was obtained 98.3, 95.5, 96 and 89% of desulfurization for DBT, 4-MDBT, 4,6-DMDBT and BT respectively, after 2 h of reaction at 60°C. This catalyst demonstrated ability to be reused and recycled for three consecutive ODS cycles without losing activity.(Mohammad Ali Rezvani, Shojaie, & Loghmani, 2012) When the trilacunar $H_7SiV_3W_9O_{40}$-TiO_2 catalyst was used, the addition of formic acid together with the oxidant H_2O_2 was needed to the ODS process. This last organic acid reacts with H_2O_2 to produce a peracid, which can efficiently convert organic sulfur to sulfones without forming a substantial amount of residual product. The results obtained were very similar to those found with $(Bu_4N)_7H_3[P_2W_{18}Cd_4(Br)_2O_{68}]$-$TiO_2$ catalyst, i.e. 98% of desulfurization for DBT, 96% for 4-MDBT, 95% for 4,6-DMDBT and 93% for BT was achieved after 2 h at 70 °C. The catalyst was recycled for three consecutive cycles without decrease of efficiency. In the last example, the use of $[PMo_{10}V_2O_{40}]^{5-}$-TiO_2 as catalyst also originated very positive result of desulfurization for each sulfur compound, after 2 h of reaction at 70 °C: 98, 96, 97 and 94% for DBT, 4-MDBT, 4,6-DMDBT and BT, respectively.(M. A. Rezvani, Shojaei, & Zonoz, 2014)

Heterogenization of POMs With Organic Counter-Cations

A simple strategy to prepare POM-based heterogeneous catalysts is the modification of POMs with appropriate organic units. Different organic groups, such as surfactants with different carbon-chain lengths and ionic-liquids, have been applied to obtain POMs-based hybrid catalysts with high efficiency and ability of recovery.(Zhou et al., 2014)

Surfactant encapsulated POMs have been used for oxidative desulfurization, and this family of materials is considered one of the most efficient heterogeneous POMs. (Junhua Xu, Zhao, Ji, & Song, 2013) Several types of POMs have been used in the preparation of surfactant encapsulated POMs to be applied in oxidative desulfurization, namely decatungstates $[W_{10}O_{32}]^{4-}$, Keggin $[XM_{12}O_{40}]^{n-}$, Anderson $[XM_6O_{24}]^{n-}$ and Lindqvist $[M_6O_{19}]^{n-}$.

Hybrid Decatungstates POMs

Li and co-workers reported the application of surfactant-type decatungstates $Q_4(W_{10}O_{32})$ [Q=$(CH_3)_3NC_{16}H_{33}$, $(CH_3)_3NC_{14}H_{29}$, $(CH_3)_3NC_{12}H_{25}$ and $(CH_3)_3NC_{10}H_{21}$] in the oxidation of a model oil using H_2O_2 as oxidant.(X. Jiang et al., 2009) The length of the carbon chain of the quaternary ammonium cations has an important role on the catalytic performance: longer carbon chain revealed better catalytic performance. In the optimized conditions, the desulfurization of DBT oil (1000 ppm) achieved 99,6% after 0.5 h at 60ºC, BT (1000 ppm) reached 99.6% at 3.25 h and for 4,6-DMDBT it was attained 99.4% after 1.25 h, using the catalyst $[(CH_3)_3NC_{16}H_{33}]_4W_{10}O_{32}$. This catalyst could be recycled several times without any significant loss of catalytic activity, thus decatungstates with quaternary ammonium cations with long carbon chain revealed to be efficient catalysts in ODS.

Hybrid Keggin-Type and Derivatives POMs

The phosphotungstic acid is one of the most studied POMs in ODS systems, owing to its excellent redox properties and commercial availability, as could be found in the previous topics of this chapter. Li and co-workers prepared the catalyst $[(C_{18}H_{37})_2N(CH_3)_2][PW_{12}O_{40}]$ assembled in emulsion, in 2004.(C. Li et al., 2004) This catalyst could oxidize completely an oil containing 4,6-DMDBT (450 ppm) after 80 min, under mild conditions (T = 30ºC), using H_2O_2 as oxidant in stoichiometric amounts. Interestingly, when the temperature was increased to 60ºC the oxidative desulfurization was completed in 45 min. This catalyst could also be easily separated from the reaction media by demulsification and sedimentation.

Li and co-workers synthesized different hybrid POMs based on quaternary ammonium cations and phosphotungstic acid, one year later.(C. Li et al., 2005) These heterogeneous catalysts were investigated in the ODS of model oils containing various sulfur compounds. The hybrid POMs behaved both as catalysts and as emulsifying agents, instead of only simple surfactants. BT, DBT and 4,6-DMDBT could be completely oxidized to its corresponding sulfones after 1.5 h at 60 ºC, using $[(C_{18}H_{37})_2N(CH_3)_2]_3[PW_{12}O_{40}]$ as catalyst. Besides, this catalyst could be recycled for three ODS cycles maintaining the catalytic performance. The ability to form metastable emulsions and consequently the capability of recovering the catalyst from reaction medium depended mainly on the alkyl chains of the cation.

Wang and collaborators also promoted the electrostatic interaction of phosphotungstic acid with different cationic surfactants based in various alkyl chains.(Nisar, Zhuang, & Wang, 2011). The authors investigated the catalytic efficiency of these hybrid POMs on the oxidative desulfurization of DBT model oil. The amphiphilic units with similar chemical behaviors presented similar catalytic efficiencies. In 2006, Luo *et al.* prepared a combined amphiphilic catalyst and optimized the molar ratio between phosphotungstic acid and the quaternary ammonium cation trimethyloctadecylammonium (ODA) for the ODS process using a DBT model oil (3000 ppm in *n*-octane). (D. Huang, Wang, Yang, & Luo, 2006) The best results were attained with a molar ratio 1:1 between the ODA cation and the POM, obtaining complete conversion of DBT just after 20 min at 50ºC. Remarkably, the ODS process could also be shortened to 10 min when the temperature was increased to 70ºC. The heterogenization of phosphotungstic acid was also achieved using an ion-exchange reaction between [3-(acryloyamino)propyl]dodecyldimethyl ammonium (APDDAB) located within the porous of poly(acrylamide) microgels (PAM) and the phosphotungstic acid in aqueous solution.(H. Li, Zhang, Zhang, Zhou, & Hu, 2009) This catalyst was tested in ODS of a model oil with DBT in decalin (2000 ppm): after 24 h of reaction the sulfur content in oil was less than 25 ppm. This heterogeneous catalyst formed by microspheres was separated by decantation and reused in three cycles with similar catalytic performances.

Rafiee and co-workers synthesized several organic-inorganic POMs composed by sulfonated pyridinum cations and Keggin tungstophosphate anion ($[PW_{12}O_{40}]^{3-}$), which were used in catalytic oxidation of sulfur containing model oil (methyl phenyl sulfide, thiophene and dibenzothiophene).(Rafiee & Eavani, 2013) Under the optimized conditions, the catalyst revealed good catalytic activity and selectivity. Besides, it could be separated by filtration and recycled five times with high catalytic activity between cycles.

The quaternary ammonium cation $[(C_{18}H_{37})N^+(CH_3)_3]$ was also used as surfactant in the synthesis of a series of emulsion catalysts using various Keggin-type mono-substituted POMs: $PW_{11}M(H_2O)O_{39}$ with M=Ti, Mn, Fe, Co, Ni and Cu. These catalysts were employed in the ODS process of a BT model oil in decalin.(Zhang et al., 2010) The emulsion catalyst with the Keggin POM $[PW_{12}O_{40}]^{3-}$ revealed to be inactive in the oxidation of BT using H_2O_2 as oxidant at 30ºC. On the other hand, the hybrid catalyst based on the mono-lacunar $[PW_{11}O_{39}]^{7-}$ is more active and could completely oxidize BT at the same conditions. However, the catalytic activity of $[(C_{18}H_{37})N(CH_3)_3]_5Na_2[PW_{11}O_{39}]$ decreased considerably when the lacunary site was coordinated with transition metal cations. Some years before, Li *et al.* also reported the preparation and utilization of an amphiphilic Keggin-type lacunary catalyst $[(C_{18}H_{37})N^+(CH_3)_3]_4[H_2NaPW_{10}O_{36}]$ in oxidative desulfurization of model oils of various sulfur-compounds, such as BT, 5-methylbenzothiophene (5-MBT), DBT and 4,6-DMDBT (Figure 11).(Lü et al., 2006) Each one of these compound was investigated separately on the respective model oils in decalin (1000 ppm). Total desulfurization was achieved for 4,6-DMDBT and DBT just in 20 min, at 40ºC. In the case of 5-MDBT were needed 90 min and for BT the complete desulfurization was only attained after 120 min of reaction.

Figure 11. Schematic representation of the oxidative desulfurization of a model oil (or diesel) containing BT in water/oil emulsion droplets.
(Lü et al., 2006)

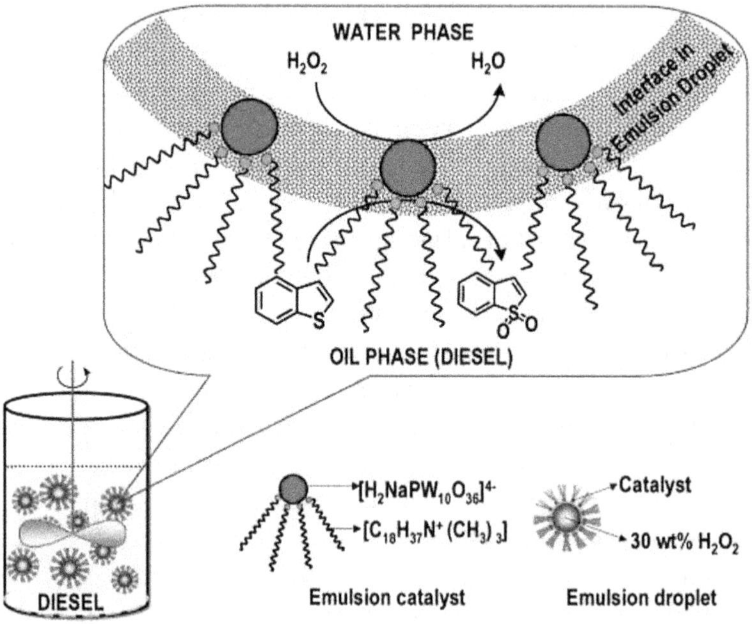

Li and co-workers used an emulsion system on the oxidative desulfurization of a DBT model oil (in decalin), composed by an amphiphilic Keggin-type derivative POM catalyst with a mixed-addenda, $[(C_{18}H_{37})N(CH_3)_3]_5[PV_2Mo_{10}O_{40}]$.(Lu et al., 2007) In this case, the molecular oxygen was used as oxidant, the isobutyl aldehyde as sacrificial agent, and the ODS process occurred in the presence of acetonitrile as extraction solvent. In the optimized conditions, DBT was completely oxidized and efficiently removed from the decalin phase after 4 h. This is an interesting ODS result mainly due to the use the green oxidant O_2; however, the activity of the own sacrificial agent is not described in this work.

In 2008, Li *et al.* used a peroxophosphomolybdate with different alkyl carbon-chain $Q_3\{PO_4[MoO(O_2)_2]_4\}$ ($Q=[(C_4H_9)_4N)]^+$, $[C_{14}H_{29}N(CH_3)_3]^+$ and $[C_{16}H_{35}NC_5H_5]^+$) to study the oxidative desulfurization of model oil with BT, DBT and 4,6-DMDBT, in the presence of an IL (1-butyl-3-methylimidazolium tetrafluoroborate ($[BMI]BF_4$)) as extraction solvent and H_2O_2 as oxidant.(He et al., 2008) The catalysts with shorter alkyl carbon-chain exhibited higher catalytic activity. In fact, the peroxophosphomolybdate with shorter alkyl chain allowed an oxidative desulfurization of DBT from 1000 ppm to 8 ppm in 3 h (reaction temperature: 70°C). Furthermore, this system was recycled for four cycles with only a slight decrease of activity. Huo and co-workers also prepared peroxo-polyoxometalates with different quaternary ammonium cations that were used in the oxidative desulfurization of DBT oil with O_2 as oxidant, a few years later.(C. Jiang et al., 2011) In this case, a peroxotitanium-polyoxometalate ($[PW_{10}Ti_2O_{38}(O_2)_2]^{7-}$) was used, and the catalyst with longer carbon chain ($[(C_{18}H_{37})N^+(CH_3)_3]$) revealed the best performance. This catalyst, allowed the decrease of sulfur level of the DBT model oil from 500 ppm to 1 ppm in 6 h at 90ºC.

Hybrid Anderson-Type POMs

The Anderson-type POMs are considerably less explored than the Keggin-type. The application of Anderson-type POMs is reduced but these compounds have gained an increased interest due to their catalytic performance under aerobic conditions, mainly when associated with quaternary ammonium cations with long carbon-chains. Furthermore, these compounds can behave as heterogeneous catalyst, capable to be removed from ODS system and recycled.

Anderson-type POMs were used by Li and collaborators in the oxidative desulfurization of BT, DBT and 4,6-DMDBT.(Lu, Zhang, Jiang, & Li, 2010) Using $[(C_{18}H_{37})_2N(CH_3)_2]IMo_6O_{24}$ as catalyst and molecular oxygen as oxidant, the sulfur compounds mentioned could be completely oxidized without the use of any sacrificial agent, after 8 h at 90ºC. Interestingly, when the same POM was prepared with shorter carbon-chain the catalytic activity decreased. Three years later, Lü *et al.* prepared a series of Anderson-type catalysts ($Q_4NiMo_{6-x}W_xO_{24}H_6$ (x=0, 2, 4, 6)) that were used on deep desulfurization of sulfur multicomponent model oil (BT, 5-MBT, DBT, 4-MDBT or 4,6-DMDBT; 500 ppm of each) using H_2O_2 as oxidant. (Lü, Ren, Wang, et al., 2013) The conversion of DBT was clearly improved with the increase of the amount of molybdenum ions in the Anderson-type POMs: the catalyst $[(C_4H_9)_4N]_4NiMo_6O_{24}H_6$ allowed a DBT conversion of 98% after 3 h at 30ºC. The same research group reported the use of a B-type Anderson polyoxomolybdate $[(C_{18}H_{37})_2N(CH_3)_2]_3Co(OH)_6Mo_6O_{18}$ in an ODS process.(Lü, Ren, Liao, et al., 2013) In this case, the molecular oxygen was used as oxidant agent and the model oil was composed by BT, DBT or 4,6-DMDBT (500 ppm of each) in decalin. This catalyst reveled to be efficient in aerobic oxidative desulfurization without the use of any sacrificial agent: DBT (500 ppm) was completely oxidized after 7 h, at 80ºC.

Hybrid Lindqvist -Type POMs

The Lindqvist-type POMs containing lanthanide bridged centers have gained notable interest in ODS systems due to their high catalytic efficiency using H_2O_2 as oxidant (see also section 1.2). Some example of these type of POMs combined with quaternary ammonium cations with long carbon-chains to prepare active heterogeneous catalyst for ODS, can also be found in the literature.

In recent years, a growing effort has been developed to use ionic liquids cations as counter-cation parts to create hybrid POMs.(Zhou et al., 2014) There are few examples of application these hybrid POMs in ODS processes. Most of these cases are based in the combination of the POMs based catalysts with ILs as extraction solvents, resulting in efficient catalytic systems behaving as homogeneous catalysts that can be reused for several times. However, when these hybrid POMs are used as heterogeneous catalysts (without ionic liquids in the system) the catalytic activity is usually very low.(W. Huang et al., 2010; W. Zhu et al., 2011)

Oxidative Desulfurization Studies Using Liquid Fuel Transportation

Liquid fuels consist in a complex mixture of organic liquids based on hydrocarbons containing sulfur, nitrogen and oxygen. The sulfur content of heavy crude oil can vary from 0.03 to 7.89 mass%, depending on the type and source of crude oils. (Soleimani et al., 2007) Sulfur compounds in liquid fuels exist in various forms and can be classified into four main groups: mercaptans, sulfides, disulfides and thiophenes (Figure 1). The current industrial method for removal of sulfur from fuels is the hydrodesulfurization (HDS). The actual requirements for very low sulfur content (10 ppm) in liquid motor fuels demand the production of low-sulfur fuel oil (< 10 ppm), without the presence of refractory sulfur compounds like 4,6-dimethyldibenzothiophene and other alkyl-substituted thiophene derivatives. The elimination of these compounds from liquid fuels has turned the actual HDS technology to operate under extreme severe conditions which implies a high economical cost of the process, increasing the price of the transportation fuels.

The oxidative desulfurization (ODS) is currently considered one of the most promising alternative or complementary process to the HDS offering several advantages, such as mild reaction conditions, high selectivity and economic viability. In addition to the operational advantages of ODS relatively to HDS, it is also very efficient for the desulfurization of the refractory sulfur compounds, because these are simply oxidized in more polar molecules. Consequently, these molecules can be easily removed from liquid fuels by conventional separation processes such as solvent extraction or adsorption. (Mjalli, Ahmed, Al-Wahaibi, Al-Wahaibi, & AlNashef, 2014) The H_2O_2 has been successfully used as oxidant for ODS process with liquid fuels. Furthermore, H_2O_2 has been preferentially chosen due to its environmental benign properties, but this oxidant needs to be activated in the presence of a catalyst. Among different oxidation systems, the H_2O_2/POMs is still poorly explored but has demonstrated high efficiency. When POMs are supported in a solid material (sections 1.1 to 1.6) or associated to a suitable cation to form an insoluble material (section 1.7), the ODS system H_2O_2/POMs gains the possibility to be reused and/or recycled, and in this case a higher amount of liquid fuel can be desulfurized using smaller amounts of reagents, decreasing the cost of the desulfurization process.

In the first topic of this chapter various studies reporting the use diverse heterogeneous ODS systems based POMs to desulfurize different model oils containing a single refractory sulfur compound or a mixture of compounds dissolved in a non-polar solvent were described. These were used as model oils to simulate the liquid fuels, however, a liquid fuel is composed of about 75% of saturated hydrocarbons (primarily paraffins including *n*, *iso*, and cycloparaffins), and 25% aromatic hydrocarbons (including naphthalenes and alkylbenzenes).((ATSDR). 1995) The presence of this wide variety of hydrocarbons containing sulfur, nitrogen and oxygen has a strong influence on the ODS process. An efficient catalytic ODS process using model oil can behave differently when a liquid fuel is used, because various other organic molecules susceptible to be oxidized are also present. Therefore, the selectivity of the catalyst to oxidize the sulfur compounds is an essential characteristic in liquid fuels systems.

A few examples comparing the heterogeneous catalytic performance of the POM based catalyst using the model oil and the liquid fuel are already reported in the literature. The first work reporting the application of a heterogeneous POM in ODS system in a liquid fuel was presented by Li and co-workers in 2006.(Lü et al., 2006) An amphiphilic Keggin-type lacunary compound, $[(C_{18}H_{37})N^{+}(CH_3)_3]_4[H_2NaPW_{10}O_{36}]$, was used as catalyst to desulfurize prehydrotrated (500 ppm) and straight-run diesels (6000 ppm), in the presence of H_2O_2 as oxidant, at 30ºC. The sulfones obtained were removed by a polar organic solvent after 2 hours. The sulfur quantity of the prehydrotrated diesel as lowered from 500 to 0.1 ppm after oxidation and extraction, while the sulfur level of a straight-run diesel decreased from 6000 to 30 ppm after oxidation and extraction. These results were similar to those obtained with the model oils (see 1.7.2 section and Figure 11) and confirm the efficiency of the amphiphilic Keggin-type lacunary catalyst for the ODS process.

A small number of works using heterogeneous POMs based catalysts have been reported three years later, and two of them also using amphiphilic POMs. Wang and collaborators explored the phosphomolybdic acid ($H_3PMo_{12}O_{40}$) modified with tetramethyl ammonium chloride (TMAC), dodecyl trimethyl ammonium chloride (DTAC) and hexadecyl trimethyl ammonium chloride (HTAC).(Qiu et al., 2009) The authors used these catalysts to treat a straight-run diesel in the presence of H_2O_2, at 70ºC. The reaction products were removed after oxidation using liquid-liquid extraction with dimethylformamide. The authors referred that some components in diesel were also oxidized along with the oxidation of sulfur-containing compounds, decreasing the recovery rate of diesel. The best catalyst was the POM with longer alkyl chain $[HTAC]_2H[PMo_{12}O_{40}]$ achieving 93.1% of desulfurization after 2 h, with a recovery rate of diesel of 84.2%. In fact, the authors mentioned that increasing the extraction times, the desulfurization rates increase but the recovery rates of diesel decrease. Li *et al.* also reported the application of the surfactant decatungstate $Q_4(W_{10}O_{32})$ ($Q=(CH_3)_3NC_{16}H_{33}$, $(CH_3)_3NC_{14}H_{29}$, $(CH_3)_3NC_{12}H_{25}$ and $(CH_3)_3NC_{10}H_{21}$) in real diesel (1078 ppm) using the same operational conditions as for model oils presented in 1.7.1.(X. Jiang et al., 2009) The oxidant used was also H_2O_2 at 60ºC, and the extraction of oxidized products was also performed with dimethylformamide for five times. The sulfur content of the real diesel could be lowered from 1078 to 4 ppm. In 2009, Yan and co-workers investigated a heterogeneous catalyst based on Keggin-type POM $H_3PW_{12}O_{40}$ supported on the surface of mesoporous silica modified with Ag^+.(X.-m. Yan et al., 2009) This catalyst was used to desulfurize a straight-run diesel (1800 ppm) using H_2O_2 as oxidant. The sulfur compounds concentration was reduced to 228 ppm after 4 h of reaction. Apparently the authors did not perform an extraction of oxidized products after oxidation. This silica composite revealed a lower performance using straight-run diesel than the model oils (presented in 1.2). However, this silica catalyst was recycled using this liquid fuel for three ODS cycles and only a slightly decrease of activity was noticed.

In 2011, a distinct silica catalyst based in the ionic liquid Keggin-type $(BMI)_3[PW_{12}O_{40}]$ (BMI=1-butyl-3-methyl imidazolium) was used for desulfurization of real diesel (530 ppm) in the presence of H_2O_2 oxidant. The oxidized compounds were also removed by liquid extraction (one extraction) after oxidation with equivalent volumes of diesel and dimethylformamide. The best sulfur removal was 97.9% after 3 h at 80ºC, using a molar ratio O/S of 5. In the same year, an ionic liquid emulsion system with $[(C_8H_{17})_3NCH_3]_2W_6O_{19}$ was used by Li and collaborators for desulfurization of a real gasoline (415 ppm), using H_2O_2 as oxidant and the ionic liquid (1-octyl-3-methylimidazolium hexafluorophosphoric) as extraction solvent, at 60ºC (Figure 12). The real gasoline was desulfurized for four ODS cycles during 1 h each cycle, and the sulfur content decreased from 415 to 25 ppm, with a total sulfur removal of 94%.(Ding et al., 2011)

Li and Abdalla used a mono-lacunar $[PW_{11}O_{39}]^{7-}$ polyanion supported in mesoporous MCM-41 to treat a FCC diesel (974 ppm) with H_2O_2 as oxidant ($H_2O_2/S = 4$) during 3 h at 70ºC.(Abdalla & Li, 2012) After the oxidation step, the oxidized products were extracted from diesel with methanol for five cycles. This POM based heterogeneous ODS process allowed a reduction of sulfur from 974 to 27.3 ppm, corresponding to 97.2% of removal. However, the results obtained with FCC diesel were inferior to the previous study performed in model oils (section 1.2) where complete desulfurization was achieved after 1 h. Furthermore, in model oils the extraction of oxidized products after oxidation was performed in a single cycle, which demonstrates the higher difficulty to remove oxidized products from liquid fuel media.

Figure 12. Schematic representation of the desulfurization of real gasoline using an ionic liquid emulsion system with $[(C_8H_{17})_3NCH_3]_2W_6O_{19}$ in the presence of H_2O_2 oxidant.
(Ding et al., 2011)

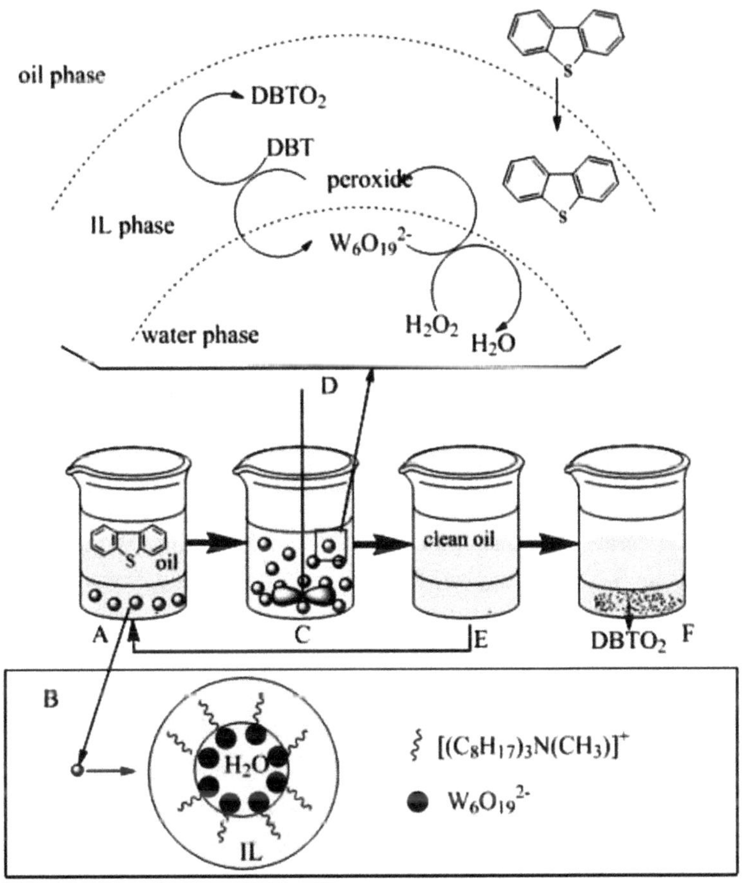

Figure 13. Oxidative Desulfurization process of prehydrotreated diesel (500 ppm).
(Y. Liu et al., 2013)

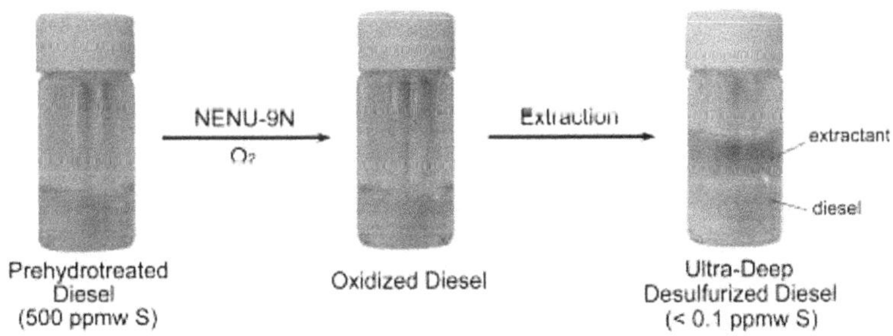

The study of ODS in liquid fuels using other solid supports to immobilize POMs as active centers from heterogeneous catalysts was carried out more recently. One of these works used the Metal-Organic Framework NENU-9 ($[(CH_3)_4N]_2\{[Cu_2(BTC)_{4/3}(H_2O)_2]_6\}$) to accommodate the POM $H_5PV_2Mo_{10}O_{40}$.(Y. Liu et al., 2013) This heterogeneous catalyst was used to desulfurize a prehydrotreated diesel (500 ppm), at 80°C in the presence of isobutyraldehyde as sacrificial agent and molecular oxygen as oxidant. The oxidative stage of the process occurred during 1.5 h followed by the liquid extraction of the oxidized products with acetonitrile. The sulfur quantity decreased from 500 ppm to 0.1 ppm (Figure 13). The results obtained with the prehydrotreated diesel were similar to those previous described using the DBT model oil in decaline (section 1.1).

Hydrotalcite was also used to support the Keggin-type POM, $[PMo_{12}O_{40}]^{3-}$ leading to the composite material $Mg_9Al_3(OH)_{24}[PMo_{12}O_{40}]$.(F. L. Yu & Wang, 2013) This heterogeneous catalyst was explored to desulfurize a diesel with 493 ppm of sulfur using a biphasic diesel/acetonitrile system (H_2O_2 utilized as oxidant with molar ratio O/S of 20, at 60°C). The sulfur content of diesel decreased from 493 to 107 ppm after 3 h of oxidation (78.30% of sulfur removal) using equal volume of diesel and acetonitrile. However, the duplication of the volume of acetonitrile relatively to the volume of diesel clearly increased the desulfurization: only 34.8 ppm remained in real diesel after de 3 h of oxidation. Furthermore, the liquid extraction performed with acetonitrile after oxidation allowed lowering the sulfur content to 9.12 ppm.

Activated carbon was used as support material to immobilize a POM, and used as heterogeneous catalyst for desulfurization of gasoline (280 ppm), for the first time in 2014.(Xiao et al., 2014) This catalyst possess as active center the Keggin-type phosphotungstic acid $H_3[PW_{12}O_{40}]$, and the oxidant used was H_2O_2 in the presence of cetyltriethylammonium as emulsion agent: after 2 h at 90°C, only 32% of the initial sulfur in gasoline has been removed. The authors compared this result with the identical system replacing the gasoline by a thiophene model oil. However, using this last simulated gasoline the desulfurization was 90% (section 1.3). This difference of performance was justified by the strong influence of gasoline composition, i.e. the large amounts of alkylated aromatic hydrocarbons, alkenes and cycloalkenes inhibit the desulfurization of sulfur compounds. The authors suggest that the inhibition occurs due to the competition for the catalytic active centers between the sulfur compounds and other active alkyl chain in alkylated aromatic hydrocarbons, and also the active pi-electrons in alkenes.

Xu and co-workers used a hybrid catalyst formed by a peroxo-POM $[C_{16}H_{33}N(CH_3)_3]_3[PO_4\{MoO(O_2)_2\}_4]$ ($C_{16}PMo(O_2)_2$) dispersed in agarose matrix to treat a diesel oil (974 ppm).(Jian Xu et al., 2014) The oxidation was performed during 6 h, using H_2O_2 as oxidant (molar ratio of H_2O_2 to oil was 6:1) at 70°C, and at the end the diesel was simply filtered to give a sulfur free diesel (4 ppm). In fact, the authors referred that agarose has the ability to absorb the oxidative products from the reaction system. The desulfurization performance of this catalyst using diesel and DBT model oil (presented in section 1.6) was similar.

FUTURE RESEARCH DIRECTIONS

The detailed approach performed in this chapter between the various heterogeneous POMs catalytic systems for ODS processes applied to model oils and to liquid fuels, allow-us to propose some lines that needed to be explored in this area of research to improve activity, selectivity, robustness of the catalytic systems and also the viability for their application in refineries.

The reported studies indicate that the Keggin-type and Lindqvist-type incorporating bridged lanthanides are the most suitable catalysts for ODS systems using H_2O_2 oxidant. The Lindqvist-type POMs have demonstrated high oxidative desulfurization performance in the presence of ionic liquids. The ionic liquids can be present in the system as extraction liquid phase receiving sulfur compounds from fuels or even more interesting, can be present as functional groups on the surface of solid support materials (such as silica) or even in the cavities of MOF materials. When ionic liquids are present as a liquid phase, these can also behave as a co-catalyst immobilizing the homogeneous active species and allowing the recyclability of the ODS systems for various cycles. On the other hand, when ionic liquids are present on the surface of a solid support can promote the stability of the POM immobilized, avoiding leaching occurrence. Some examples could be found in the literature reporting the application of the ionic liquid as an extraction solvent with co-catalyst properties, but unfortunately scarce examples are described using heterogeneous catalysts functionalized with ionic liquids.

The Keggin-type POMs $[PM_{12}O_{40}]^{n-}$ (M = W, Mo and V) have been largely explored in ODS systems using model and liquid fuels, probably as consequence of their easy availability (most of these compounds are commercially available). However, from the comparative analysis of the ODS it is possible to find that the lacunar derivatives are more active, probably due to the presence of available non-coordinated oxygen atoms in the lacunary region of the POM structure.

The Anderson-type POM structures are also of high interest to be used in ODS systems because these compounds have revealed high capacity to catalyze under aerobic conditions without the use of co-catalysts or sacrificial agents. A small number of examples were reported recently however, this is a subject that need to be explored to enlarge the application of molecular oxygen in ODS based POMs catalytic systems.

The application of ultrasound irradiation also necessitate to be investigated using the heterogeneous POMs catalytic ODS systems, given that the few examples already reported demonstrated that the use of ultrasound irradiation avoids the use of costly phase transfer agent.

The main effort required in this area is the application of novel heterogeneous POMs based ODS systems with liquid fuels, because the literature has demonstrated that the efficiency of these systems in liquid fuels can be different from that of the model oils. The optimization and the investigation of the principal parameters responsible for the desulfurization efficiency need to be carried out also using the complex real liquid fuels.

CONCLUDING REMARKS

A myriad of heterogeneous POMs based nanocatalysts successfully applied in the oxidative desulfurization of model oils and liquid fuels, was reported minutely. However, much more work has been developed using model oils and only a few examples report the comparison of efficiency between model oils and liquid fuels. In fact, the desulfurization of liquid fuels is much more difficult than with model oils. In liquid fuels sulfur compounds co-exist in various forms (mercaptans, sulfides, disulfides and thiophenes) and their selective oxidation is a challenge as consequence of the complex mixture of hydrocarbons (75% saturated hydrocarbons and 25% of aromatic hydrocarbons), which has a strong influence in the ODS process. This feature has been detected mainly using heterogeneous catalysts based in POM supported in some specific supports (mainly using activated carbon and silica supports). In these cases, the authors defend that competitive oxidation reactions may occur with other alkylated aromatic hydrocarbons, alkenes and cycloalkenes compounds decreasing the oxidation of the sulfur ones. On the other hand, successful examples in the oxidative desulfurization of liquid fuels could also be here described. Some of these examples used hybrid POM catalysts containing quaternary ammonium cations with different carbon-chain lengths that behave as amphiphilic POMs and also ionic-liquids. The presence of an extraction solvent during the oxidation stage of the system also demonstrated to have a positive contribution to achieve higher sulfur removal from simulated and real liquid fuels. In addition, the oxidative desulfurization of liquid fuels was increased when a final extraction of oxidized products was performed to achieve ultra-low sulfur fuels. This indicates that ODS systems with liquid fuels are more difficult to oxidize selectively, and also the removal of oxidized products is more difficult than in clean model fuel. This is probably due to the strong interaction between the complex mixture of molecules present in liquid fuels. The H_2O_2 is a successful oxidant used in the presence of heterogeneous POMs catalysts, based on various structures (decatungstates, Keggin, Anderson and Lindqvist-type) supported in different solid supports (MOFs, silica, carbon materials, LDHs, etc). The high reactivity of this oxidant with supported POMs or hybrid POMs must be associated to the oxidative mechanism involved. This is well described in the literature and is based on the formation of a peroxometalate active intermediate, by the interaction of the POM with the H_2O_2. Some few examples already reported the use of molecular oxygen as oxidant. The most successful cases were verified with Anderson-type POMs as catalysts, because when Keggin-type derivatives were used the presence of aldehydes as sacrificial agent was required.

In conclusion, the heterogeneous POMs based catalysts are promising catalysts able to produce ultra-low sulfur liquid fuels under sustainable conditions. However, this is still considered an unexplored area of research that need to improve the design of catalysts, as well as optimize desulfurization procedures and conditions to better understand the crucial parameters that domain the desulfurization of fuel liquids.

ACKNOWLEDGMENT

This work was partly financed by FEDER (Fundo Europeu de Desenvolvimento Regional) through COMPETE (Programa Operacional Factores de Competitividade) and by national funds through the FCT (Fundação para a Ciência e a Tecnologia) within the projects FCOMP-01-0124-FEDER-020658 (FCT ref. PTDC/EQU-EQU/121677/2010), FCOMP-01-0124-FEDER-013026 (FCT ref. PTDC/CTM/100357/2008) and Requimte (UID/QUI/50006/2013).

REFERENCES

Abdalla, Z. E. A., & Li, B. S. (2012). Preparation of MCM-41 supported (Bu4N)(4) H-3(PW11O39) catalyst and its performance in oxidative desulfurization. *Chemical Engineering Journal*, *200*, 113–121. doi:10.1016/j.cej.2012.06.004

Abdalla, Z. E. A., Li, B. S., & Tufail, A. (2009). Direct synthesis of mesoporous (C19H42N)(4)H-3(PW11O39)/SiO2 and its catalytic performance in oxidative desulfurization. *Colloids and Surfaces a-Physicochemical and Engineering Aspects*, *341*(1-3), 86-92.

Armatas, G. S. (2013). Heterogeneous Polyoxometalate-Containing Mesoporous Catalysts. In S. L. Suib (Ed.), *New and Future Developments in Catalysis* (pp. 311–342). Amsterdam: Elsevier. doi:10.1016/B978-0-444-53876-5.00014-3

Armatas, G. S., Bilis, G., & Louloudi, M. (2011). Highly ordered mesoporous zirconia-polyoxometalate nanocomposite materials for catalytic oxidation of alkenes. *Journal of Materials Chemistry*, *21*(9), 2997–3005. doi:10.1039/c0jm03395b

ATSDR. (1995). *Toxicological profile for fuel oils.* Atlanta, GA: U.S. Department of Health and Human Services, Public Health Service.

Balula, S. S., Granadeiro, C. M., Barbosa, A. D. S., Santos, I. C. M. S., & Cunha-Silva, L. (2013). Multifunctional catalyst based on sandwich-type polyoxotungstate and MIL-101 for liquid phase oxidations. *Catalysis Today*, *210*, 142–148. doi:10.1016/j.cattod.2012.12.003

Chamack, M., Mahjoub, A. R., & Aghayan, H. (2014). Cesium salts of tungsten-substituted molybdophosphoric acid immobilized onto platelet mesoporous silica: Efficient catalysts for oxidative desulfurization of dibenzothiophene. *Chemical Engineering Journal*, *255*, 686–694. doi:10.1016/j.cej.2014.06.054

Chen, Y., & Song, Y. F. (2014). Immobilization of LaW10 onto Ionic-Liquid-Modified Mesoporous Silica: Deep Desulfurization with Zero-Order Reaction Kinetics. *ChemPlusChem*, *79*(2), 304–309. doi:10.1002/cplu.201300323

Chen, Y., Yan, D., & Song, Y.-F. (2014). Tris(hydroxymethyl)aminomethane modified layered double hydroxides greatly facilitate polyoxometalate intercalation. *Dalton Transactions (Cambridge, England)*, *43*(39), 14570–14576. doi:10.1039/C4DT01827C PMID:25191827

Chen, Y., Zhao, S., & Song, Y. F. (2013). An efficient heterogeneous catalyst based on highly dispersed Na(7)H(2)LaW(10)O(36 center dot)32H(2)O nanoparticles on mesoporous silica for deep desulfurization. *Applied Catalysis A, General*, *466*, 307–314. doi:10.1016/j.apcata.2013.06.030

Coronado, E., & Gómez-García, C. J. (1998). Polyoxometalate-Based Molecular Materials. *Chemical Reviews*, *98*(1), 273–296. doi:10.1021/cr970471c PMID:11851506

Cui, X., Yao, D., Li, H., Yang, J., & Hu, D. (2012). Nano-magnetic particles as multifunctional microreactor for deep desulfurization. *Journal of Hazardous Materials*, *205–206*(0), 17–23. doi:10.1016/j.jhazmat.2011.11.063 PMID:22244972

Ding, Y., Zhu, W., Li, H., Jiang, W., Zhang, M., Duan, Y., & Chang, Y. (2011). Catalytic oxidative desulfurization with a hexatungstate/aqueous H2O2/ionic liquid emulsion system. *Green Chemistry*, *13*(5), 1210–1216. doi:10.1039/c0gc00787k

Du, D. Y., Qin, J. S., Li, S. L., Su, Z. M., & Lan, Y. Q. (2014). Recent advances in porous polyoxometalate-based metal-organic framework materials. *Chemical Society Reviews*, *43*(13), 4615–4632. doi:10.1039/C3CS60404G PMID:24676127

Farsani, M. R., & Yadollahi, B. (2014). Synthesis, characterization and catalytic performance of a Fe polyoxometalate/silica composite in the oxidation of alcohols with hydrogen peroxide. *Journal of Molecular Catalysis A Chemical*, *392*, 8–15. doi:10.1016/j.molcata.2014.05.001

Férey, G., Mellot-Draznieks, C., Serre, C., Millange, F., Dutour, J., Surblé, S., & Margiolaki, I. (2005). A Chromium Terephthalate-Based Solid with Unusually Large Pore Volumes and Surface Area. *Science*, *309*(5743), 2040–2042. doi:10.1126cience.1116275 PMID:16179475

García-Gutiérrez, J. L., Fuentes, G. A., Hernández-Terán, M. E., Murrieta, F., Navarrete, J., & Jiménez-Cruz, F. (2006). Ultra deep oxidative desulfurization of diesel fuel with H2O2 catalyzed under mild conditions by polymolybdates supported on Al2O3. *Applied Catalysis A, General*, *305*(1), 15–20. doi:10.1016/j.apcata.2006.01.027

Granadeiro, C. M., Barbosa, A. D. S., Ribeiro, S., Santos, I. C. M. S., de Castro, B., Cunha-Silva, L., & Balula, S. S. (2014). Oxidative catalytic versatility of a trivacant polyoxotungstate incorporated into MIL-101(Cr). *Catalysis Science & Technology*, *4*(5), 1416–1425. doi:10.1039/c3cy00853c

Granadeiro, C. M., Barbosa, A. D. S., Silva, P., Paz, F. A. A., Saini, V. K., Pires, J., ... Cunha-Silva, L. (2013). Monovacant polyoxometalates incorporated into MIL-101(Cr): Novel heterogeneous catalysts for liquid phase oxidation. *Applied Catalysis A, General*, *453*(0), 316–326. doi:10.1016/j.apcata.2012.12.039

He, L., Li, H., Zhu, W., Guo, J., Jiang, X., Lu, J., & Yan, Y. (2008). Deep Oxidative Desulfurization of Fuels Using Peroxophosphomolybdate Catalysts in Ionic Liquids. *Industrial & Engineering Chemistry Research*, *47*(18), 6890–6895. doi:10.1021/ie800857a

Hill, C. L., & Prosser-McCartha, C. M. (1995). Homogeneous catalysis by transition metal oxygen anion clusters. *Coordination Chemistry Reviews*, *143*(0), 407–455. doi:10.1016/0010-8545(95)01141-B

Hu, X., Lu, Y., Dai, F., Liu, C., & Liu, Y. (2013). Host–guest synthesis and encapsulation of phosphotungstic acid in MIL-101 via "bottle around ship": An effective catalyst for oxidative desulfurization. *Microporous and Mesoporous Materials*, *170*(0), 36–44. doi:10.1016/j.micromeso.2012.11.021

Huang, D., Wang, Y. J., Yang, L. M., & Luo, G. S. (2006). Chemical Oxidation of Dibenzothiophene with a Directly Combined Amphiphilic Catalyst for Deep Desulfurization. *Industrial & Engineering Chemistry Research, 45*(6), 1880–1885. doi:10.1021/ie0513346

Huang, W., Zhu, W., Li, H., Shi, H., Zhu, G., Liu, H., & Chen, G. (2010). Heteropolyanion-Based Ionic Liquid for Deep Desulfurization of Fuels in Ionic Liquids. *Industrial & Engineering Chemistry Research, 49*(19), 8998–9003. doi:10.1021/ie100234d

Jia, S.-Y., Zhang, Y.-F., Liu, Y., Qin, F.-X., Ren, H.-T., & Wu, S.-H. (2013). Adsorptive removal of dibenzothiophene from model fuels over one-pot synthesized PTA@ MIL-101(Cr) hybrid material. *Journal of Hazardous Materials, 262*(0), 589–597. doi:10.1016/j.jhazmat.2013.08.056 PMID:24095999

Jiang, C., Wang, J., Wang, S., Guan, H., Wang, X., & Huo, M. (2011). Oxidative desulfurization of dibenzothiophene with dioxygen and reverse micellar peroxotitanium under mild conditions. *Applied Catalysis B: Environmental, 106*(3–4), 343–349. doi:10.1016/j.apcatb.2011.05.038

Jiang, X., Li, H., Zhu, W., He, L., Shu, H., & Lu, J. (2009). Deep desulfurization of fuels catalyzed by surfactant-type decatungstates using H2O2 as oxidant. *Fuel, 88*(3), 431–436. doi:10.1016/j.fuel.2008.11.010

Juan-Alcaniz, J., Gascon, J., & Kapteijn, F. (2012). Metal-organic frameworks as scaffolds for the encapsulation of active species: state of the art and future perspectives. *Journal of Materials Chemistry, 22*(20), 10102-10118.

Juan-Alcaniz, J., Ramos-Fernandez, E. V., Lafont, U., Gascon, J., & Kapteijn, F. (2010). Building MOF bottles around phosphotungstic acid ships: One-pot synthesis of bi-functional polyoxometalate-MIL-101 catalysts. *Journal of Catalysis, 269*(1), 229–241. doi:10.1016/j.jcat.2009.11.011

Julião, D., Gomes, A. C., Pillinger, M., Cunha-Silva, L., de Castro, B., Gonçalves, I. S., & Balula, S. S. (2015). Desulfurization of model diesel by extraction/oxidation using a zinc-substituted polyoxometalate as catalyst under homogeneous and heterogeneous (MIL-101(Cr) encapsulated) conditions. *Fuel Processing Technology, 131*(0), 78–86. doi:10.1016/j.fuproc.2014.10.030

Katsoulis, D. E. (1998). A survey of applications of polyoxometalates. *Chemical Reviews, 98*(1), 359–387. doi:10.1021/cr960398a PMID:11851510

Kholdeeva, O. A., Maksimchuk, N. V., & Maksimov, G. M. (2010). Polyoxometalate-based heterogeneous catalysts for liquid phase selective oxidations: Comparison of different strategies. *Catalysis Today*, *157*(1–4), 107–113. doi:10.1016/j.cattod.2009.12.016

Komintarachat, C., & Trakarnpruk, W. (2006). Oxidative Desulfurization Using Polyoxometalates. *Industrial & Engineering Chemistry Research*, *45*(6), 1853–1856. doi:10.1021/ie051199x

Kwon, T., Tsigdinos, G. A., & Pinnavaia, T. J. (1988). Pillaring of layered double hydroxides (LDH's) by polyoxometalate anions. *Journal of the American Chemical Society*, *110*(11), 3653–3654. doi:10.1021/ja00219a048

Lefebvre, F. (2013). *Polyoxometalates Encapsulated in Inorganic Materials: Applications in Catalysis. In New and Future Developments in Catalysis: Hybrid Materials* (pp. 265–288). Composites, and Organocatalysts.

Li, C., Gao, J., Jiang, Z., Wang, S., Lu, H., Yang, Y., & Jing, F. (2005). Selective Oxidations on Recoverable Catalysts Assembled in Emulsions. *Topics in Catalysis*, *35*(1-2), 169–175. doi:10.100711244-005-3821-5

Li, C., Jiang, Z., Gao, J., Yang, Y., Wang, S., Tian, F., ... Han, C. (2004). Ultra-Deep Desulfurization of Diesel: Oxidation with a Recoverable Catalyst Assembled in Emulsion. *Chemistry (Weinheim an der Bergstrasse, Germany)*, *10*(9), 2277–2280. doi:10.1002/chem.200305679 PMID:15112217

Li, H., Zhang, P., Zhang, L., Zhou, T., & Hu, D. (2009). Composite microspheres with PAM microgel core and polymerisable surfactant/polyoxometalate complexes shell. *Journal of Materials Chemistry*, *19*(26), 4575–4586. doi:10.1039/b823075g

Li, K., Yang, X., Guo, Y., Ma, F., Li, H., Chen, L., & Guo Yihang, Y. (2010). Design of mesostructured H3PW12O40-titania materials with controllable structural orderings and pore geometries and their simulated sunlight photocatalytic activity towards diethyl phthalate degradation. *Applied Catalysis B: Environmental*, *99*(1-2), 364–375. doi:10.1016/j.apcatb.2010.07.012

Lissner, E., de Souza, W. F., Ferrera, B., & Dupont, J. (2009). Oxidative Desulfurization of Fuels with Task-Specific Ionic Liquids. *ChemSusChem*, *2*(10), 962–964. doi:10.1002/cssc.200900171 PMID:19795435

Liu, L., Zhang, Y., & Tan, W. (2014). Ultrasound-assisted oxidation of dibenzothiophene with phosphotungstic acid supported on activated carbon. *Ultrasonics Sonochemistry*, *21*(3), 970–974. doi:10.1016/j.ultsonch.2013.10.028 PMID:24326220

Liu, Y., Liu, S., Liu, S., Liang, D., Li, S., Tang, Q., ... Zheng, Z. (2013). Facile Synthesis of a Nanocrystalline Metal–Organic Framework Impregnated with a Phosphovanadomolybdate and Its Remarkable Catalytic Performance in Ultradeep Oxidative Desulfurization. *ChemCatChem*, *5*(10), 3086–3091. doi:10.1002/cctc.201300378

Lü, H., Gao, J., Jiang, Z., Jing, F., Yang, Y., Wang, G., & Li, C. (2006). Ultra-deep desulfurization of diesel by selective oxidation with [C18H37N(CH3)3]4[H2NaPW10O36] catalyst assembled in emulsion droplets. *Journal of Catalysis*, *239*(2), 369–375. doi:10.1016/j.jcat.2006.01.025

Lu, H., Gao, J., Jiang, Z., Yang, Y., Song, B., & Li, C. (2007). Oxidative desulfurization of dibenzothiophene with molecular oxygen using emulsion catalysis. *Chemical Communications*, (2): 150–152. doi:10.1039/B610504A PMID:17180229

Lü, H., Ren, W., Liao, W., Chen, W., Li, Y., & Suo, Z. (2013). Aerobic oxidative desulfurization of model diesel using a B-type Anderson catalyst [(C18H37)2N(CH3)2]3Co(OH)6Mo6O18·3H2O. *Applied Catalysis B: Environmental*, *138–139*(0), 79–83. doi:10.1016/j.apcatb.2013.02.034

Lü, H., Ren, W., Wang, H., Wang, Y., Chen, W., & Suo, Z. (2013). Deep desulfurization of diesel by ionic liquid extraction coupled with catalytic oxidation using an Anderson-type catalyst [(C4H9)4N]4NiMo6O24H6. *Applied Catalysis A, General*, *453*(0), 376–382. doi:10.1016/j.apcata.2012.12.047

Lu, H., Zhang, Y., Jiang, Z., & Li, C. (2010). Aerobic oxidative desulfurization of benzothiophene, dibenzothiophene and 4,6-dimethyldibenzothiophene using an Anderson-type catalyst [(C18H37)2N(CH3)2]5 [IMo6O24]. *Green Chemistry*, *12*(11), 1954–1958. doi:10.1039/c0gc00271b

Maciuca, A.-L., Ciocan, C.-E., Dumitriu, E., Fajula, F., & Hulea, V. (2008). V-, Mo- and W-containing layered double hydroxides as effective catalysts for mild oxidation of thioethers and thiophenes with H2O2. *Catalysis Today*, *138*(1–2), 33–37. doi:10.1016/j.cattod.2008.04.031

Mizuno, N., & Misono, M. (1998). Heterogeneous Catalysis. *Chemical Reviews*, *98*(1), 199–218. doi:10.1021/cr960401q PMID:11851503

Mjalli, F. S., Ahmed, O. U., Al-Wahaibi, T., Al-Wahaibi, Y., & AlNashef, I. M. (2014). Deep oxidative desulfurization of liquid fuels. *Reviews in Chemical Engineering*, *30*(4), 337–378. doi:10.1515/revce-2014-0001

Nie, Y., Dong, Y., Bai, L., Dong, H., & Zhang, X. (2013). Fast oxidative desulfurization of fuel oil using dialkylpyridinium tetrachloroferrates ionic liquids. *Fuel*, *103*(0), 997–1002. doi:10.1016/j.fuel.2012.07.071

Nisar, A., Lu, Y., Zhuang, J., & Wang, X. (2011). Polyoxometalate Nanocone Nanoreactors: Magnetic Manipulation and Enhanced Catalytic Performance. *Angewandte Chemie International Edition*, *50*(14), 3187–3192. doi:10.1002/anie.201006155 PMID:21370361

Nisar, A., Zhuang, J., & Wang, X. (2011). Construction of Amphiphilic Polyoxometalate Mesostructures as a Highly Efficient Desulfurization Catalyst. *Advanced Materials*, *23*(9), 1130–1135. doi:10.1002/adma.201003520 PMID:21360767

Nogueira, L. S., Ribeiro, S., Granadeiro, C. M., Pereira, E., Feio, G., Cunha-Silva, L., & Balula, S. S. (2014). Novel polyoxometalate silica nano-sized spheres: Efficient catalysts for olefin oxidation and the deep desulfurization process. *Dalton Transactions (Cambridge, England)*, *43*(25), 9518–9528. doi:10.1039/C3DT53444H PMID:24829931

Okada, T., Miyamoto, K., Sakai, T., & Mishima, S. (2014). Encapsulation of a polyoxometalate into an organosilica microcapsule for highly active solid acid catalysis. *ACS Catalysis*, *4*(1), 73–78. doi:10.1021/cs400923q

Omwoma, S., Chen, W., Tsunashima, R., & Song, Y.-F. (2014). Recent advances on polyoxometalates intercalated layered double hydroxides: From synthetic approaches to functional material applications. *Coordination Chemistry Reviews*, *258–259*(0), 58–71. doi:10.1016/j.ccr.2013.08.039

Pandiyarajan, C., Pandikumar, A., & Ramaraj, R. (2013). Photoelectrocatalytic performance of a titania-keggin type polyoxometalate-gold nanocomposite modified electrode in methanol oxidation. *Nanotechnology*, *24*(43), 435401. doi:10.1088/0957-4484/24/43/435401 PMID:24077520

Pires, J., Pinto, M. L., Granadeiro, C. M., Barbosa, A. D. S., Cunha-Silva, L., Balula, S. S., & Saini, V. K. (2014). Effect on selective adsorption of ethane and ethylene of the polyoxometalates impregnation in the metal-organic framework MIL-101. *Adsorption*, *20*(4), 533–543. doi:10.100710450-013-9592-6

Proust, A., Thouvenot, R., & Gouzerh, P. (2008). Functionalization of polyoxometalates: Towards advanced applications in catalysis and materials science. *Chemical Communications*, (16): 1837–1852. doi:10.1039/b715502f PMID:18401495

Qiu, J., Wang, G., Zeng, D., Tang, Y., Wang, M., & Li, Y. (2009). Oxidative desulfurization of diesel fuel using amphiphilic quaternary ammonium phosphomolybdate catalysts. *Fuel Processing Technology*, *90*(12), 1538–1542. doi:10.1016/j.fuproc.2009.08.001

Rafiee, E., & Eavani, S. (2013). Organic–inorganic polyoxometalate based salts as thermoregulated phase-separable catalysts for selective oxidation of thioethers and thiophenes and deep desulfurization of model fuels. *Journal of Molecular Catalysis A Chemical*, *380*(0), 18–27. doi:10.1016/j.molcata.2013.09.009

Rezvani, M. A., Shojaei, A. F., & Zonoz, F. M. (2014). Anatase titania-vanadium polyphosphomolybdate as an efficient and reusable nano catalyst for the desulphurization of gas oil. *Journal of the Serbian Chemical Society*, *79*(9), 1099–1110. doi:10.2298/JSC131026015A

Rezvani, M. A., Shojaie, A. F., & Loghmani, M. H. (2012). Synthesis and characterization of novel nanocomposite, anatase sandwich type polyoxometalate, as a reusable and green nano catalyst in oxidation desulfurization of simulated gas oil. *Catalysis Communications*, *25*(0), 36–40. doi:10.1016/j.catcom.2012.04.007

Rhule, J. T., Hill, C. L., Judd, D. A., & Schinazi, R. F. (1998). Polyoxometalates in Medicine. *Chemical Reviews*, *98*(1), 327–358. doi:10.1021/cr960396q PMID:11851509

Ribeiro, S., Barbosa, A. D. S., Gomes, A. C., Pillinger, M., Goncalves, I. S., Cunha-Silva, L., & Balula, S. S. (2013). Catalytic oxidative desulfurization systems based on Keggin phosphotungstate and metal-organic framework MIL-101. *Fuel Processing Technology*, *116*, 350–357. doi:10.1016/j.fuproc.2013.07.011

Ribeiro, S., Barbosa, A. D. S., Gomes, A. C., Pillinger, M., Gonçalves, I. S., Cunha-Silva, L., & Balula, S. S. (2013). Catalytic oxidative desulfurization systems based on Keggin phosphotungstate and metal-organic framework MIL-101. *Fuel Processing Technology*, *116*(0), 350–357. doi:10.1016/j.fuproc.2013.07.011

Ribeiro, S., Granadeiro, C. M., Silva, P., Almeida Paz, F. A., de Biani, F. F., Cunha-Silva, L., & Balula, S. S. (2013). An efficient oxidative desulfurization process using terbium-polyoxometalate@MIL-101(Cr). *Catalysis Science & Technology*, *3*(9), 2404-2414.

Salavati, H., Tavakkoli, N., & Hosseinpoor, M. (2012). Preparation and characterization of polyphosphotungstate/ZrO 2 nanocomposite and their sonocatalytic and photocatalytic activity under UV light illumination. *Ultrasonics Sonochemistry*, *19*(3), 546–553. doi:10.1016/j.ultsonch.2011.09.001 PMID:21986517

Soleimani, M., Bassi, A., & Margaritis, A. (2007). Biodesulfurization of refractory organic sulfur compounds in fossil fuels. *Biotechnology Advances*, *25*(6), 570–596. doi:10.1016/j.biotechadv.2007.07.003 PMID:17716849

Thanasilp, S., Schwank, J. W., Meeyoo, V., Pengpanich, S., & Hunsom, M. (2013). Preparation of supported POM catalysts for liquid phase oxydehydration of glycerol to acrylic acid. *Journal of Molecular Catalysis A Chemical*, *380*, 49–56. doi:10.1016/j.molcata.2013.09.023

Wang, R., Yu, F., Zhang, G., & Zhao, H. (2010). Performance evaluation of the carbon nanotubes supported Cs2.5H0.5PW12O40 as efficient and recoverable catalyst for the oxidative removal of dibenzothiophene. *Catalysis Today*, *150*(1–2), 37–41. doi:10.1016/j.cattod.2009.10.001

Wang, R., Zhang, G., & Zhao, H. (2010). Polyoxometalate as effective catalyst for the deep desulfurization of diesel oil. *Catalysis Today*, *149*(1–2), 117–121. doi:10.1016/j.cattod.2009.03.011

Wang, X., Chen, W., & Song, Y.-F. (2014). Directional Self-Assembly of Exfoliated Layered Europium Hydroxide Nanosheets and Na9EuW10O36·32H2O for Application in Desulfurization. *European Journal of Inorganic Chemistry*, *2014*(17), 2779–2786. doi:10.1002/ejic.201400122

Wang, X.-S., Huang, Y.-B., Lin, Z.-J., & Cao, R. (2014). Phosphotungstic acid encapsulated in the mesocages of amine-functionalized metal-organic frameworks for catalytic oxidative desulfurization. *Dalton Transactions (Cambridge, England)*, *43*(31), 11950–11958. doi:10.1039/C4DT01043D PMID:24967570

Wee, L. H., Bonino, F., Lamberti, C., Bordiga, S., & Martens, J. A. (2014). Cr-MIL-101 encapsulated Keggin phosphotungstic acid as active nanomaterial for catalysing the alcoholysis of styrene oxide. *Green Chemistry*, *16*(3), 1351–1357. doi:10.1039/C3GC41988F

Xiao, J., Wu, L., Wu, Y., Liu, B., Dai, L., Li, Z., ... Xi, H. (2014). Effect of gasoline composition on oxidative desulfurization using a phosphotungstic acid/activated carbon catalyst with hydrogen peroxide. *Applied Energy*, *113*(0), 78–85. doi:10.1016/j.apenergy.2013.06.047

Xie, Y. (2006). Photoelectrochemical reactivity of a hybrid electrode composed of polyoxophosphotungstate encapsulated in titania nanotubes. *Advanced Functional Materials*, *16*(14), 1823–1831. doi:10.1002/adfm.200500695

Xu, J., Li, H., Wang, S., Luo, F., Liu, Y., Wang, X., & Jiang, Z. (2014). Ultra-deep desulfurization via reactive adsorption on peroxophosphomolybdate/agarose hybrids. *Chemosphere*, *111*(0), 631–637. doi:10.1016/j.chemosphere.2014.05.033 PMID:24997975

Xu, J., Zhao, S., Ji, Y., & Song, Y.-F. (2013). Deep Desulfurization by Amphiphilic Lanthanide-Containing Polyoxometalates in Ionic-Liquid Emulsion Systems under Mild Conditions. *Chemistry (Weinheim an der Bergstrasse, Germany)*, *19*(2), 709–715. doi:10.1002/chem.201202595 PMID:23161825

Yamase, T. (1998). Photo- and Electrochromism of Polyoxometalates and Related Materials. *Chemical Reviews*, *98*(1), 307–326. doi:10.1021/cr9604043 PMID:11851508

Yamase, T. (2005). Anti-tumor, -viral, and -bacterial activities of polyoxometalates for realizing an inorganic drug. *Journal of Materials Chemistry, 15*(45), 4773-4782.

Yan, X., Su, G., & Xiong, L. (2009). Oxidative desulfurization of diesel oil over Ag-modified mesoporous HPW/SiO2 catalyst. *Journal of Fuel Chemistry and Technology*, *37*(3), 318–323. doi:10.1016/S1872-5813(09)60022-0

Yan, X. M., Mei, P., Xiong, L., Gao, L., Yang, Q., & Gong, L. (2013). Mesoporous titania-silica-polyoxometalate nanocomposite materials for catalytic oxidation desulfurization of fuel oil. *Catalysis Science & Technology*, *3*(8), 1985–1992. doi:10.1039/c3cy20732c

Yan, X. M., Mei, Z., Mei, P., & Yang, Q. (2014). Self-assembled HPW/silica-alumina mesoporous nanocomposite as catalysts for oxidative desulfurization of fuel oil. *Journal of Porous Materials*, *21*(5), 729–737. doi:10.100710934-014-9819-2

Yu, F., Zheng, P. Q., Long, Y. X., Ren, Y. P., Kong, X. J., Long, L. S., ... Zheng, L. S. (2010). Polyoxometalate-based metal-organic frameworks as heterogeneous catalysts for selective oxidation of ethylbenzene. *European Journal of Inorganic Chemistry*, *2010*(28), 4526–4531. doi:10.1002/ejic.201000491

Yu, F. L., & Wang, R. (2013). Deep Oxidative Desulfurization of Dibenzothiophene in Simulated Oil and Real Diesel Using Heteropolyanion-Substituted Hydrotalcite-Like Compounds as Catalysts. *Molecules (Basel, Switzerland)*, *18*(11), 13691–13704. doi:10.3390/molecules181113691 PMID:24284486

Zhang, Y., Lü, H., Wang, L., Zhang, Y., Liu, P., Han, H., ... Li, C. (2010). The oxidation of benzothiophene using the Keggin-type lacunary polytungstophosphate as catalysts in emulsion. *Journal of Molecular Catalysis A Chemical*, *332*(1–2), 59–64. doi:10.1016/j.molcata.2010.08.021

Zheng, H., Sun, Z., Chen, X., Zhao, Q., Wang, X., & Jiang, Z. (2013). A micro reaction-controlled phase-transfer catalyst for oxidative desulfurization based on polyoxometalate modified silica. *Applied Catalysis A, General*, *467*(0), 26–32. doi:10.1016/j.apcata.2013.06.026

Zhou, Y., Chen, G., Long, Z., & Wang, J. (2014). Recent advances in polyoxometalate-based heterogeneous catalytic materials for liquid-phase organic transformations. *RSC Advances*, *4*(79), 42092–42113. doi:10.1039/C4RA05175K

Zhu, Q.-L., & Xu, Q. (2014). Metal-organic framework composites. *Chemical Society Reviews, 43*(16), 5468-5512.

Zhu, W., Huang, W., Li, H., Zhang, M., Jiang, W., Chen, G., & Han, C. (2011). Polyoxometalate-based ionic liquids as catalysts for deep desulfurization of fuels. *Fuel Processing Technology*, *92*(10), 1842–1848. doi:10.1016/j.fuproc.2011.04.030

Chapter 5
The Relation of Ni/ZnO Nano Structures With Properties of Reactive Adsorption Desulfurization

Hui Ge
Institute of Coal Chemistry, Chinese Academy of Sciences, China

Weixing Wang
School of Chemistry and Chemical Technology, Southwest Petroleum University, China

Lichun Huang
Ordos Dalu Institute of Coal Chemical Industry, China

Mingxing Tang
Institute of Coal Chemistry, Chinese Academy of Sciences, China

Zhenyu Ge
Institute of Coal Chemistry, Chinese Academy of Sciences, China

ABSTRACT

Ni/ZnO nano-sorbent systems have been extensively used in the reactive adsorption desulfurization (RADS) of gasoline steams, especially in China, to meet the more rigorous regulation on the sulfur content. The apparent advantage of RADS is that most of the olefins are kept in the product with low consumption of hydrogen and little loss of octane. The authors discussed in this chapter the relation of catalytic properties with components and structures of Ni/ZnO sorbent. Based on detailed characterization and reaction results, they revealed the dynamic change of Ni/ZnO sorbents during RADS, the mechanisms of desulfurization, and the sulfur transfer and sulfur adsorption. Apart from the RADS of Ni/ZnO nano-sorbent for cleaner gasoline production, they also presented other potential applications.

DOI: 10.4018/978-1-7998-2146-5.ch005

INTRODUCTION

The desulfurization is one of the most important processes in the petroleum refining industry. The conventional technology is hydrodesulfurization (HDS) where molybdenum or tungsten supported on alumina support promoted with cobalt or nickel is used as a catalyst (Babich and Moulijn 2003). Recently, many governments make more and more stringent regulations controlling the sulfur content in transport fuels. For example, in the Europe V standard, the sulfur content in gasoline and diesel is limited to below 10 ppm. The HDS process faces a serious challenge to achieving this objective. Thus other desulfurization technologies have been vividly investigated. Among them, reactive adsorption desulfurization (RADS) emerges as a perspective field, where a sorbent is used to remove the sulfur impurity from oil stream and keeps the sulfur in the sorbent by chemical reaction (Ge et al. 2016). The great challenge for RADS technology is to develop sorbent with high desulfurization activity and high sulfur capacity. Nano Ni/ZnO sorbent is found as an ideal candidate for this objective.

Ni/ZnO was considered one of the most excellent sorbents for the RADS. The metallic Ni can strongly adsorb sulfur, meantime, the sulfur is difficult to be removed. In the Ni/ZnO system, the Ni captures sulfur from the sulfur-containing compound in feed forming NiS_x, then the sulfur is transferred to the ZnO component in the presence of hydrogen, and the Ni active center is regenerated, while the hydrocarbon portion of the sulfur-containing compound is released back into the process stream (Tawara et al. 2000, 2001a). Due to the dual identities of Ni/ZnO, namely, both as a catalyst and as sorbent, it has been called different names in references, such as "catalyst", "sorbent", or "adsorbent" etc.. In this chapter, for the consistency, we adopt the term "sorbent". To RADS sorbent, there are three key specifications, one is the activity of desulfurization, another is the sulfur capacity, namely possible adsorbed sulfur quantity when sulfur content is kept below some content, and the third is the regeneration ability. Many methods have been used to improve the reaction properties of RADS sorbents. By controlling the nano morphology and size of Ni/ZnO sorbent, the activity and sulfur capacity can be remarkably improved (Zhang et al. 2012a). Addition of Mn forms the new phase of $ZnMnO_3$, which enhances activity and renewability (Zhang et al. 2013). Substitution ZnO component with MnO increases the sulfur capacity (Tang et al. 2015). Introduction of a second metal, such as Co, Fe, Cu, Silver, can inhibit the hydrogenation to olefins (Khare 1999, Gyanesh 2001).

Ryzhikov et al. (2008) observed that NiO/ZnO can be reduced directly in situ and shows the better RADS of thiophene than the pre-reduced counterpart. They suggested that the H_2 pretreatment results in the formation of Ni-Zn alloy and agglomeration of nano particles, leading the decrease of activity. However, the pre-reduction to NiO/ZnO-SiO_2-Al_2O_3 sorbent can improve the desulfurization capability (Fan et al. 2010), which is attributed to the additives of alumina and silica stabilizing the particles (Wen et al. 2012, Meng et al. 2013). Decreasing the size of ZnO nano particles leads to efficient contact between Ni and ZnO particles and enhances the desulfurization ability and sulfur adsorption capacity (Zhang et al. 2012a). But small ZnO particles are not stable under the calcination or reaction conditions, the sinter of particles lead a noticeable drop of activity (Bezverkhyy et al. 2008). Thus structure additives, such as diatomite, perlite, attapulgite, silica sol, pseudo-boehmite or their mixtures are added to reinforce the texture of nano sorbent (Shangguan et al. 2013, Zhou et al. 2013).

Babich and Moulijn (2003) proposed a RADS mechanism of Ni/ZnO sorbent where H_2S as a sulfur carrier transfers the sulfur from Ni surface to ZnO bulk, which has been accepted by a lot of authors. But DFT calculation showed that the direct sulfur transfer from Ni site to Zn site has lower energy barrier, suggesting the new mechanism that sulfur may be transferred through the interface between Ni and ZnO phases (Zhang et al. 2012b). The accumulation of sulfur in the sorbent leads to a gradual deactivation of the sorbent. Regeneration at above 500 °C can recover the RADS activity at least partly. However, the formation of new species, such as Ni_2SO_4, $ZnSO_4$, NiAl or AlZn spinel, influence the performance of regenerated sorbent (Wen et al. 2012, Meng et al. 2012).

It notes that desulfurization can apparently impact the quality of fuel product. For gasoline, deep HDS leads the loss of octane number owing to the saturation to olefins. Thus a process of recovery to octane number is usually followed, such as isomerization or addition of high octane-number compounds, MTBE. This is especially true for the fluidized catalytic cracker (FCC) gasoline, which contains large olefins and refractory sulfur compounds, such as thiophene and its derivatives. In modern refining, there is a trend to get more light oil fractions from heavy fractions. The ratio of diesel/gasoline is expected to decrease. Thus the ratio of FCC gasoline has increased. For the deep desulfurization of FCC gasoline, avoiding hydrogenation to olefins is necessary.

The RADS of Ni/ZnO nano sorbent is the preferred alternative for the desulfurization of gasoline owing to the low activity for olefin saturation. Conoco Philips Petroleum Co. was the first one using the nano-Ni/ZnO sorbent for the RADS of gasoline in the refinery. They termed this technology "S-Zorb", illustrating that it is originated from the older Z-Sorb technique for the removal of H_2S gas (Johnson et al. 2001). The first commercial S-Zorb unit was run at Borger refinery in April of 2001. It was demonstrated to possess some valuable features like low hydrogen consumption and low loss in octane number (Khare 1999, Gyanesh 2001). It was then bought out by China Petroleum & Chemical Corp. (Sinopec) and have been improved continually. With the latest sorbent and circulated fluidized bed technology, the clean gasoline can be produced with sulfur content below 10 ppm and little octane number loss (Qiu et al. 2013). By the end of 2014, over thirty units have been built and operated (Jia et al. 2014). Recently the fix bed technique of RADS has also been developed.

In Ni/ZnO RADS processes, two regenerations play important roles, one is auto-regeneration where the Ni active sites are continually regenerated by the sulfur transfer from the Ni surface to the ZnO component. Another is that the deactivated Ni/ZnO sorbent is regenerated periodically by oxidation and then H_2 reduction. Understanding the two regenerations and making them run effectively warrant the excellent performance of Ni/ZnO nano sorbent for RADS. Thus we try to illustrate these two regeneration processes based on our understanding and knowledge. Based on the characterizations and theory researches from us and many others, we can now reveal the structures of the active components of Ni/ZnO sorbent, building the relation of structures with the adsorption and reaction processes. And we deduced the adsorption, reaction and regeneration mechanisms.

RELATION OF STRUCTURES OF Ni/ZnO SORBENT WITH REACTIVE ADSORPTION DESULFURIZATION PROPERTIES

Until now, Ni/ZnO is the most effective sorbent for RADS. In this section, we focus on the sulfur transfer on the Ni/ZnO, following the process from adsorption of sulfur compounds, desulfurization on Ni surface, sulfur transfer to ZnO and regeneration of Ni active sites. We bring readers insight into the intrinsic structures and properties of Ni/ZnO, as well as the reaction mechanisms, which are supported by the characterizations and theory calculations.

Characterizations of Ni/ZnO Nano-Sorbents

1. XRD

XRD can obtain crystal information of the main components of Ni/ZnO sorbent. Comparison of XRD patterns between calcined and reduced NiO/ZnO (Ryzhikov et al. 2008), it is found that the calcined sample contains the peaks characteristic of the hexagonal ZnO. The additional peak (at $2\theta = 43.2^{o}$) corresponds to NiO. A similar additional peak, but which is slightly displaced to $2\theta= 43.7^{o}$, is observed for the sample reduced in H_2 at 360 ^{o}C for 6 h, this is suggested as the features of

Ni-Zn alloy (Homs et al. 2006). To verify this postulation, a reduction treatment for a shorter time, which can reduce Ni but avoid the formation of alloy, in this case, the peak is situated at $2\theta = 44.1\ ^{\circ}$ which is close to metallic Ni. This illustrates that the species obtained after a 6 hour treatment is truly metallic Ni-Zn alloy.

Using the on-site synchrotron radiation XRD technology, the reaction between reduced Ni/ZnO (containing NiZn alloy) and thiophene was followed *(Bezverkhyy et al. 2009)*. In the beginning, the alloy was observed to decompose, producing metallic Ni particles. The cell parameter (a =3.68 Å) of Ni, which is larger than the 3.55 Å of pure cubic Ni, indicates that Ni phase is dilated. After decomposition of the NiZn alloy, the dilated Ni particles remained unchanged throughout the ZnO sulfidation but yielded the hexagonal Ni_3C phase after ZnO is completely transformed to ZnS. Base on this observation, the authors tentatively proposed that the expansion of lattice is due to the permeation of carbon.

After the thiophene RADS, a complex XRD pattern was observed (*Ryzhikov et al. 2008*). It contains the peaks characteristic of sphalerite, wurtzite-2H, metallic Ni and Ni_3S_2 (heazlewoodite). ZnO was completely sulfided, while nickel was only partial sulfided. And after reaction, Ni^0 adopts the hexagonal structure rather than the cubic one in the initial sample.

We investigated the NiO/ZnO system for the RADS of diesel (*Huang et al. 2010a*). Figure 1 showed the change of crystal phases during the reaction. The NiO was found to completely be reduced to metallic Ni after 2h on stream. The ratio of $I_{28.6(ZnS)}/I_{36.3(ZnO)}$ grows from 26 to 122 h, indicating that the ZnS content in the sorbent increases monotonously with the time on stream. The peak of Ni_3S_2 appears after 98 h on stream, which corresponds the apparent deactivation of the sorbent. It was found that the reduction resulted in the increase of crystallite sizes of ZnO and Ni species. However, during the RADS process, the crystallite sizes of ZnO and Ni phases were basically stable.

We also compared the XRD patterns between catalyst precursors and calcined catalysts which were prepared by co-precipitation (Figure 2). Two poorly defined phases of $Ni(OH)_2{\cdot}2H_2O$ and $Zn_5(CO_3)_2(OH)_6$ were detected in the precursor catalyst without calcination. The degree of crystallinity of this sample is noticeably low, indicating that the main component is an amorphous phase. The Ni precursor has a structure similar to that of α-nickel hydroxide. For the catalyst calcined at 260 °C, the XRD pattern is similar to that of precursor catalyst, suggesting that little precursors are decomposed. For the catalyst calcined at 280 °C, NiO phase with NaCl-type structure is distinguishable. ZnO phase with the wurtzite type structure began to form at 300 °C. The low-intensity peaks imply the existence of a considerable amount of amorphous phase. Higher calcination temperatures resulted in the increase of the crystallinities of NiO and ZnO. It appears that the optimal calcination temperature is 350 °C. In fact, this temperature is sufficient for the formation of metal oxides, and still avoids serious sintering that could cause loss of surface area and increase of crystallite size.

*Figure 1. XRD patterns of the calcined adsorbent (a) and used adsorbents after the RADS of diesel oil for different reaction times on stream (b, c, d, e, f, g, and h represent 2, 6, 12, 26, 75, 98, and 122 h on stream, respectively; *, NiO; Δ, Ni; +, Ni_3S_2; o, ZnO; #, ZnS).RADS conditions:350 °C, 6 MPa, H_2/diesel = 500, LHSV = 3.20 h^{-1}.*

(Huang et al. 2010a. Copyright 2010 Elsevier.)

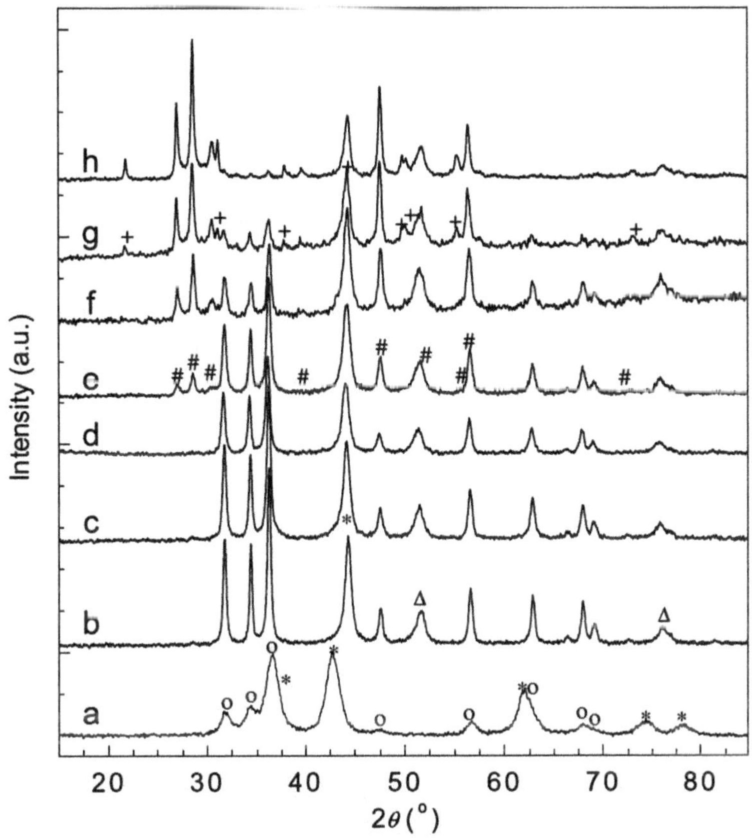

*Figure 2. The XRD patterns of the catalyst with different calcination temperature: a) catalyst precursor; (b) 260 °C; (c) 280 °C; (d) 300 °C; (e) 350 °C; (f) 400 °C; (g) 450 °C. &, $Ni(OH)_2{\cdot}H_2O$; #, $Zn_5(CO_3)_2(OH)_6$; o, ZnO; *, NiO.*
(Huang et al. 2017. Copyright 2017 Wiley.)

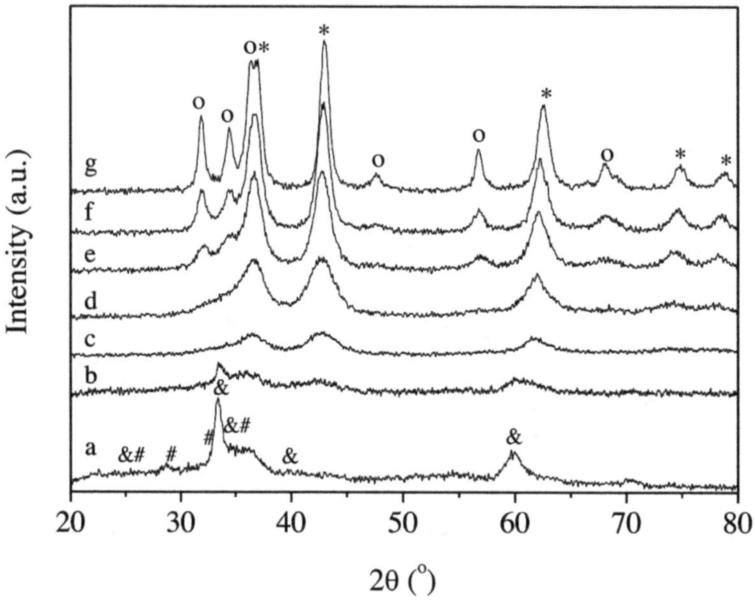

2. TGA

We (*Huang et al. 2017*) conducted the TGA experiments to study the temperature effect on the catalyst precursor which was obtained by co-precipitation method using $Ni(NO_3)_2{\cdot}6H_2O$ and $Zn(NO_3)_2{\cdot}6H_2O$ solutions and the precipitation agent of $(NH_4)_2CO_3$ (1M). Figure 3 shows the curves of differential thermal gravimetric (DTA) and mass spectra (MS) analysis of TGA. The catalyst precursor exhibits 2 stages of weight loss: the first one below 220 °C with a broad endothermic effect accounts for a weight loss of 5.2% which is due to the loss of moisture; the second one between 220 and 350 °C with a large endothermic effect can be ascribed to the decomposition of the catalyst precursor to NiO and ZnO that accounts for a weight loss of 21.7%. The peak temperature of weight loss in the second stage locates at 320 °C and the endothermic peak presents at 323 °C respectively. As shown in MS analysis, H_2O and little CO_2 are detected below 220 °C. A large amount of H_2O and CO_2 are released between 220 and 350 °C. Above 350 °C, little weight changes are observed. The total weight loss of the precursor (28.9%) is very near to that calculated for the thermal transformation of the precursor into NiO and ZnO.

Figure 3. TG/DTA and MS analysis curves of catalyst precursor.
(Huang et al. 2017. Copyright 2017 Wiley.)

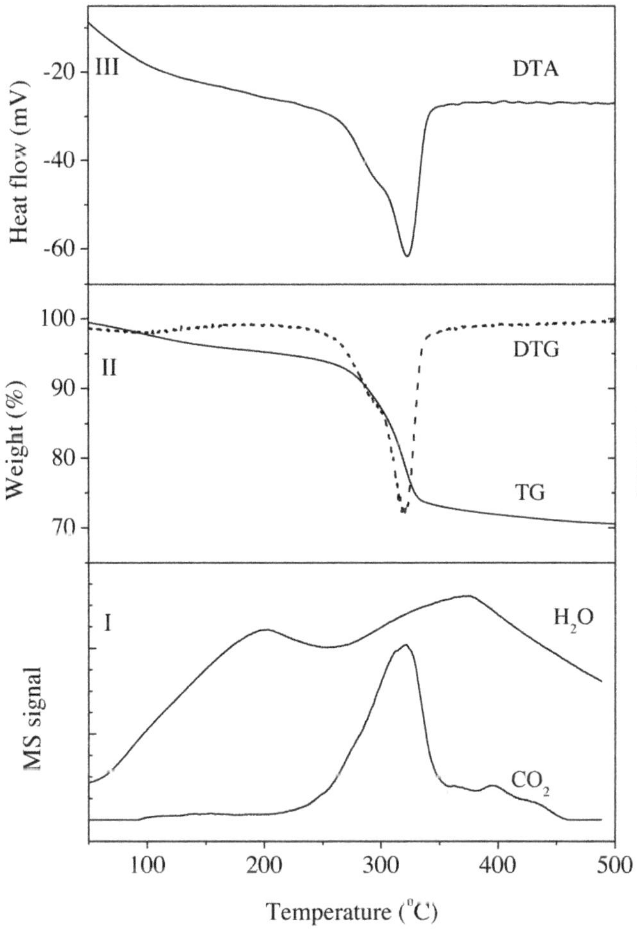

Bezverkhyy et al. (2008) conducted the thermal gravimetric analysis (TGA) to study the kinetics of reduced Ni/ZnO sorbent reacted with thiophene/H_2. It was observed that the weight change follows a complex pattern in which at least three domains can be distinguished. At the beginning, the conversion increased rapidly up to about 0.07-0.1, suggesting that the metallic Ni surface is fast sulfided by the thiophene HDS. Then an S-shaped domain was followed up to conversion $X \approx 0.25$. This shape is characteristic of a process controlled by nucleation and growth of a new phase. The induction time corresponds to the formation of ZnS nuclei which grow on the surface of ZnO particles and cover them progressively. Using the kinetics equation, the activation energy was calculated as 94 kJ/mol, which suggests a two-dimensional growth of the nuclei. At this stage, the sulfur transfer lead to the regeneration of the Ni surface. After the end of the surface sulfidation of ZnO (conversion ca. 0.25), the reaction rate decreases. It is postulated that at this third stage the thiophene HDS on NiS_x surface and sulfur diffusion from the surface to ZnO bulk have comparable rates and their contributions vary with temperature due to the difference in activation energies.

3. TEM

The instability of Ni/ZnO sorbent can be observed by comparing the unreduced sample before and after RADS reaction. TEM showed that the initial sample consisted of strongly agglomerated shapeless particles of about 10-30 nm (*Bezverkhyy et al. 2008*). It was found that the sulfidation significantly changed textural properties of the solid: particle size was visibly increased and the smaller particles, observed in the initial sample, disappeared. This observation is in line with a decrease of BET surface area after sulfidation. The shape of the particles was also changed after the reaction. An irregular corrugated structure of the facets and a strong variation of contrast throughout a particle pointed out to a polycrystalline nature and a complex morphology of the particles in sulfide solids.

4. XPS

XPS spectra, as a surface-sensitive technology, can give the valent states of surficial species. We compared the RADS under hydrogen or nitrogen for Ni/ZnO sorbent, respectively (*Huang et al. 2011*). Figure 4(I) shows the Zn 2p3/2 XPS spectra of the spent Ni/ZnO adsorbents after diesel RADS in different atmospheres. For the calcined NiO/ZnO, the binding energy of the spent sample after RADS in hydrogen is shifted to a higher value by 0.4 eV, compared with that in nitrogen. It confirms that the state of ZnO in the calcined sample changes little after RADS in nitrogen; however, it is transferred to ZnS after RADS in hydrogen. The broadening of Zn 2p3/2 peak in the spectra of the reduced adsorbent and especially the spent samples after RADS in hydrogen is also observed. This can be partially ascribed to the reduction of ZnO. Moreover, the reduction of ZnO may also shift the Zn 2p3/2 binding energy to a lower value, possibly due to the formation of Ni-Zn alloy. To act as the active desulfurization species, Ni should be liberated from the Ni-Zn alloy after the sulfidation of Zn counterpart in the alloy during the desulfurization process. Figure 4(II) shows the C 1s XPS spectra of the spent adsorbents. The main peak centered at 284.8 eV with two wide components of low intensity at 286.3 and 288.6 eV is present. The peak at 284.8 eV is from the carbonaceous contamination of the adsorbent. The peaks at 286.3 and 288.6 eV are assigned to species containing

Figure 4. Zn $2P_{3/2}$ (I) and C 1S (II) XPS spectra of the spent adsorbents after the RADS in different atmospheres: (a) calcined, spent in N_2; (b) reduced, spent in N_2; (c) calcined, spent in H_2; (d) reduced, spent in H_2.
(Huang et al. 2011. Copyright 2011 Elsevier.)

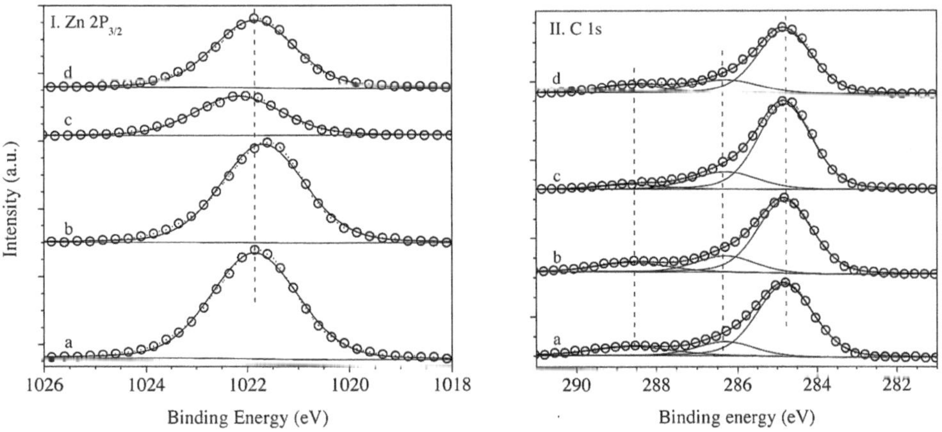

C-O and C=O bonds, respectively. However, the formation of Ni_3C at 283.5 eV is not found here *(Bezverkhyy et al. 2009)*.

5. XAS

Using the in-situ XAS technique, we studied the phase transformation during RADS reaction of diesel oil *(Huang et al. 2010a)*. Figure 5 shows the sulfur K-edge XANES spectra of the standard samples NiS, Ni_3S_2, and ZnS and the Ni/ZnO sorbent at different times on stream in RADS of diesel. The relative concentrations of each sulfur species are evaluated by least-squares fitting (LSF) of the spectra with those of standard samples; combined with the total sulfur content in the sorbents, the contents of each sulfur species are estimated (Figure 5). In the beginning, the sulfur content in Ni_3S_2 is higher than that of ZnS. The content of ZnS increases steadily with the reaction time on stream, while the content of Ni_3S_2 keeps relatively stable. The content of Ni_3S_2 in the sorbent changes little with reaction time, and the steady-state activity of RADS is attained. After 75 h, the content of Ni_3S_2 increases gradually, accompanied by the slow decrease of the desulfurization efficiency of Ni/ZnO. But the Ni_3S_2 phase cannot be detected using XRD, until the apparent deactivation of the sorbent. This illustrates that the early formed Ni_3S_2 is amorphous. Different from the thiophene RADS, the Ni_3C was not observed. It was proposed that the metallic Ni should be responsible for keeping the high HDS activity of the diesel. Ni_3S_2 corresponds to the final state of Ni species and has low HDS activity for desulfurization.

Figure 5. Sulfur K-edge XANES of the standard samples (a, NiS; b, Ni_3S_2; c, ZnS) and the used Ni/ZnO catalysts after the RADS of diesel oil for different times on stream (d, e, f, g, h, i, and j represent 2, 4, 12, 26, 75 98, and 122 h on stream, respectively).
(Huang et al. 2010a. Copyright 2010 Elsevier.)

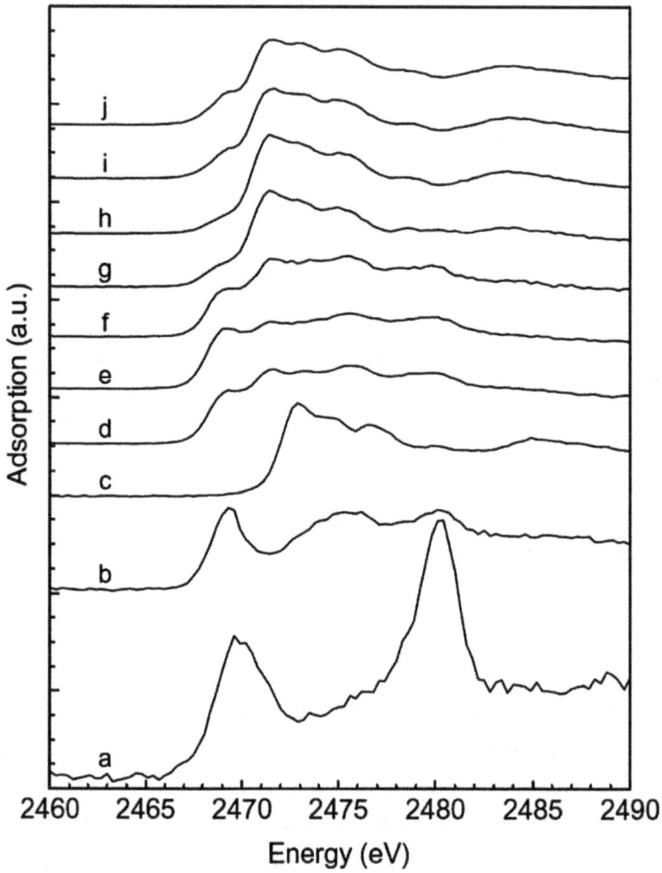

To get a better insight into the evolution of Ni/ZnO adsorbent structure and the transfer of sulfur species from dibenzothiophene (DBT) in the feed oil to the adsorbent during RADS over Ni/ZnO, the desulfurization of the model oil containing DBT was carried out in an in situ XAS cell. Figure 7(I) shows the Ni spectra of the calcined NiO/ZnO adsorbent during RADS of the model oil in hydrogen. Ni species in the fresh calcined adsorbent is NiO; after carrying out the RADS for 40 min on stream, its white line intensity is greatly decreased and the spectrum becomes quite similar to that of metallic Ni, suggesting that most of NiO are quickly converted to Ni in the reductive reaction atmosphere. During the whole RADS period, metallic Ni is the dominant Ni species and bulk NiS_x species is indiscernible even after

Figure 6. Amounts of different sulfur species adsorbed in NiO/ZnO samples collected after RADS of diesel oil with the reaction time on stream: (a) Ni_3S_2; (b) ZnS; (c) total sulfur.
(Huang et al. 2010a. Copyright 2010 Elsevier.)

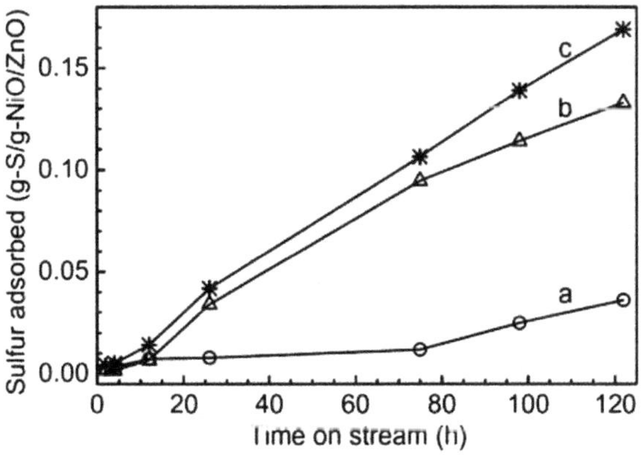

Figure 7. In situ Ni (I) and Zn (II) K-edge XANES spectra of the calcined NiO/ZnO adsorbent during the RADS of the model oil in hydrogen. The desulfurization of model oil containing DBT was carried out at 350 °C and 3.0 MPa in the stream of H_2-He mixture and the XAS spectra were collected in the transmission mode at 40 (i), 80 (ii), 120 (iii), 160 (iv), 200 (v), 240 (vi), 280 (vii), 320 (viii), 400 (ix), 480 (x), and 660 min (xi) on stream. References: (a) NiO; (b) Ni; (c) Ni_3S_2; (d) NiS; (e) Zn; (f) $3Zn(OH)_2 2ZnCO_3 xH_2O$; (g) ZnO; (h) ZnS; and (o) the fresh calcined Ni/ZnO.
(Huang et al. 2011a. Copyright 2011 Elsevier.)

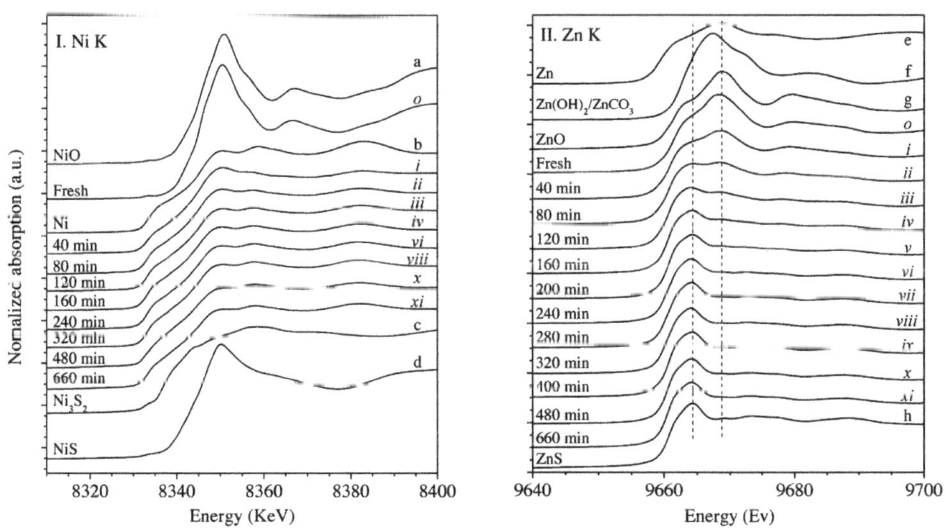

660 min on stream, suggesting that a complete sulfidation of Ni is difficult under current reaction atmosphere. These XAS results indicate that the main Ni-containing components in the adsorbents during RADS are NiO, Ni, and Ni_3S_2. The content of NiO decreases successively with a time of stream (TOS), while that of Ni increases at first, and then remains stable until 480 min on stream and thereafter decreases with TOS. On the other side, the content of Ni_3S_2 remains almost unchanged before 120 min on stream due to the balance between sulfidation and regeneration; after that, it increases gradually with the TOS for RADS.

Figure 7(II) presents the in situ Zn K-edge XANES spectra. A change in the chemical state of Zn is observed during RADS. Compared with the spectra of the reference samples, it was found that the intensity of the dominant peak attributed to ZnO decreases gradually along with TOS, meanwhile, the peak attributed to ZnS becomes distinct and its intensity increases incessantly. The gradual evolvement of the XANES spectra with increasing TOS indicates a smooth progression from ZnO to ZnS. As shown in Fig. 7(II), the main Zn components in the adsorbent are ZnO and ZnS; the content of ZnS increases successively, accompanied by a gradual decrease of ZnO content with TOS. It should be mentioned that the sulfidation rate of ZnO to ZnS decreases with TOS and becomes very low at the later period of the RADS reaction, properly due to the difficulty of sulfur diffusion into the bulk of unreacted ZnO through a previously formed ZnS layer.

The in-situ K-edge EXAFS spectra of corresponding samples are collected also (Figure 8). The Ni spectra of the calcined NiO/ZnO adsorbent during RADS are shown in Figure 8(I). After conducting the desulfurization for 40 min on stream, the spectrum feature of the adsorbent is changed significantly, compared with that of the fresh adsorbent; the intensity of Ni-O and Ni-Ni (NiO) peaks are strongly attenuated. A broad feature between the Ni-O and Ni-Ni peaks attributed to Ni-Ni (metallic Ni) and Ni-S (Ni_3S_2) emerges. It seems that metallic Ni is still the main phase of Ni species even after 660 min on stream, confirming that it is difficult to form bulk NiS_x at current reaction atmosphere.

Figure 8(II) shows the in situ Zn K-edge EXAFS spectra. Compared with the fresh adsorbent, the intensity of Zn-O and Zn-Zn peaks attributed to ZnO is attenuated and a peak at 2 Å attributed to ZnS is emerged after conducting the desulfurization for 40 min on stream. The ZnS islands formed at an early period may involve a distortion of the regular wurtzite structure, which leads to a longer Zn-S bond than the standard one. These ZnS islands are then developed eventually to bulk sulfide by the migration of sulfide ions through the oxide lattice. After RADS for 240 min on stream, the main phase in the adsorbent turns to be ZnS, suggesting that the sulfidation of zinc species is easier than that of nickel species.

Figure 8. In situ Ni (I) and Zn (II) K-edge EXAFS spectra of the calcined NiO/ZnO adsorbent during the RADS of the model oil in hydrogen. The spectra were collected in the transmission mode at 40 (i), 80 (ii), 120 (iii), 160 (iv), 200 (v), 240 (vi), 320 (vii), 480 (viii), and 660 min (ix) on stream. References: (a) NiO; (b) Ni; (c) Ni_3S_2; (e) Zn; (f) $3Zn(OH)_2 2ZnCO_3 xH_2O$; (g) ZnO; (h) ZnS; and (o) the fresh calcined Ni/ZnO.

(Huang et al. 2011. Copyright 2011 Elsevier.)

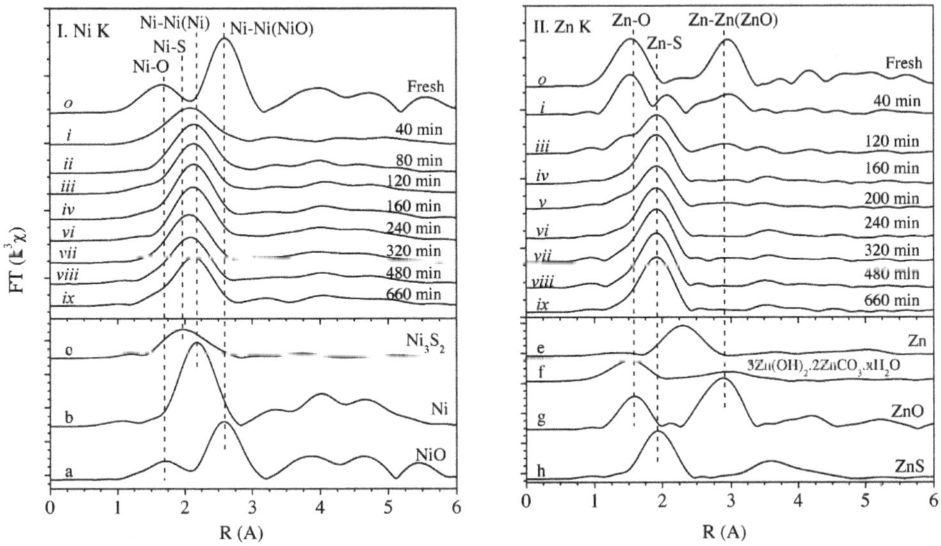

Key Factors Determining the Activity and Capacity for RADS

1. Effect of the Chemical State of Ni Active Components

Mangnus et al (1995) reported that the surface of Ni particles was sulfided at lower temperatures than NiO particles and the sulfiding rate of the core of the particle was determined by solid-state diffusion. We (Huang et al 2011) observed that in an inert nitrogen atmosphere the reduced adsorbent exhibits higher desulfurization activity than the calcined one, indicating that metallic Ni is more active than NiO in the RADS reaction. When the RADS is conducted in hydrogen, an induction time is observed over the calcined adsorbent, suggesting that NiO in the adsorbent is first reduced to Ni in situ and the desulfurization activity is then acquired. Meanwhile, when the reduction is performed in the reaction atmosphere, the agglomeration of Ni and the formation of Ni-Zn alloy may be suppressed due to the presence of a sulfur-containing compound. These results suggest that the metallic nickel among the several nickel species (NiO, Ni, and Ni_3S_2) detected in the adsorbent may be the principal active sites for the RADS over Ni/ZnO.

2. Effect of Hydrogen on RADS

To examine the effect of hydrogen on the sulfur transfer, we (*Huang et al. 2011*) treated the mechanical mixture of Ni_3S_2-ZnO under different atmosphere (Figure 9). In the presence of nitrogen, the formation of Ni and ZnS only occurs at a high temperature of 600 °C by the solid reaction. With H_2 treatment, the sulfur exchange has apparently appeared at 300 °C, the complete formation of Ni and ZnS is observed at 400 °C. This evidenced that hydrogen facilitates the sulfur transfer from Ni_3S_2 to ZnO.

When the RADS was carried out in an inert nitrogen atmosphere (Huang et al 2011), we found that the Ni/ZnO adsorbent shows little desulfurization capability, only trance amount of ZnS was formed in N_2 (both calcined and reduced). And the amounts of carbon deposited are much greater than that in hydrogen (Table 1); moreover, in nitrogen the calcined adsorbent deposits more carbonaceous materials than the pre-reduced one. When the RADS was carried out in hydrogen, however, almost all ZnO was converted to ZnS in the spent adsorbent (both calcined and reduced). Since ZnO alone does not exhibit any actual RADS capacity, it is proposed that hydrogen takes an important role in promoting the mobility of sulfur from NiS_x to ZnO. And the hydrogen can also enhance the desulfurization of the sulfur compound on Ni surface by hydrolysis. The coke deposited in the adsorbent is reduced greatly when the RADS is carried out in hydrogen, suggesting that the presence of hydrogen can effectively suppress the coke accumulation.

3. Effect of Residual Sodium on RADS

The alkali ions may remain in sorbent if they are not removed by sufficient washing. We studied the effect of Na ions in Ni/ZnO sorbent on the RADS properties (Huang et al. 2010b). As shown in figure 10, the precipitate or the calcined catalyst samples were washed for different times to get a series of adsorbents with different residual sodium contents. The adsorbents prepared are denoted as C*x*-*y*, where *x* is the washing times of the precipitate before drying and *y* is the washing times of the sample after calcination. The adsorbents C0-0 and C1-0 with high sodium content (10.59 wt% and 2.70 wt%, respectively) exhibit little capability in the RADS of diesel fuel. When the residual sodium content is decreased to 0.38 wt%, the adsorbent C2-0 shows certain capability for the RADS; the sulfur content in diesel fuel is reduced from 560 ppm to 300 ppm. When the residual sodium content is further reduced to 0.03 wt% by washing the precipitate for 4 times, the adsorbent C4-0 exhibits

*Figure 9. XRD patterns of the mechanical Ni_3S_2-ZnO mixture treated under nitrogen (I) and hydrogen (II) for 24 h at different temperatures: (a) untreated; (b) 300 °C; (c) 400 °C; (d) 500 °C; (e) 600 °C. o, ZnO; *, Ni_3S_2; +, Ni; &, Ni–Zn; Δ, h-ZnS; and ^, c-ZnS.*
(Huang et al. 2011. Copyright 2011 Elsevier.)

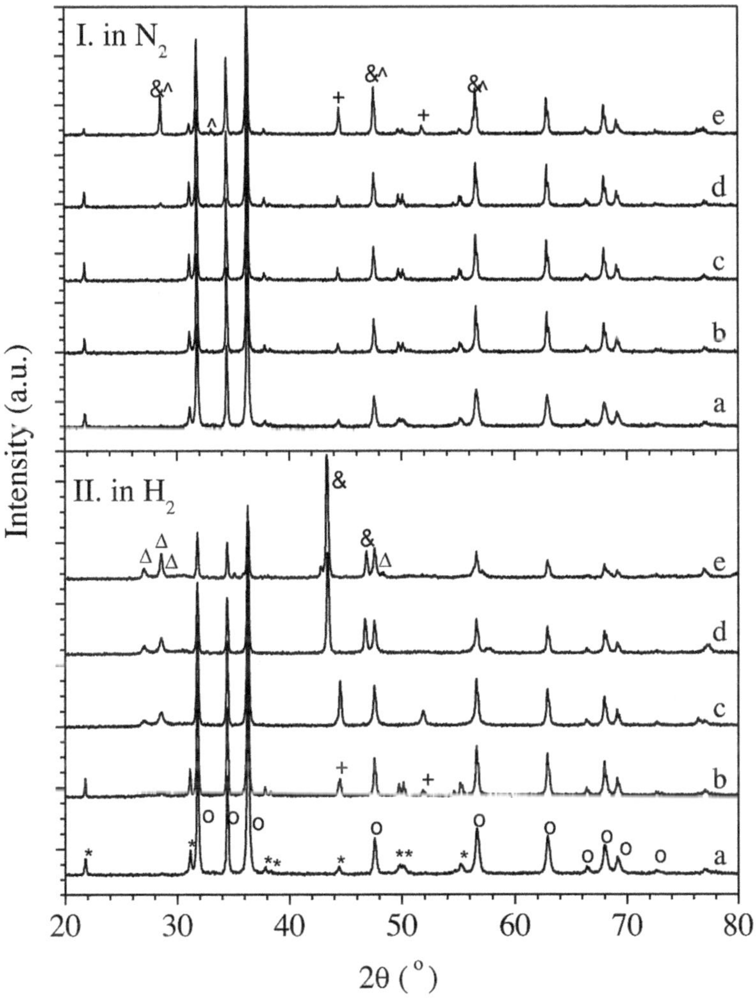

Table 1. Elemental S and C analysis of the spent adsorbents and their desulfurization capacities for the RADS in different atmospheres

Adsorbents	Elemental Content (%)		Capacity (g-S/g-NiO/ZnO)	
	C	S	Breakthrough	Total (57 h)
Calcined, spent in N_2	5.7	0.5	–	0.003
Reduced, spent in N_2	4.2	1.4	–	0.015
Calcined, spent in H_2	1.2	23.2	0.245	0.343
Reduced, spent in H_2	1.5	24.4	0.116	0.307

Note: The desulfurization capacity is estimated by the difference between the sulfur content in oil feed and that in the oil effluent after RADS.

excellent capability for the RADS of diesel fuel; the sulfur content in diesel fuel is reduced to a very low value (about 0 ppm in first 40 h and less than 10 ppm in next 30 h). It was evidenced that the residual sodium in Ni/ZnO has an apparent negative effect on its reaction performance. The extent of this negative effect is relevant to the content of residual sodium. With the decrease of the residual sodium content by washing the adsorbent Ni/ZnO with deionized water, the RADS capability can be recovered substantially.

Figure 10. Total sulfur concentration in the outflow diesel fuel after RADS over the Ni/ZnO adsorbents with different residual sodium contents: (a) C0-0; (b) C1-0; (c) C2-0; (d) C4-0; (e) C1-4; and (f) C2-4; the dashed line is the total sulfur concentration in the feed diesel fuel (560 ppm).
(Huang et al. 2010b. Copyright 2010 Elsevier.)

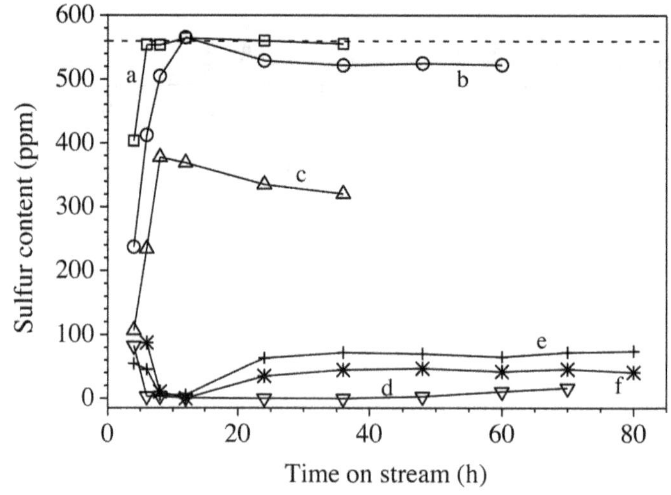

Figure 11. Crystallite sizes of NiO(200) (a) and ZnO(100) (b) in the calcined NiO/ZnO adsorbents and Ni(111) (c) and ZnO(100) (d) in the reduced Ni/ZnO adsorbents with different residual sodium contents (solid symbols for samples C0-0, C1-0, C2-0, and C4-0; open symbols for C1-4 and C2-4).
(Huang et al. 2010b. Copyright 2010 Elsevier.)

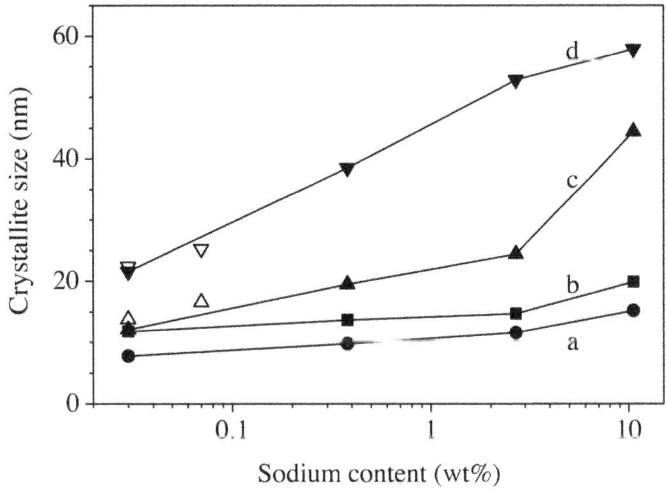

The crystallite size of ZnO and NiO in the adsorbent estimated from the XRD patterns is correlated with the content of residual sodium, as shown in Figure 11. The presence of the residual sodium in NiO/ZnO adsorbent leads to an increase in the crystallite size of zinc and nickel oxides. Moreover, the diffraction peaks of NiO and ZnO overlap at 2θ of 36-38° in C2-0 and C4-0. All these indicate that the crystallite size of ZnO and NiO increases with the residual sodium content in the adsorbent.

4. Effect of Naphthalene on RADS

The RADS activity is influenced by the hydrogenation reaction of aromatics. Our results showed that the selectivity of RADS over Ni/ZnO was apparently changed by the existence of naphthalene, although the activities were hardly influenced (Figure 11). The presence of naphthalene caused inhibition of the hydrogenation ability of aromatic cycles of DBT. This suggests that, in the hydrogenation (HYD) route of RADS, DBT and nickel may interact mainly through conjugated π-electrons. And the competition adsorption of naphthalene leads to decreased interaction, resulting in low HYD selectivity. However, it was found that even the weight concentration of naphthalene is about 10 times higher than that of DBT in the liquid phase, the conversion of DBT decreased only slightly. This is contrary to the conventional

Figure 12. Inhibition of the RADS of DBT in the presence of naphthalene: a), DBT alone; b), DBT + naphthalene (1:10).
(Huang et al. 2017. Copyright 2017 Wiley.)

(I) 0.1 MPa (II) 0.4 MPa

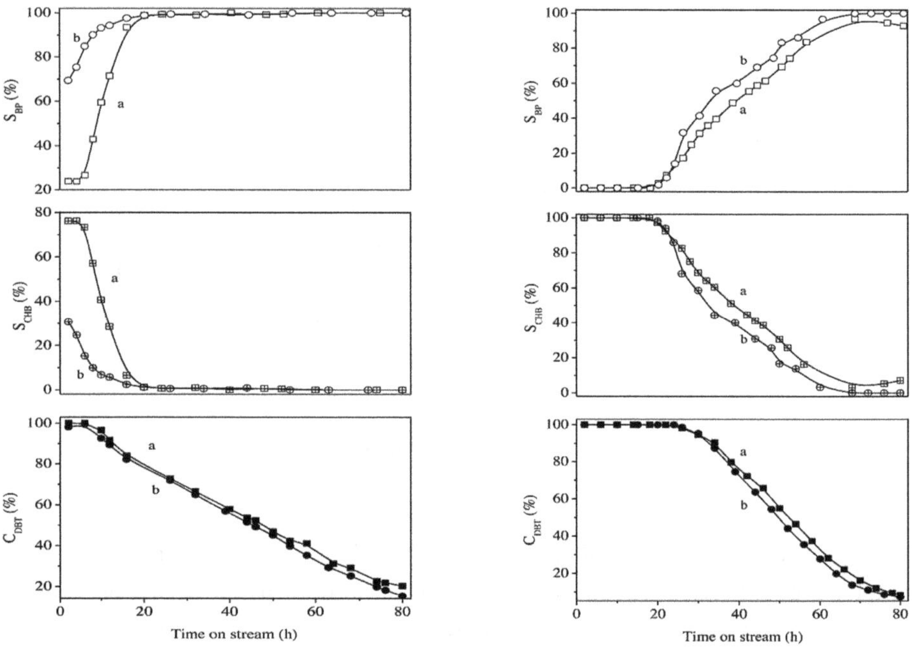

desulfurization, most studies indicated that the influence of aromatics on HDS was not negligible. We suggest that the favored direct desulfurization (DDS) may compensate for the HYD inhibition. The naphthalene changes the mode of adsorption of the DBT molecule on the Ni active phase. We also observed that the HYD/DDS ratio varied with the conversion of DBT. This illustrated that the sulfide states of Ni surface not only dominate the activity but also influence the selectivity of RADS.

Figure 13. The mechanism for the RADS of DBT over the Ni/ZnO adsorbent.
(Huang et al. 2011. Copyright 2011 Elsevier.)

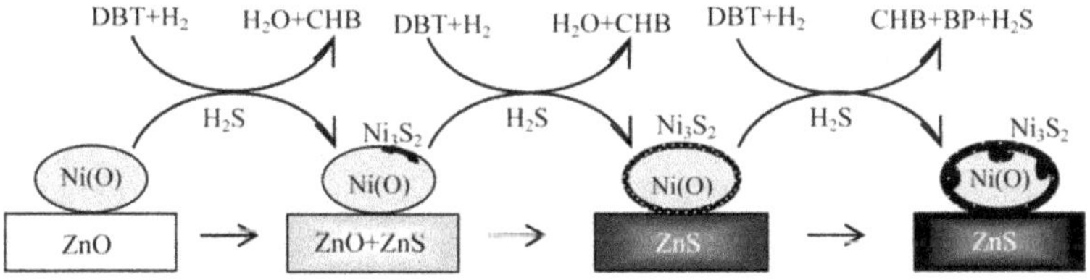

The Adsorption and Reaction Mechanism of Ni/ZnO Nano-Sorbent

1. The Process of Reactive Adsorption Desulfurization

According to Babich and Moulijn (2003), for Ni/ZnO sorbent, thiophene is firstly adsorbed on Ni active sites and then decomposed through hydrogenolysis. The sulfur atoms remained on the Ni surface are then transferred to ZnO to form ZnS where H_2S is proposed as the sulfur carrier.

Using the DMol3 software package, Zhang et al. (2012b) investigated the adsorption and desulfurization of thiophene over a Zn_3NiO_4 cluster. They observed that thiophene is favored to adsorb on Ni site. It then decomposed on the Ni site to form nickel sulfide. Two reaction pathways of the reduction of sulfide nickel were compared. Formation of H_2S by the attack of the H_2 molecule passed an energy barrier of 0.859 eV. And sulfur is directly transferred from Ni site to Zn site in the cluster with a barrier of 0.634 eV. This result provides a clue that direct sulfur transfer from Ni to ZnO may be the more favored path.

Ling et al (2013) investigated the adsorption and reaction of H_2S on ZnO ($10\bar{1}0$). It is observed that the dissociative adsorption of H_2S molecule is preferred. The most stable configuration for SH adsorption on the ZnO ($10\bar{1}0$) surface is that the S atom is bonded to two adjacent Zn atoms via the bridge bond mode. The S atom bridging a Zn-O bond is the most stable configuration for single S atom adsorption. The H atom is preferentially adsorbed on the O-top site. In the coadsorption structure of SH and H, the adsorption sites are the same as the sites of SH and H adsorption separately. But in the coadsorption of S and 2H, the most stable structure is that the S atom is adsorbed on two adjacent Zn atoms via the bridge bond mode, and two H atoms are located on the O-top sites. The surface adsorbed hydrogen atoms can further combine to form H_2 by the dehydrogenation processes or react with oxygen to form H_2O. It is observed that the H_2O-forming via H_2S-ZnO interaction is the most probable reaction route. An oxygen vacancy is generated on the surface after the formation of H_2O, and the H_2S can directly absorb on the vacancy.

2. Mechanism of RADS

As for the thiophene RADS, Tawara et al. (2000, 2001a) proposed that in NiO/ZnO system the NiO is firstly reduced to metallic Ni, which serves as the active site for adsorption and hydrogenolysis of thiophene. The sulfur species remained on Ni surface is transferred to ZnO forming ZnS by H_2S under a hydrogen atmosphere, and the Ni active surface is regenerated.

Based on the reaction results, we (Huang et al 2011) proposed a three-stage process of Ni/ZnO sorbent for DBT RADS. At the beginning, the Ni regenerating rate is basically equal to its poisoning rate and the reaction activity is kept at a stable state. In the second stage, the Ni regeneration rate is lower than the Ni poisoning rate; the sulfur transfer from Ni_3S_2 to ZnO is slowed down owing to the decrease of the active surface of ZnO. As a result, the content of Ni_3S_2 is increased gradually at the expense of metallic Ni phase. Thus a gradual deactivation occurs. In the last stage, most of ZnO is converted to ZnS and the sulfur transfer rate is access zero; the DBT conversion is mainly ascribed to the HDS catalyzed by Ni_3S_2. At this stage sulfur removed from DBT is mainly released to the effluent as H_2S.

Similarly, we (Huang et al 2010a) proposed a three-stage mechanism of Ni/ZnO sorbent for diesel RADS. Firstly, the organic sulfur compounds in diesel oil are decomposed on the adsorbent Ni/ZnO surface and Ni_3S_2 is formed from the sulfidation of the surface Ni (Eq. (1)). Secondly, Ni_3S_2 is regenerated by H_2 via the equilibrium reaction (Eq. (2)). Lastly, H_2S is accepted by ZnO to form ZnS (Eq. (3)). Hence, the active Ni species can be continually obtained by shifting the equilibrium of the reaction (Eq.(2)) to the right side.

$$2C_xH_yS + 3Ni + H_2 \rightarrow Ni_3S_2 + 2C_xH_{y+2} \tag{1}$$

$$Ni_3S_2 + 2H_2 \leftrightarrow 3Ni + 2H_2S \tag{2}$$

$$ZnO + H_2S \leftrightarrow ZnS + H_2O \tag{3}$$

Although the mechanism of sulfur transfer from Ni phase to ZnO phase by H_2S has been extensively accepted. But DFT study shows that the sulfur direct transfer from Ni to Zn site goes through lower energy barrier than the transfer by H_2S (Zhang et al. 2012c). And the Ni (111) surface has the highest adsorption energy to sulfur among Ag (111), Au (111), Cu (111), Ir (111), Ni (111), Pd (111) and Pt (111) surfaces (Alfonso 2008). This suggests that the sulfur on the Ni surface is difficult to be desorbed. Marécot found that Ni catalyst poisoned by sulfur lost the hydrogenation activity and cannot be reversed by hydrogen treatment. Thus we propose that RADS mechanism need further investigation. The S transfer directly at the interface and/or indirectly by H_2S may exist simultaneously, they may compete and play different roles at different stages.

Brief Summary

Some important features can be drawn from the discussion above. The pre-reduction of NiO/ZnO sorbent may not be a requirement. The reduction of NiO can be achieved under reaction condition, which avoids the formation of Ni-Zn alloy and the agglomeration of active particles. The metallic Ni is sulfurized to Ni_3S_2 when contacting with sulfur impurity in fuels, however the sulfur atoms of Ni_3S_2 is continually transferred to ZnO forming ZnS through H_2S or interface S transfer, leading to that the Ni active centers are auto-regenerated.

During RADS period, a balance between sulfur deposition and sulfur transfer is built, the ratio of Ni active centers is kept constant, and RADS reaction is in a stable state. With the accumulation of ZnS, the sulfur transfer and sulfur diffusion in ZnO bulk are inhibited, and the content of Ni_3S_2 increases with the expanse of Ni active sites. The sorbent begins deactivated until a quick drop of activity. At this time the Ni surface is totally covered by Ni_3S_2, and most of ZnO is transformed into ZnS. The H_2S formed by HDS on Ni_3S_2 surface is released directly into effluent gas.

Carbon species originated from the decomposition of an organic compound may dissolve into Ni phase and slows down the sulfur poison to Ni bulk. Ni and ZnS are formed obviously when the Ni_3S_2-ZnO mixture is exposed to hydrogen at 300 °C, suggesting the important role of hydrogen in the auto-regeneration of Ni active centers. Regeneration of the pertly deactivated sorbent may be required if the auto-regeneration cannot meet the requirement of RADS.

Large of fundamental researches have helped us understand the auto-regeneration process of Ni/ZnO sorbent a lot. But some important issues are still necessary to be solved. For example, why is the adsorbed sulfur transferred to the ZnO component rather than diffuse into the Ni bulk? Does the carbon permeation during RADS reaction retard the sulfidation of Ni bulk? Is the formation of Ni carbide is beneficial for the RADS reaction? Whether is the sulfur transferred by H_2S or directly through the interface? Answer to these problems will helpful for the better application of nano Ni/ZnO RADS sorbent.

Table 2. 4th generation of S-Zorb process for RADS of gasoline (adapted from Hou and Zhuang 2013)

	Unit1(Ferndale)	Unit2(Western)
Pressure /MPa	1.64	2.5
Temperature /°C	429	440
LHSV /h^{-1}	3.5	2.8
H_2/oil mol/mol)	0.5	0.42
S content in feed /ppm	1000	840
S content in product /ppm	13	8
desulfurization degreee /%	98.7	99
RON loss	1	1.6

COMMERCIAL Ni/ZnO SORBENT: RELATION OF STRUCTURES WITH INDUSTRIAL APPLICATION

S-Zorb Technique Based on Ni/ZnO for RADS of Gasoline

1. The History of S-Zorb Technique

S-Zorb technique was firstly developed by Conoco Philips Petroleum Co. for desulfurization of gasoline and diesel by circulating fluidized bed. Owing to the squeezing and attrition between sorbent particles during the operation, the structural additive is needed to add into the sorbent to increase the strength. In the test period, it was found that the attrition resistance of sorbent was noticeably enhanced with expanded perlite replacing diatomite as structural additive. First industrial unit of S-Zorb was built at Borger refinery in September of 1999 and started operation in April of 2001 using the 1st generation sorbent.

The 2nd generation sorbent was developed by Süd-Chemie AG, which could produce clean gasoline with sulfur content less than 30 ppm. And the 3rd generation S-Zorb sorbent was invented by Engelhard Co. New expanded perlite was used and the preparation process was optimized. But this sorbent is inferior to the followed sorbent of Süd-Chemie AG (4th generation) in desulfurization activity and attrition resistance. It was so given up before commercial application. A new mineral was used in the 4th generation S-Zorb sorbent. The strength of attrition resistance was largely increased. Owing to the excellent desulfurization activity and attrition resistance, this S-Zorb sorbent has been commercially used to produce clean gasoline since 2002 (Table 2). The 5th generation sorbent was manufactured with a simpler method, there was only one key step-spray drying. The cost was decreased and the preparation efficiency was improved remarkably. The test showed that the 5th generation sorbent had a very high activity of desulfurization and high ability to keep octane number. However, it produced much more fine powders than the 4th generation sorbent, influencing the stable run of the unit. Thus it has not been commercially applied.

In 2007, the S-Zorb technique was bought out by China's Sinopec, which began developing new generation sorbent. Based on the concept formulation provided by Conoco Phillips Co., the Research Institute of Petroleum Processing of Sinopec developed the FCAS sorbents and applied them successfully. Thirty-four S-Zorb units have been built in the refineries of Sinopec by September 2017; the total processing capacity has reached over 40 Mt/y. And another five units are under design. The RADS technique of Sinopec has been licensed to two other refineries in China. Meantime there are other six refineries abroad producing deep desulfurized gasoline with S-Zorb units; they are Borger refinery, 0.25 Mt/y; Ferndale refinery, 0.85 Mt/y; Lake Charles refinery, 1.57 Mt/y; Wood River refinery, 1.30 Mt/y; Pasadena Refining Systems refinery, 1.60 Mt/y; Western Yorktown (Giant) refinery, 1.20 Mt/y.

2. The Circulating Fluidized Bed Process of S-Zorb

Owing to the continual deactivation of RADS sorbent, the S-Zorb technique uses the circulating fluidized bed process to produce clean gasoline with low sulfur content and high content olefins, where the partly deactivated sorbent is continually regenerated by oxidation and reduction. The temperature and pressure at the inlet are controlled at about 420 °C and 2-3 MPa with a temperature rising about 22 °C and a pressure drop about 0.17 MPa, the LHSV and the mole ratio of H_2/oil is about 4 $h^{-1}$and <0.3, respectively. At this condition, the sulfur capacity of a sorbent is about 8 wt%. The H_2 consumption compared with feed is about 0.2 wt%, and the loss of (RON+MON)/2 are about 0.3.

The regenerator is run at about 525 °C and 0.1 MPa with 0.2 v/v% O_2 as a regeneration gas. The cycle time of sorbent is 20 min/per cycle (Hou and Zhuang, 2013). Usually, the sulfur in sorbent is not totally burnt out for the cause of energy efficiency and the protection to the sorbent. The supplement of sorbent is needed due to the abrasion loss. The FCAS sorbent is made with at least two additives to increase the strength and the resistance to the volume expansion from ZnO to ZnS. It is especially fitful to treat the fluid catalytic cracking (FCC) gasoline fraction, which usually contain high contents of olefins and refractory thiophene analogs. The commercial application shows that Z-Sorb process can decrease the sulfur content in FCC gasoline from 1000 to 10 ppm with little loss of octane number.

3. The Regeneration of the Sorbent

During the RADS reaction, the sorbent is transformed into a mixture of sulfides and gradually deactivated. To extend the lifetime of sorbent, the deactivated sorbent is usually regenerated in a two-step process: sulfides are firstly calcined to obtain oxides which are then treated in H_2 to reduce the Ni oxide. As for S-Zorb process, the $ZnAl_2O_4$ and Zn_2SiO_4 spinel analog are formed in the regenerated sorbent, which decreases remarkably the content of the active component ZnO. Furthermore, the sorbent is easily crushed with the $ZnAl_2O_4$ and Zn_2SiO_4 in the sorbent. The $ZnSO_4$ and $NiSO_4$ were found in regenerated sorbent also by XRD and XPS (Meng et al. 2013, Qiu et al. 2013). The formation of $ZnSO_4$ should be avoided. But the $NiSO_4$ seems to have some positive effect due to its Lewis acidity. The formation of $Zn_3O(SO_4)_2$ was observed in Wang et al.'s (2010) study.

Fix Bed Techniques for RADS of Gasoline

The fix bed technique affords the advantages of low cost and easy maintenance. A fix bed unit of RADS with the ability of 0.5 Mt/y started up successfully at Huizhou refinery of China National Offshore Oil Corporation (CNOOC) at the end of 2013, which was co-developed by Beijing Haishunde Titanium Catalyst Co. Ltd. and CNOOC. The produced gasoline can meet national V standard of China, corresponding to the Europe V standard. The YDCADS technique developed by Dalian Institute of Chemical Physics and Shaanxi Yanchang Petroleum (Group) Corp. Ltd. was run in a pilot plant in July, 2013. It decreases the sulfur content in gasoline from 100 ppm to below 10 ppm. The loss of octane number is lower than 0.8 units and the hydrogen consumption is less than 0.2 wt%.

The Comparison of RADS Technology With HDS Technology

RADS technology has good activity for the deep desulfurization of gasoline. Because the hydrogenation of olefins is inhibited, the hydrogen consumption can be largely decreased. And the octane number has a little drop. This brings the merits of cost. But the circulating fluidized bed process is complicated; the operation control of the unit is more difficult. Meanwhile, there is a requirement for the high strength and resistance to abrasion for Ni/ZnO sorbent. Although fix-bed technology has been developed, the sorbent is still needed to regenerate frequently.

HDS technology can be used for deep desulfurization of versatile feeds. But the hydrogenation to olefin results in the loss of octane number of gasoline. This can be solved with followed octane number recovering. But the cost is higher. On the other hand, the fix bed technology is simpler, and the catalysts usually have a longer life.

Both technologies play important roles in the deep desulfurization of petroleum refining. The RADS shows its merit to deep desulfurization of FCC gasoline, while HDS is more fitful for desulfurization of diesel and other heavy oil streams.

FUTURE VISION

RADS Application for Petroleum Oil Streams Other Than Gasoline

Tawara et al. (2001a and 2001b) investigated the RADS sorbent for the ultra-deep desulfurization of kerosene of the fuel cell (FC). A 13% Ni/ZnO sorbent was certified to decrease the average sulfur content of kerosene to less than 0.03 ppm for 1 year. They predicted that 1-year operation of petroleum feed FC cogenerations and 1110 thousand-km running of FC vehicles are possible by applying this sorbent.

Ni/ZnO sorbent can be effective for producing low-sulfur diesel. We (Huang et al. 2010) observed that the sulfur content in the diesel oil is reduced from the original 560.0 ppm to a very low value after an induced period over a NiO/ZnO sorbent with 36.6 wt% Ni content. However, the sulfur content increased slowly after 12 h on stream, until to 120.0 ppm at 122 h on stream. This primary result illustrate that for diesel RADS the sulfur capacity and desulfurization ability are necessary to be largely improved if the commercial application can be run.

RADS can also be applicable for deep desulfurization in the production of chemicals. The sulfur level can be dropped to very low (<0.03 ppm) level, and the process treating the H_2S gas pollutant can be avoided due to the sulfur adsorption of ZnO. This is especially favored for the small chemical plants. We have applied the RADS fix-bed technique for ultra-deep desulfurization of chemical feeds in three chemical enterprises. The sulfur content can be reduced below 0.1 ppm with the feed containing dozens of ppm of sulfur.

Suppression of Methanation and the RADS of Coal Coke Gas

Although Ni/ZnO is an efficient RADS sorbent for deeply removing organic sulfur compounds from liquid petroleum fuels such as gasoline and diesel, its application in the desulfurization of industrial gases, such as syngas and coke oven gas, is hindered due to the methanation function of metal Ni. Metal Ni can catalyze the reaction of H_2 and CO (CO_2) to CH_4 during desulfurization, which causes not only the loss of H_2 and CO (CO_2) but also the temperature runaway in the reactor, leading to the sintering of sorbent. Therefore, exploring methods or routes to suppress the

Figure 14. Schematic illustration for SMSI in Ni/ZnO systems reduced at 350, 450 and 500 °C. The color bar depicts the degree of the negative charge state of the Ni particles

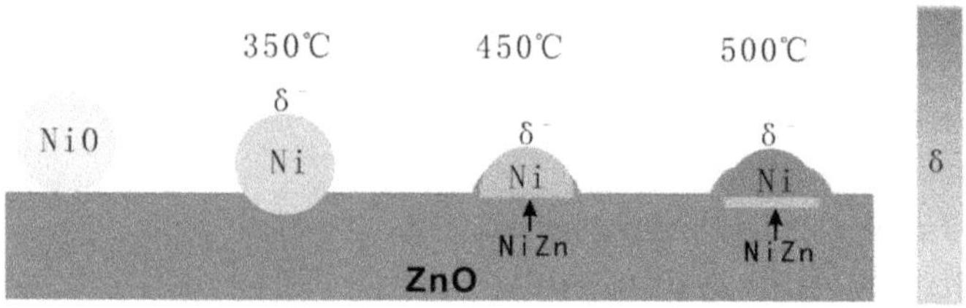

methanation activity of Ni/ZnO RADS sorbent is the prerequisite for expanding its application to the desulfurization of coke oven gas or syngas.

We (Wang et al. 2017) have made some efforts to deal with this issue by modulating the interaction between Ni and ZnO particles in the Ni/ZnO. We prepared a NiO/ZnO system by co-precipitation, which was then reduced in H_2 at different temperatures (350, 400, 450 and 500 °C). H_2 and CO chemisorption experiments combined with XRD, SEM, TEM analyses verified that reduction treatment in H_2 could induce the strong metal-support interaction (SMSI) in Ni/ZnO system and the degree of SMSI strengthened as the reduction temperature rose. TEM, EPR, and XPS studies revealed that the generation of SMSI was attributed to the geometric decoration of Ni particles by ZnO and electron transfer from ZnO to Ni atoms as well as the chemical interaction between Ni and ZnO leading to the formation of a NiZn alloy. Figure 14 illustrates the generation mechanism of SMSI between Ni and ZnO particles.

The effect of SMSI on the methanation performance of Ni/ZnO was investigated by comparing the methanation performance of Ni/γ-Al_2O_3 and Ni/ZnO samples reduced at different temperatures (Figure15). The Ni/γ-Al_2O_3 sample was prepared by impregnation and reduced at 500 °C, containing the same Ni content as the Ni/ZnO sample. As figure 15 shows, the methanation reaction over Ni/γ-Al_2O_3 starts at 200 °C and its CO conversion reaches 100% at 300 °C, while methanation over the Ni/ZnO sample reduced at 350 °C starts at 280 °C and the corresponding CO conversion at 300 °C is only about 7.0%, showing that the methanation activity of Ni/γ-Al_2O_3 is much higher than that of Ni/ZnO, which means that SMSI exhibits a remarkable suppression effect on the methanation activity. It was found that the methanation activity of Ni/ZnO greatly relates to the reduction temperature and the higher reduction temperature the weaker methanation activity. In situ FTIR spectra studies revealed that CO adsorption on the Ni particles in Ni/γ-Al_2O_3 in both linear and bridge forms, while CO adsorption on the Ni particles in Ni/ZnO only in linear form. The suppression of methanation arose from the fact that SMSI changed the

Figure 15. CO conversion curves for Ni/γ-Al_2O_3 (a) and Ni/ZnO samples reduced at 350 °C (b), 400 °C (c), 450 °C (d) and 500 °C (e); CH_4 selectivity curves for Ni/γ-Al_2O_3 (f) and Ni/ZnO samples reduced at 350 °C (g), 400 °C (h), 450 °C (i) and 500 °C (j) (Wang et al. 2017, Copyright 2017, The Royal Society of Chemistry).

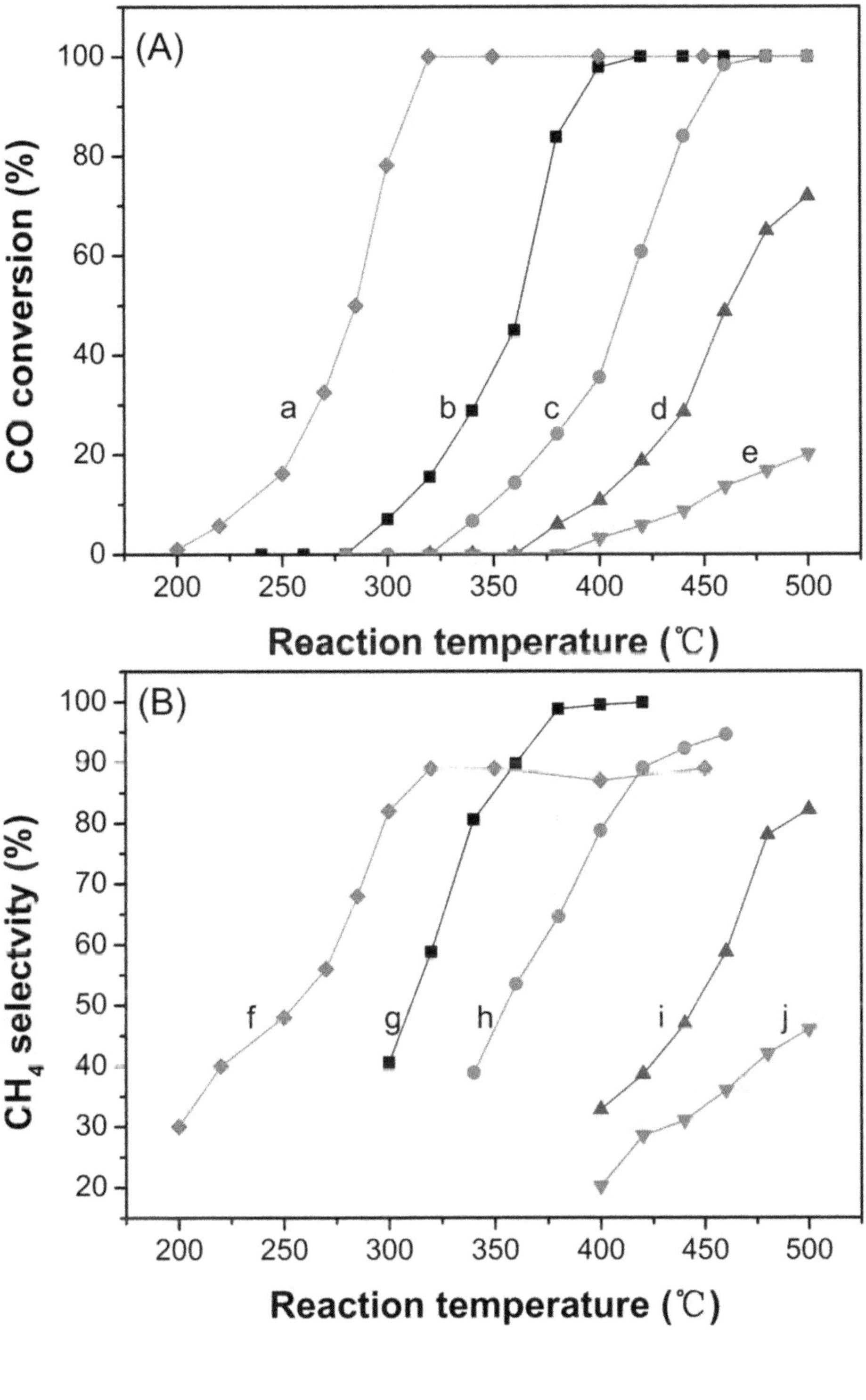

Figure 16. Desulfurization and methanation activities over Ni/ZnO-COP samples reduced at different temperature
(Wang 2018)

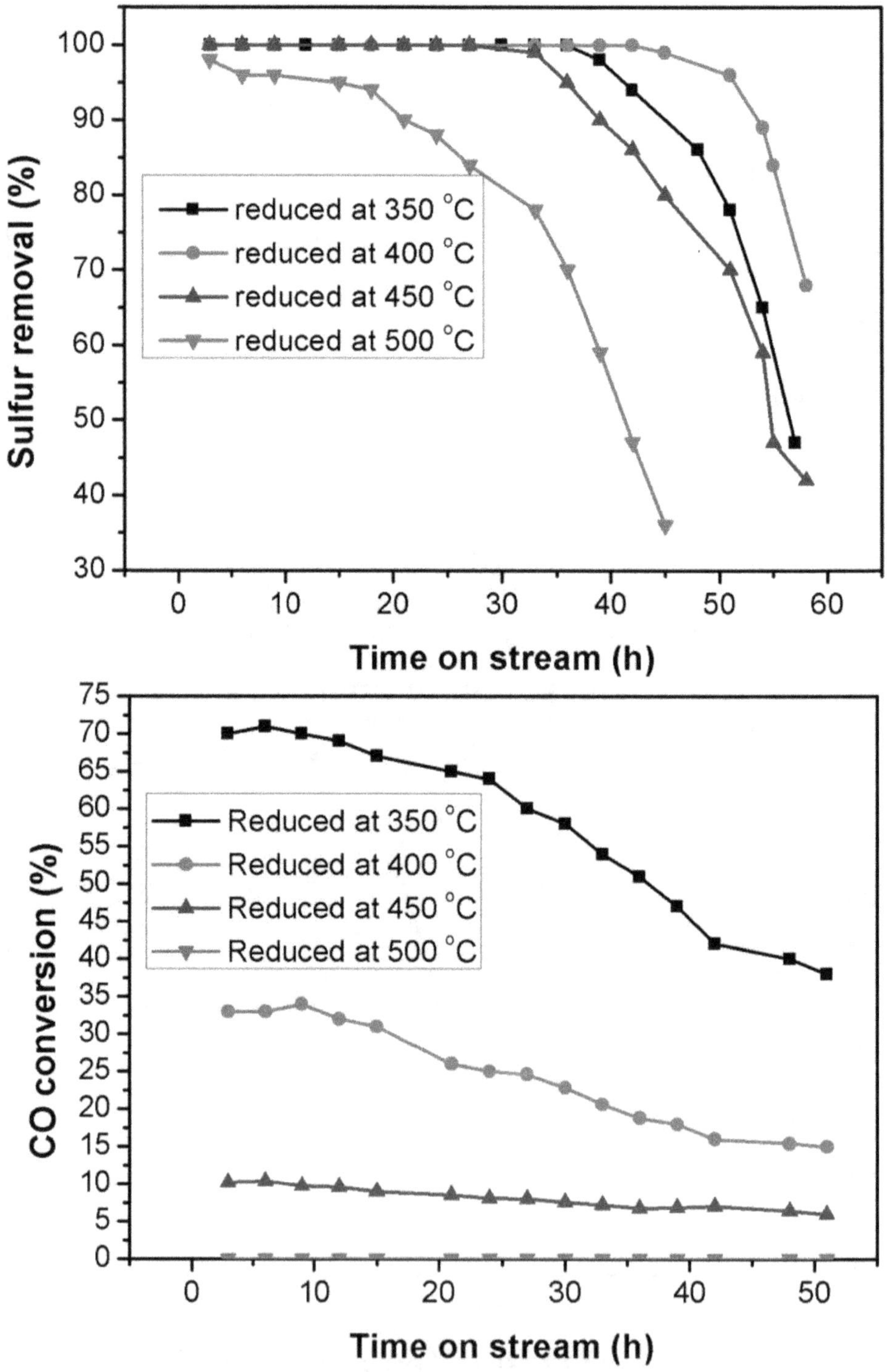

geometric, chemical and electronic properties of Ni particles, and then adsorption manner of CO was changed.

To verify whether Ni/ZnO RADS sorbent has potential application in the desulfurization of coke oven gas and syngas, the methanation and desulfurization behaviors over Ni/ZnO sorbent were investigated simultaneously (Wang, 2018). Figure 16 demonstrates that the CO conversion decreases sharply as the reduction temperature rises and activity disappears when the reduction temperature rises to 500 °C, indicating that SMSI has a strong suppression effect on methanation. Meanwhile, the time on stream for 100% sulfur removal prolongs as the reduction temperature rises from 350 to 400 °C, but this time shortens when the sorbent reduced at a higher temperature, suggesting that the low SMSI state facilitates the improvement of desulfurization performance, but the desulfurization activity is suppressed when SMSI is too strong. This work showed that the side reaction methanation during desulfurization could be suppressed remarkably through the introduction of SMSI in Ni/ZnO sorbent, suggesting a potential application for the desulfurization of coke oven gas and syngas.

CONCLUSION

The fundamental research reveals the micro-regeneration process of Ni/ZnO sorbent. During the RADS reaction, the Ni active center is auto-regenerated continually by the sulfur transfer from Ni surface to the ZnO particles. Thus the Ni catalytic centers do not deactivate by sulfur poison, desulfurization activity is kept. The deactivation does develop after long reaction time due to the inhibition of sulfur transfer by the accumulation of ZnS on the surface of ZnO; this can be partly recovered by the regeneration of oxidation and reduction, namely conventional regeneration. Owing to the deactivation by the sinter of the sorbent is difficult to recover, structure reagent, such as alumina, silica or perlite, are usually added into the sorbent. The macro regeneration, namely the periodic deactivation-regeneration of commercial Ni/ZnO sorbent, guarantees the stable usage of sorbent in ultra-deep desulfurization of gasoline. The synergy coordination of the micro-regeneration and macro regeneration features the relation of RADS properties with the dynamic structures of Ni/ZnO sorbent.

ACKNOWLEDGMENT

The authors are acknowledged of the sponsorship from the National Natural Science Foundation of China (21473231,21603256) for the fundamental research of RADS of Ni/ZnO sorbent.

REFERENCES

Alfonso, D. R. (2008). First-principles studies of H_2S adsorption and dissociation on metal surfaces. *Surface Science*, *602*(16), 2758–2768. doi:10.1016/j.susc.2008.07.001

Babich, I. V., & Moulijn, J. A. (2003). Science and technology of novel processes for deep desulfurization of oil refinery streams: A review. *Fuel*, *82*(6), 607–631. doi:10.1016/S0016-2361(02)00324-1

Bezverkhyy, I., Ryzhikov, A., Gadacz, G., & Bellat, J. P. (2008). Kinetics of thiophene reactive adsorption on Ni/SiO_2 and Ni/ZnO. *Catalysis Today*, *130*(1), 199–205. doi:10.1016/j.cattod.2007.06.038

Bezverkhyy, I., Safonova, O. V., Afanasiev, P., & Bellat, J. P. (2009). Reaction between thiophene and Ni nanoparticles supported on SiO_2 or ZnO: In situ synchrotron X-ray diffraction study. *The Journal of Physical Chemistry C*, *113*(39), 17064–17069. doi:10.1021/jp903488p

Fan, J., Wang, G., Sun, Y., Xu, C., Zhou, H., Zhou, G., & Gao, J. (2010). Research on reactive adsorption desulfurization over Ni/ZnO-SiO_2-Al_2O_3 adsorbent in a fixed-fluidized bed reactor. *Industrial & Engineering Chemistry Research*, *49*(18), 8450–8460. doi:10.1021/ie100923v

Ge, H., Tang, M., & Wen, X. (2016) Ni/ZnO Nano Sorbent for Reactive Adsorption Desulfurization of Refinery Oil Streams. In Applying Nanotechnology to the Desulfurization Process in Petroleum Engineering (pp. 216-239). IGI Global.

Gyanesh, K. (2001). Sorbent composition, process for producing and use in desulfurization. *PCT Int. Appl*. WO 0114051.

Homs, N., Llorca, J., & de la Piscina, P. R. (2006). Low-temperature steam-reforming of ethanol over ZnO-supported Ni and Cu catalysts: The effect of nickel and copper addition to ZnO-supported cobalt-based catalysts. *Catalysis Today*, *116*(3), 361–366. doi:10.1016/j.cattod.2006.05.081

Hou, X., & Zhuang, J. (Eds.). (2013). *Technology handbook of S-Zorb unit for desorption desulfurization of FCC gasoline*. Beijing, China: China Petrochemical Press.

Huang, L., Ge, H., Yan, L., Wang, G., Qin, Z., & Wang, J. (2017). Competitive reactive adsorption desulfurization of dibenzothiophene and hydrogenation of naphthalene over Ni/ZnO. *Canadian Journal of Chemical Engineering*, *9999*, 1–8.

Huang, L., Qin, Z., Wang, G., Du, M., Ge, H., Li, X., ... Wang, J. (2010b). A detailed study on the negative effect of residual sodium on the performance of Ni/ZnO adsorbent for diesel fuel desulfurization. *Industrial & Engineering Chemistry Research*, *49*(10), 4670–4675. doi:10.1021/ie100293h

Huang, L., Wang, G., Qin, Z., Dong, M., Du, M., Ge, H., ... Wang, J. (2011). In situ XAS study on the mechanism of reactive adsorption desulfurization of oil product over Ni/ZnO. *Applied Catalysis B: Environmental*, *106*(1-2), 26–38.

Huang, L., Wang, G., Qin, Z., Du, M., Dong, M., Ge, H., ... Wang, J. (2010a). A sulfur K-edge XANES study on the transfer of sulfur species in the reactive adsorption desulfurization of diesel oil over Ni/ZnO. *Catalysis Communications*, *11*(7), 592–596. doi:10.1016/j.catcom.2010.01.001

Jia, G. (2014). Retrieved from http://www.sinopecnews.com.cn/news/content/2014-02/17/content_1377530.shtml (in Chinese)

Johnson, B. G., Slater, P., & Kidd, D. (2001). Application of Phillips's Zorb process to distillates-meeting the challenge. AM-01-14, New Orleans, LA.

Khare, G. P. (1999). Sorbent compositions. US Pat Appl. 4948830.

Ling, L., Zhang, R., Han, P., & Wang, B. (2013). DFT study on the sulfurization mechanism during the desulfurization of H_2S on the ZnO desulfurizer. *Fuel Processing Technology*, *106*, 222–230. doi:10.1016/j.fuproc.2012.08.001

Mangnus, P. J., Poels, E. K., Van Langeveld, A. D., & Moulijn, J. A. (1995). Comparison of the sulfiding rate and mechanism of supported NiO and Ni0 particles. *Journal of Catalysis*, *137*(1), 92–101. doi:10.1016/0021-9517(92)90141-4

Marécot, P., Paraiso, E., Dumas, J. M., & Barbier, J. (1992). Deactivation of nickel catalysts by sulphur compounds: I. Benzene hydrogenation. *Applied Catalysis A*, *80*(1-2), 79–88. doi:10.1016/0926-860X(92)85109-O

Meng, X., Huang, H., & Shi, L. (2013). Reactive mechanism and regeneration performance of NiZnO/Al_2O_3-diatomite adsorbent by reactive adsorption desulfurization. *Industrial & Engineering Chemistry Research*, *52*(18), 6092–6100. doi:10.1021/ie303514y

Qiu, L., Zou, K., & Xu, G. (2013). Investigation on the sulfur state and phase transformation of spent and regenerated S zorb sorbents using XPS and XRD. *Applied Surface Science*, *266*, 230–234. doi:10.1016/j.apsusc.2012.11.156

Ryzhikov, A., Bezverkhyy, I., & Bellat, J. (2008). Reactive adsorption of thiophene on Ni/ZnO: Role of hydrogen pretreatment. *Applied Catalysis B: Environmental*, *84*(3-4), 766–772. doi:10.1016/j.apcatb.2008.06.009

Shangguan, J., Zhao, Y., Fan, H., Liang, L., Shen, F., & Miao, M. (2013). Desulfurization behavior of zinc oxide based sorbent modified by the combination of Al_2O_3 and K_2CO_3. *Fuel, 108*, 80-84.

Tang, M., Zhou, L., Du, M., Lyu, Z., Wen, X.-D., Li, X., & Ge, H. (2015). A novel reactive adsorption desulfurization Ni/MnO adsorbent and its hydrodesulfurization ability compared with Ni/ZnO. *Catalysis Communications*, *61*, 37–40. doi:10.1016/j.catcom.2014.11.005

Tawara, K., Nishimura, T., & Iwanami, H. (2000). Ultra-deep hydrodesulfurization of kerosene for fuel cell system (Part 2) Regeneration of sulfur-poisoned nickel catalyst in hydrogen and finding of auto-regenerative nickel catalyst. *Sekiyu Gakkaishi*, *43*(2), 114–120. doi:10.1627/jpi1958.43.114

Tawara, K., Nishimura, T., Iwanami, H., Nishimoto, T., & Hasuike, T. (2001). New hydrodesulfurization catalyst for petroleum-feed fuel cell vehicles and cogenerations. *Industrial & Engineering Chemistry Research*, *40*(10), 2367–2370. doi:10.1021/ie000453c

Tawara, K., Nishimura, T., Iwanami, H., Nishimoto, T., & Hasuike, T. (2001b). Ultra-deep hydrodesulfurization of kerosene for fuel cell system (Part 3) Development and evaluation of Ni/ZnO catalyst. *Sekiyu Gakkaishi*, *44*(1), 43–51. doi:10.1627/jpi1958.44.43

Wang, M. Z., & Ruan, Y. J. (2010). Study on process conditions of adsorption desulfurization reaction of FCC gasoline. *Petroleum Refinery Engineering*, *40*(9), 5–10.

Wang, W. (2018). *Study on the methanation and desulfurization behaviors on Ni/ZnO* (PhD dissertation). University of Chinese Academy of Sciences.

Wang, W., Li, X., Zhang, Y., Zhang, R., Ge, H., Bi, J., & Tang, M. (2017). Strong metal–support interactions between Ni and ZnO particles and their effect on the methanation performance of Ni/ZnO. *Catalysis Science & Technology*, *7*(19), 4413–4421. doi:10.1039/C7CY01119A

Wen, Y., Wang, G., Wang, Q., Xu, C., & Gao, J. (2012). Regeneration characteristics and kinetics of Ni/ZnO-SiO_2-Al_2O_3 adsorbent for reactive adsorption desulfurization. *Industrial & Engineering Chemistry Research*, *51*(10), 3939–3950. doi:10.1021/ie202730w

Zhang, S., Zhang, Y., Huang, S., Wang, P., & Tian, H. (2012b). Mechanistic investigations on the adsorption of thiophene over Zn_3NiO_4 bimetallic oxide cluster. *Applied Surface Science*, *258*(24), 10148–10153. doi:10.1016/j.apsusc.2012.06.096

Zhang, Y., Yang, Y., Hana, H., Yang, M., Wang, L., Zhang, Y., ... Li, C. (2012a). Ultra-deep desulfurization via reactive adsorption on Ni/ZnO: The effect of ZnO particle size on the adsorption performance. *Applied Catalysis B: Environmental*, *119-120*, 13–19. doi:10.1016/j.apcatb.2012.02.004

Zhang, Y., Yang, Y., Lin, F., Yang, M., Liu, T., Jiang, Z., & Li, C. (2013). Improvement of adsorptive desulfurization performance of Ni/ZnO adsorbent by doping with Mn additive. *Chinese Journal of Catalysis*, *34*(1), 140–145. doi:10.1016/S1872-2067(11)60513-5

Zhou, G., Sun, W., Jiang, W., & Zhou, H. (2013). Deep desulfurization of hydro-gasoline by adsorption in H_2 atmosphere over Ni/ZnO-SiO_2-Al_2O_3. *Petroleum Processing and Petrochemicals*, *44*(7), 17–21.

Chapter 6
Coordination Polymers and Polymer Nanofibers for Effective Adsorptive Desulfurization

Tendai O. Dembaremba
https://orcid.org/0000-0003-4937-249X
Nelson Mandela University, Port Elizabeth, South Africa

Adeniyi S. Ogunlaja
https://orcid.org/0000-0003-2479-4928
Nelson Mandela University, Port Elizabeth, South Africa

Zenixole R. Tshentu
https://orcid.org/0000-0002-5340-8005
Nelson Mandela University, Port Elizabeth, South Africa

ABSTRACT

Desulfurization of fuel oils is an essential process employed in petroleum refineries to reduce the sulfur content to levels mandated for environmental protection. Hydrodesulfurization (HDS), which is currently being employed, is limited in treating refractory organosulfur compounds and only reduces the sulfur content in fuels to a range of 200-500 ppmS. In this chapter, several scientific and technological advances reported in the literature for the desulfurization of fuels are reviewed and discussed. Amongst these techniques, oxidative desulfurization (ODS) and adsorptive desulfurization (ADS) are proposed as additional steps to complement HDS in meeting the mandated ultra-low sulfur levels (10 ppmS). In the ODS technique, refractory organosulfur compounds are oxidized to organosulfones, followed by solvent extraction

DOI: 10.4018/978-1-7998-2146-5.ch006

or adsorption of the organosulfones. The chemistry involved in the development and fabrication of sulfur/sulfone responsive adsorbents is also discussed. The use of molecular imprinted polymers (MIPs) and coordination polymers (CPs) for the selective adsorption of organosulfone compounds (in ODS) and/or organosulfur (in ADS) offers various properties such as imprinting effect, hydrogen bonding, π-π interactions, van der Waals forces, π-complexation, and electrostatic interactions. CPs, in particular metal organic frameworks (MOFs), have been reported to possess suitable features to overcome most of these challenges associated with adsorptive ultra-deep desulfurization when design strategies to achieve good selectivity are strictly followed. Matching the sizes of the cavities to the critical dimensions of the sulfur containing compounds (SCCs), using suitable metal centres which allow for coordinative interaction with the SCCs and using linkers with suitable functionality as to enhance specific interaction (dispersion forces) with the SCCs were considered to be pivotal features to prioritize. The prospects for the use of MIPs and CPs for future industrial applications in desulfurization are envisaged.

INTRODUCTION

Crude oil is a complex blend containing thousands of hydrocarbons, non-hydrocarbon compounds and heavy metals (Sami *et al.* 1994). The hydrocarbons are mixed with variable quantities of sulfur-, nitrogen-, and oxygen-containing compounds. Crude oil is expected to be the source of energy in the world for several decades to a century. Crude oil reserves are distributed around all the continents (Figure 1). The Middle East has the highest reserves of over 750 thousand million barrels and the Asia Pacific region having the least reverses of just over 40 thousand million barrels of oil. Generally, crude oils are refined to separate the complex mixture into simpler fractions that can be used as fuels, lubricants, and as intermediate feedstock for petrochemical industries.

The sulfur content distribution in some crude oil reservoirs around the world is depicted in Figure 2. The sulfur content of crude oils varies appreciably from one reservoir to another. Over 70% of the world's oil reserves are classified to be of heavier and sourer composition, meaning they have a high sulfur content. The low sulfur-containing crude oils are referred to as sweet oils and are more desirable due to fewer costs during refinery processes, of which removal of sulfur is an important step. (Source: EPA, 2013; Crude oil reserve, 2013). Removal of sulfur has become important over the years since sulfur oxides produced from the combustion of fuel containing the sulfur compounds are emitted into the atmosphere, constituting serious environmental hazards such as acid rain and the generation of airborne particulate (such as smog and sulfates). Sulfur oxides also poison and deactivate catalytic converters in vehicles, while smog and particulates, on the other hand,

Figure 1. Crude oil reserves distribution around the continents of the world
(Source: EPA, 2013; Crude oil reserve, 2013)

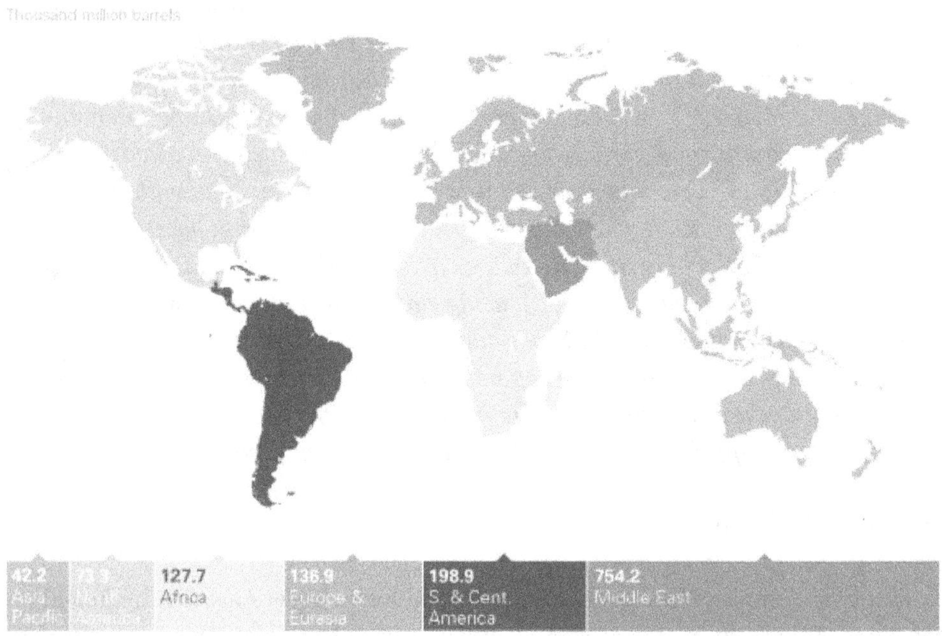

cause serious respiratory illnesses such as asthma when inhaled. The lowering and/or complete eradication of emission of sulfur oxides by motorized transport is currently being addressed through regulation of sulfur levels in diesel and gasoline by the environmental protection agencies (EPA) and other regulatory bodies.

The current process employed for the removal of sulfur compounds from crude oils is hydro-desulfurization. In this process hydrogen gas passes through a stream of crude oil under high temperature and pressure in the presence of a catalyst (such as Co–Mo/ɣ-Al_2O_3). Hydrogen sulfide gas is produced, and it is scrubbed off and converted to elemental sulfur in the Claus process. The environmental regulating agencies have mandated stringent legislation with the aim of reducing sulfur levels in fuels to an ultra-low level of between 10-15 ppm by 2015 (Figure 3) (Barbara *et al.* 2011; Song *et al.* 2003; Ogunlaja A.S. *et al.* 2013).

In South Africa, regulations regarding the introduction of Clean Fuels II (10 ppmS) were gazetted in 2012 and effective from 2017 (Clean Fuel II, 2019). In 2015, the South Africa Petroleum Industry Association (SAPIA) requested the South Africa Department of Energy (DOE) to rescind the regulations pertaining to the implementation of cleaner fuels by 2017 due to policy uncertainty generated by the delay in a decision to implement cost recovery. Hence, oil refining fleet in

Figure 2. Sulfur distribution in crude oils in the world (1 wt.% Sulfur = 10000 ppm)
(Source: EPA, 2013; Crude oil reserve, 2013)

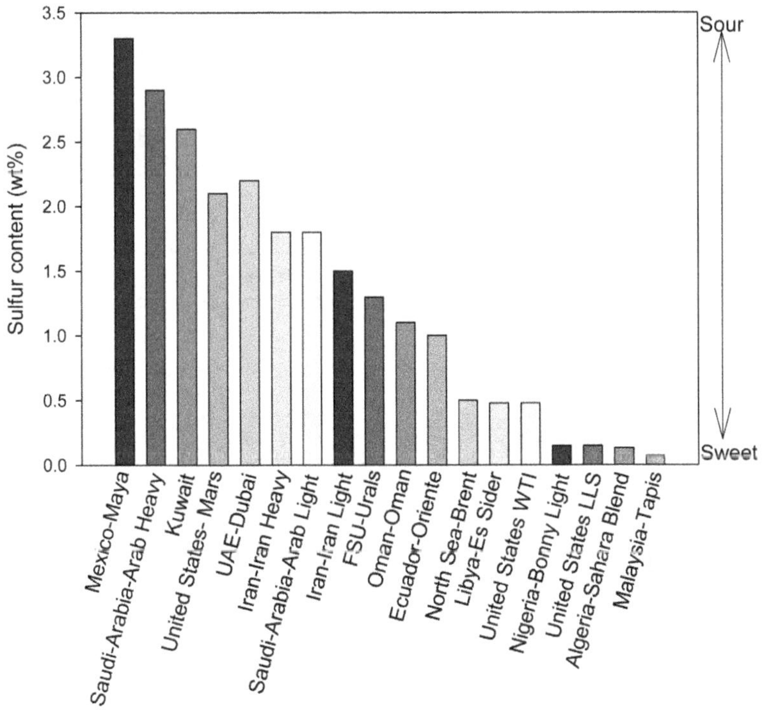

South Africa will not have progressed with the necessary upgrades to implement cleaner fuels II (10 ppmS) by July 2017. (Clean Fuel II, 2019). DOE and SAPIA are working on the issues related to cost recovery so that an informed decision can be made by the government supported by the oil industry. The top three countries with stricter sulfur limits (≤10 ppmS) include Sweden, Germany, and Japan. EU countries have continued to legislate the use of sulfur-free (<10 ppmS) fuels since 2009, with Sweden leading the way.

Investigations by researchers demonstrated that sulfur compounds remaining in diesel fuels at sulfur levels lower than 500 ppm are predominantly dibenzothiophenes and dibenzothiophenes with alkyl substituents at the 4- and/or 6-position (Song *et al.* 2006; Whitehurst *et al.* 1998; Badich *et al.* 2003, Ma *et al.* 1996; Andersson *et al.* 2006, 2007). These species are lower in hydro-desulfurization (HDS) reactivity and are termed refractory sulfur compounds. Steric hindrance and electronic factors were reported as factors responsible for the observed low HDS reactivity of DBTs and 4- and 6-substituted DBTs (Ma *et al.*1995). The HDS reactivity of sulfur species in fuel have been reported in detail (Ma *et al.* 1994, 1996); the authors employed a high-resolution gas chromatography (GC) with a sulfur-selective Flame Photometry

Figure 3. Trends in diesel sulfur fuel specifications from the United States (US), United Kingdom (UK), European Union (EU) and South Africa (SA) over the years.[a] The SA target of 10 ppmS (Clean Fuels II) meant to come into effect in 2017 was receded.

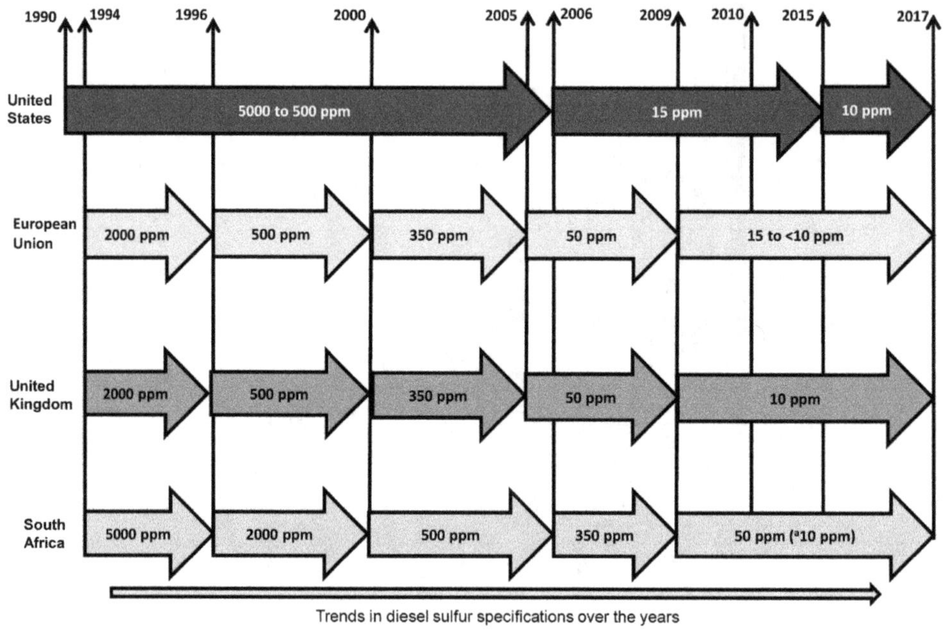

Detector (FPD) to quantify the sulfur species in diesel and gas fuel oils. It was observed that dibenzothiophenes and dibenzothiophenes having alkyl substituents at the 4 and/or 6 positions are the most resistant to hydrodesulfurization. Hydrogen sulfide (H_2S) formed during the HDS process also inhibits HDS catalyst activity by modifying the catalyst surface (Topsoe *et al.* 1996; Song *et al.* 2006). Topsoe and co-workers reported that high concentrations of H_2S are known to increase the density of Brönsted acid sites on a commercial catalyst and it is also responsible for the poisoning of mainly hydrogenolysis sites on a sulfided Co-Mo/ɣ-Al_2O_3 catalyst. This leaves the sulfur content relatively high considering the mandatory levels (Topsoe et al. 1996; Barbara et al. 2011). Pushing the HDS processing conditions (temperature, amount of hydrogen and pressure) further to produce ultra-low sulfur fuels results in the aromatic and polynuclear aromatic content falling outside the requisite specifications. Some fuel specifications that could be altered are the cetane index, polyaromatic hydrocarbons (PAH) index and T_{95} distillation temperature (Barbara *et al.* 2011). Several techniques have been developed for the elimination of sulfur in fuel to the ultra-low levels, for example, bio-desulfurization (BDS), oxidative-desulfurization (ODS) and adsorptive desulfurization.

The ODS process is carried out under mild conditions (temperature and pressure) in the presence of a catalyst (mostly transition metals in their high oxidation states) and an oxidant agent. The oxidized sulfur compounds (sulfones) are subsequently removed by extraction, adsorption, or distillation (Abdullah *et al.* 2015; Wang *et al.* 2003; Ishilara *et al.* 2005; Chica *et al.* 2006; Zhou *et al.* 2007; Kuznetsova *et al.* 2008; Prasad *et al.* 2008; Torres *et al.* 2011). The main advantage of ODS is that refractory sulfur compounds can be removed without using hydrogen, at relatively low temperatures and at atmospheric pressure. A suitable extraction solvent or adsorbent is important for the ODS process to be effectively carried out. Extraction solvents have been reported to extract a large quantity of aromatics along with sulfones hence altering fuel specification (Campos-Martin *et al.* 2010). Lyondell chemicals (Karas *et al.* 2008; Karas *et al.* 2004), EniChem/UOP (Kocal *et al.* 2001; Kocal *et al.* 2002) and Alternative Petroleum Technologies (ATP) independently patented ODS process. In the Lyondell process, solvent extraction was used for separation of sulfones, while EniChem/UOP process removed the sulfone species by adsorption. In both processes, *tert*-butylhydroperoxide was employed as the oxidant due to its solubility in fuel. Lyondell chemicals and EniChem/UOP ODS processes produced *tert*-butyl alcohol after the oxidation step, and this compound was employed as a potential octane-booster compound for gasoline. APT desulfurization technology is known as Sulfex™. It provides recyclable fluid containing a catalyst which serves the dual purpose of a catalyst and a transfer agent, thus increasing the rate of oxidation of the sulfur-containing compounds in the fuel. In the Sulfex™ process, a mixture of fluid-borne reagents and the hydrocarbon forms an emulsion at high mix rates. Once these sulfur compounds are oxidized, the fluid and fuel are separated, and the sulfur is removed from the fuel. In the Sulfex™ process, hydrogen peroxide is employed. To date, APT's laboratory, pilot plant have resulted in as much as 99.9% reductions of all sulfur compounds found in existing high sulfur fuels and middle distillates sourced from various refineries (Sulfex™, 2019).

Heterogeneous catalysts have been prepared using a variety of support materials, such as alumina, zirconia, titania, zeolites, silica, polymer microspheres and Metal-Organic Frameworks (MOFs) (Campos-Martin *et al.* 2010; Barbara *et al.* 2011; Karas *et al.* 2008). Molybdenum oxide was investigated in the catalyzed oxidation of hydro-treated fuels (Prasad *et al.* 2008) and it showed good ODS activity. The catalyst was employed as a heterogeneous catalyst, by incorporating molybdenum (Mo) unto alumina ($Mo/\gamma\text{-}Al_2O_3$) (Prasad *et al.* 2008; Abdullah *et al.* 2015). Other heterogeneous catalysts which are more stable during use have been investigated; some examples of such catalysts include Ti, Mo, V, W (Campestrini *et al.* 1988), gold (Au) (Si *et al.* 2008) and rhenium (Re) (Di-Giuseppe *et al.* 2009). Titanium catalysts gave the best results in the catalysed oxidation step during oxidative-desulfurization (Chica *et al.* 2006). However, titanium (Ti) is known to be expensive and the use of titanium in catalyzed reactions in industry, generally, would result in an increase in the cost of production. Hence, the search for a cheaper and more robust potential catalyst that can be applied industrially in the ODS process as well as a robust material with a high surface area that can selectively remove sulfones without changing the fuel specification. In that regard, vanadium lends itself useful for catalyst development.

In this chapter, we briefly give an overview of some of the desulfurization techniques that have been proposed to achieve deep desulfurization while detailing problems associated with each technique. The concept of adsorptive separation is then presented as a possible solution. A general scope on adsorptive separation is given before placing emphasis on the need for strategic design of adsorbents to achieve good selectivity. In that regard, a brief discussion on attempts that have been made to try and improve the selectivity of general inorganic adsorbents is presented. The discussion then advances to the chemistry involved in molecular imprinting of organosulfones on functional polymers and electrospinning of the polymeric matrix to produce molecularly imprinted nanofibers for selective adsorption of organosulfones from oxidized mildly hydrotreated fuels as a second step in the ODS process. Chemical interactions, apart from the imprinting effect, that can be exploited in molecularly imprinted polymers for selective extraction of organosulfones include hydrogen bonding, π-π interactions, van der Waals forces, and electrostatic interactions, and these have been discussed by employing density functional theory calculations. The theoretical calculations are instrumental in predicting suitable polymers for adsorption of organosulfones. In this chapter, the use of molecularly imprinted chitosan and polybenzimidazole, respectively, as sorption materials for the removal of organosulfones is discussed in detail. The chapter also covers desulfurization research that has been done with coordination polymers, especially metal-organic frameworks, outlining some of the achievements and shortcomings, leading to a proposal of three key design strategies that can be adopted to achieve good selectivity for sulfur-containing compounds.

METHODS OF DESULFURIZATION

The refining process employed for the elimination of sulfur in the production of cleaner fuels is known as desulfurization. The following methods of desulfurization will be described briefly; (i) hydrodesulfurization, (ii) bio-desulfurization, (iii) oxidative-desulfurization, and (iv) adsorptive desulfurization.

Figure 4. (a) General HDS reaction scheme and (b) mechanistic routes for the hydro-desulfurization of thiophene

A

$$\text{R-S-H} + H_2 \longrightarrow \text{R-H} + H_2S$$

$$\text{R-S-R'} + 2H_2 \longrightarrow \text{R-H} + H_2S + \text{R'-H}$$

$$\text{R-S-S-R'} + 3H_2 \longrightarrow \text{R-H} + 2H_2S + \text{R'-H}$$

B

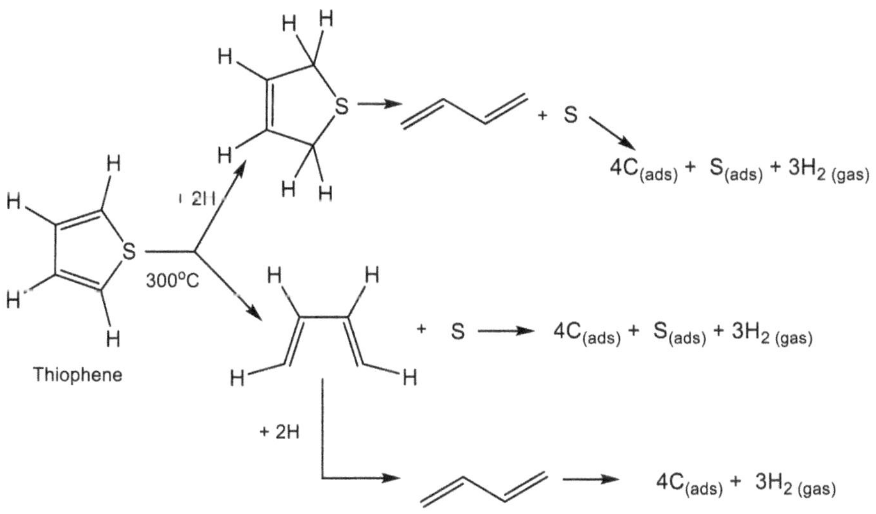

Hydro-Desulfurization

The hydro-desulfurization (HDS) process has been employed for the removal of sulfur compounds from as early as the 1930's. In the HDS process, crude oil is heated, mixed with hydrogen, and fed to a reactor packed with a pelleted CoMo/Al_2O_3 catalyst. The reactor temperatures range from 300 to 350°C in a pressure range of 15 to 90 bars to generate a mixture of the treated/desulfurized crude oil, hydrogen sulfides (H_2S) and hydrogen gas. H_2S is further recycled through a Claus process to recover the elemental sulfur (Song *et al.* 2003; Babich *et al.* 2003). HDS occurs directly through the hydrogenolysis pathway (Figure 4a).

The addition of hydrogen to sulfur compounds at higher temperatures activates carbon-sulfur (C-S) bond cleavage as shown in the reaction Figure 4b for the hydro-desulfurization of thiophene. Two mechanistic routes are presented in Figure 6, and in the first route, hydrogenation of thiophene precedes C-S bond scission, and in the second route, C-S bond scission occurs, before hydrogenation takes place. Both

Table 1. Some organosulfur compounds and their sulfur electron densities

Compound names	Structures	Sulfur electron densities
Thiophene	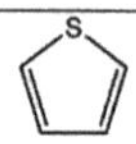	5.696
Thiophenes with methyl at the 2 or 5 position	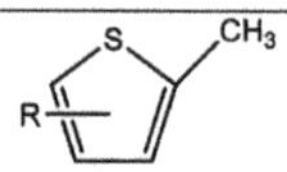	5.706-5.716
Benzothiophenes (BTs)	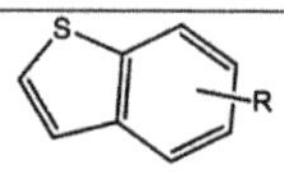	5.739
Benzothiophenes with methyl at the 2 or 7 position	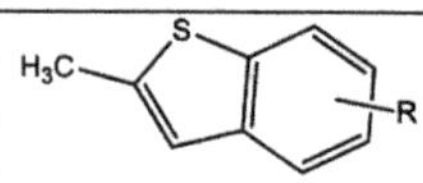	5.746
Dibenzothiophenes (DBTs)	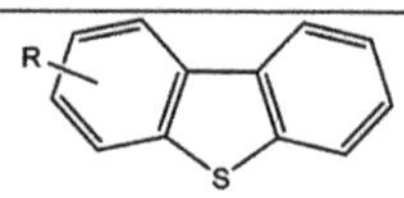	5.758
Dibenzothiophenes with methyl at 4 position	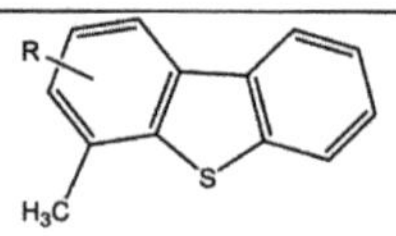	5.759
Dibenzothiophenes with methyl at the 4 and 6 position	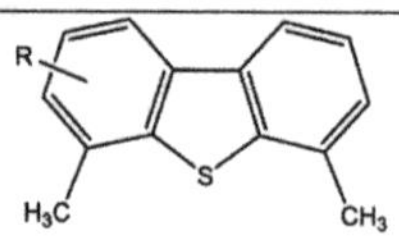	5.760
Phenanthro[4,5-*bcd*]thiophene	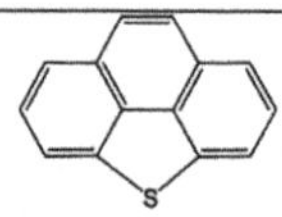	5.770
5,7-Dimethylphenanthro[4,5-*bcd*]thiophene	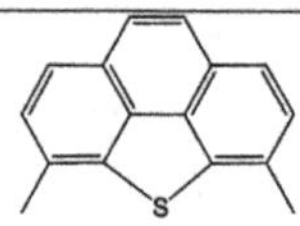	5.774

mechanistic routes produce carbon and sulfur, which can easily be absorbed into mesoporous aluminosilicate molecular sieves (Song *et al.* 2003; Song *et al.* 2003; Babich *et al.* 2003). Commercial hydrotreating catalysts are, typically, sulfides of Mo or tungsten (W) supported on Al_2O_3 and promoted by either Co or Ni.

The HDS process is however limited in treating refractory sulfur compounds such as dibenzothiophenes (DBTs) and especially the alkyl derivatives such as 4,6-dimethyldibenzothiophene (Song *et al.* 2003; Song *et al.* 2004; Schulz *et al.* 1999; Babich *et al.* 2003; Andersson *et al.* 2006, 2007; Nolte *et al.* 2013). These are the main non-reactive organosulfur compounds in the hydrogenation process. The HDS reactivity of various sulfur compounds decreases in the order of disulfides > sulfides, thiols > thiophenes > BTs (benzothiophenes), NTs (naphthothiophenes) > BNTs (benzonaphthothiophenes), DBTs (dibenzothiophenes) > DBTs with alkyl group(s) at the 4- and or 6-position(s) (Song *et al.* 2003, 2004). Table 2 lists some sulfur compounds found in crude oils, the order of reactivity of these sulfur compounds is governed by electron density of sulfur in the compounds mainly in the ODS process. The electron density is a representation of the probability of finding an electron in a specific location around an atom or molecule. In general, an electron is likely to be found in regions with high electron density than region low electron density, in so doing, influencing the reactivity of sulfur in the compounds (Song *et al.* 2003). Hence, the higher electron density results in decreased HDS activity.

The production of ultra-low sulfur fuels (10-50 ppm) using HDS process, therefore, requires the application of extreme operating conditions, *i.e.* high temperatures and high hydrogen pressures, and the use of highly active catalysts at slow velocity (Schulz *et al.* 1999; Ma *et al.* 2001; Song *et al.* 1999). Under these conditions, the aromatics and polynuclear aromatics content would fall outside the requisite specification as they tend to be hydrogenated. The other fuel requirements such as oxygen content, vapor pressure and olefin content for petrol, and cetane number, density and T95 point for diesel would also be compromised. In addition, the high cost involved because of the use of excess hydrogen gas and energy, therefore, makes the end-product expensive. Alternative processes which can complement the HDS process under mild conditions to achieve ultra-low sulfur fuels are therefore essential and oxidative-desulfurization shows potential.

Bio-Desulfurization (BDS)

Bio-desulfurization involves the use of micro-organisms (e.g. bacteria) to desulfurize fuels under biological processes.[4] Microorganisms require sulfur for their growth and biological activities. The dried weight of bacterial cells comprises 0.5-1% sulfur atoms. Sulfur generally occurs in the structure of some enzyme cofactors (such as co-enzyme A, thiamine and biotin), amino acids and proteins (cysteine,

Figure 5. Kodama pathway of dibenzothiophene (DBT) degradation

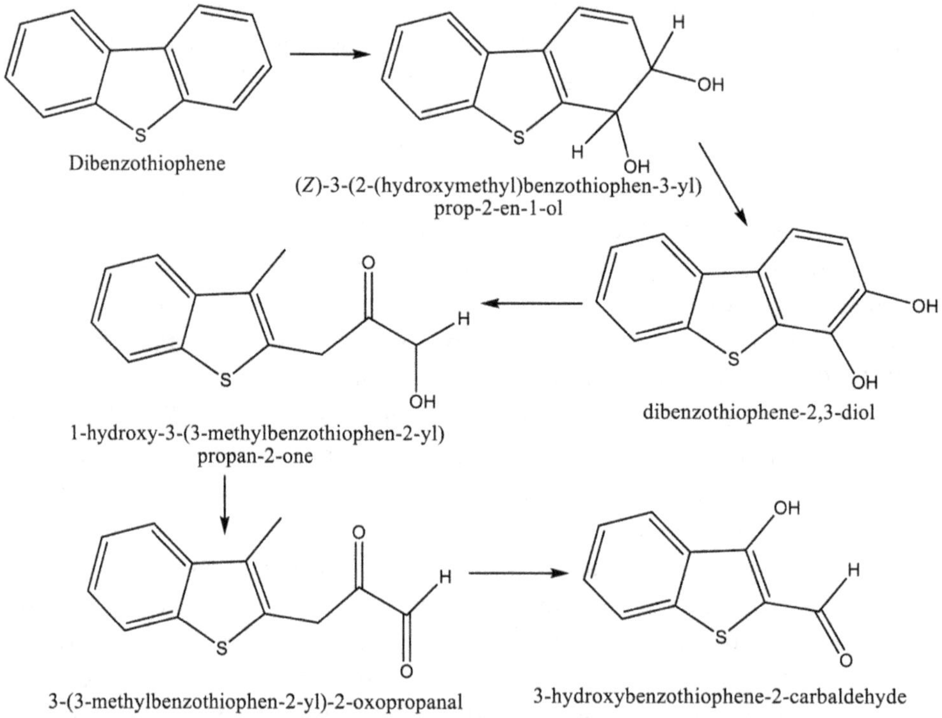

methionine, and disulfide bonds). Depending on their metabolic pathways, micro-organisms may have the ability to obtain sulfur from different sources. Some micro-organisms consume the sulfur in thiophenic compounds such as dibenzothiophene (DBT), thus reducing the sulfur content in fuel. Energy BioSystems Corporation (EBC) was initially the only commercial venture dedicated to the development of bio-desulfurization technology. EBC's concept of a bio-desulfurization process was to primarily treat diesel and to produce a value-added surfactant by-product to achieve a more economical process (Monticello *et al.* 2000; Chang *et al.* 2000; Soleimani *et al.* 2007; Chang *et al.* 1998). Bacteria that convert dibenzothiophene and alkyl sulfides are relatively well known while fewer bacteria are found for benzothiophene, and only a few bacteria detected for thiophene (Campos-Martin *et al.* 2010 and Kodame *et al.* 1973).

The two main pathways known for the utilization of dibenzothiophene by bacteria are: destructive (degradation) pathway and sulfur-specific (desulfurization) pathway.

Figure 6. The pathway of biological desulfurization of DBT relies on biocatalysts for specificity. NADH is reduced nicotinamide adenosine dinucleotide; FMN is flavin mononucleotide

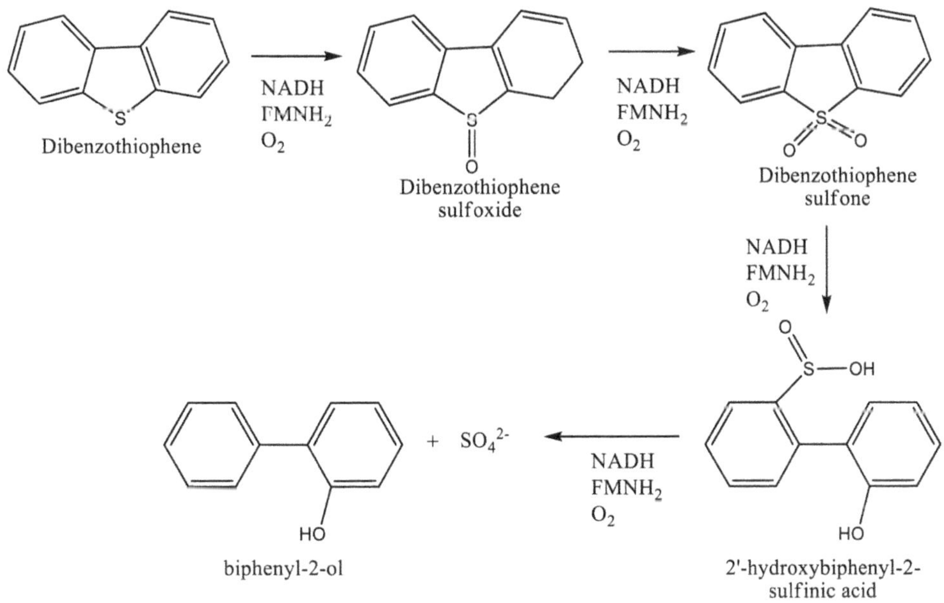

Ring Destructive Pathway

Rhodococcus erythropolis and *Pseudomonas stutzeri* bacteria strains are mostly employed for ring destructive biodesulfurization mechanistic process, also known as the 'Kodama pathway' (Figure 5) (Campos-Martin *et al.* 2010 and Kodame *et al.* 1973), involves the initial dioxygenation of the peripheral aromatic ring of dibenzothiophene, followed by cleavage of the ring which finally leads to the accumulation of 3-hydroxybenzothiophene-2-carbaldehyde as a water-soluble end product with lower carbon content than dibenzothiophene. In this pathway, no desulfurization occurs; the dibenzothiophene (organosulfur compound) is converted to a more polar compound which can be stripped off.

Sulfur-Specific (Desulfurization) Pathway

The sulfur-specific desulfurization pathway can take place in the presence of isolated bacteria strain *Brevibacterium sp.* capable of using dibenzothiophene for growth as the sole source of carbon, sulfur, and energy (Campos-Martin *et al.* 2010; Kodame *et al.* 1973). It involves a series of oxidations of dibenzothiophene

Figure 7. The oxidation reaction step in ODS process for dibenzothiophene and methyl substituted derivatives

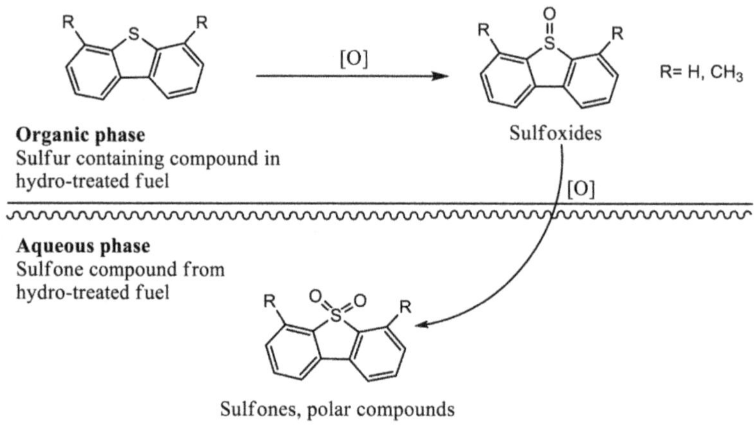

(DBT) to dibenzothiophene sulfoxide (DBTO), dibenzothiophene sulfone ($DBTO_2$), sulfinate and the hydroxybiphenyl in the presence of oxygen and water under ambient temperature and pressure (Figure 6). Enzymes also show potential in the desulfurization of DBT to $DBTO_2$ following a similar mechanistic pathway as the isolated bacteria strain (*Brevibacterium sp.*). Dibenzothiophene monooxygenase (DszC) catalyzes the conversion of dibenzothiophene (DBT) to dibenzothiophene sulfone ($DBTO_2$). The dszA gene encodes dibenzothiophene-5,5- dioxide monooxygenase (DszA) catalyzes the conversion of $DBTO_2$ into 2-hydroxybiphenyl-2-sulfinic acid. While dszB gene encodes 2-hydroxybiphenyl-2-sulfinate sulfinolyase (DszB) catalyzes the conversion of 2-hydroxybiphenyl-2-sulfinic acid into biphenyl-2-ol and sulphite (Campos-Martin *et al.* 2010; Kodame *et al.* 1973). Specific oxidative biodesulfurization of DBT degradation. catalyzes the conversion of HBPSi I nto 2-hydroxybiphenyl (2-HBP) and sulfite (Figure 5), (Campos-Martin *et al.* 2010; Kodame *et al.* 1973).

In general, the BDS does not show very deep desulfurization, the maximum sulfur content limit of 50-200 ppm sulfur at best has been achieved (Barbara *et al.* 2011). The bio-desulfurization process also suffers the following limitations: (i) production of active resting cells (biocatalysts) with a high specific activity; (ii) preparation of a biphasic system containing oil fraction, aqueous phase and biocatalyst; (iii) bio-desulfurization of a wide range of organosulfur compounds at a suitable rate; (iv) separation of desulfurized oil fraction, and recovery of the biocatalyst and its return to the bioreactor; and (v) efficient methods for treatment of wastewater (Booth *et al.* 2001).

Oxidative Desulfurization

Oxidative-desulfurization (ODS) has always been considered a complementary technique for desulfurizing fuel oils (Ogunlaja *et al.* 2013). The ODS process includes two stages: (i) oxidation of the sulfur atoms and (ii) liquid extraction of the polar products. In the ODS process, sulfur-containing compounds in fuel oils are oxidized using selective oxidants such as nitric acid and/or nitrogen oxides (Tam *et al.* 1984, 1990; Baxendale *et al.* 1946; Venturiello *et al.* 1983), organic hydroperoxides (Chica *et al.* 2006; Karas *et al.* 2004, 2008; Kocal *et al.* 2002; Mokhtar *et al.* 2014; Abdullah *et al.* 2015), peroxyacids (Zannikos *et al.* 1995; Aida *et al.* 1994), and hydrogen peroxide (Otsuki *et al.* 1999, 2000; Collins *et al.* 1997; Yazu *et al.* 2003; Paybarah *et al.* 1982). The use of hydrogen peroxide, a cheap aqueous oxidant, has been promoted for use in the oxidation of fuel oil by introducing surfactants which help to disperse oxidant in the form of small spherical droplets in the crude oil. The process is described as an emulsion catalysis system (Jiang *et al.* 2011). The ODS process generally takes place in the presence of a catalyst to produce sulfone compounds that can be preferentially extracted due to their increased relative polarity. For oxidation to take place, the oxidant needs to be in contact with fuel oils under optimum conditions, and the oxidant donates oxygen atoms to the sulfur in benzothiophenes, dibenzothiophenes, and its derivatives to form sulfoxides and/or sulfones. Some of the by-products generated by the oxidant influence the quality of the fuels; for example, *tert*-butyl alcohol, which is a potential octane booster for fuels, is generated as by-products of *tert*-butylhydroperoxide (Karas *et al.* 2008).

The oxidation of organosulfur compounds, which is the first step of the oxidative desulfurization process in hydro-treated fuel, takes place in the presence of metal catalyst(s) (Wang *et al.* 2003; Ishilara *et al.* 2005; Chica *et al.* 2006; Zhou *et al.* 2007; Kuznetsova *et al.* 2008; Prasad *et al.* 2008). A typical catalysed oxidation of hydro-treated fuel using vanadium as catalyst was conducted under batch and continuous flow processes at 40°C by using *tert*-butylhydroperoxide (*t*-BuOOH) as oxidant (Ogunlaja *et al.* 2013). The oxidation process showed the formation of organosulfones which tends to be polar in nature, thus miscible in aqueous solvents as shown in Figure 7. This chemistry has led to the development of the ODS system which is employed for the desulfurization of sulfur in fuels.

A practical industrial flow system for oxidative desulfurization of hydro-treated fuel (Figure 8), where catalyst in the hydro-treated is extracted and washed, un-used oxidant are also separated in the separator section before further oil processing can take place. Extraction of the oxidized compounds (sulfones and sulfoxides) has been carried out by contacting the oxidized oil with a non-miscible solvent which is selective for the polar oxidized sulfur compounds (Hulea *et al.* 2001; Anisimov *et*

Figure 8. Schematics of the oxidative-desulfurization (ODS) process

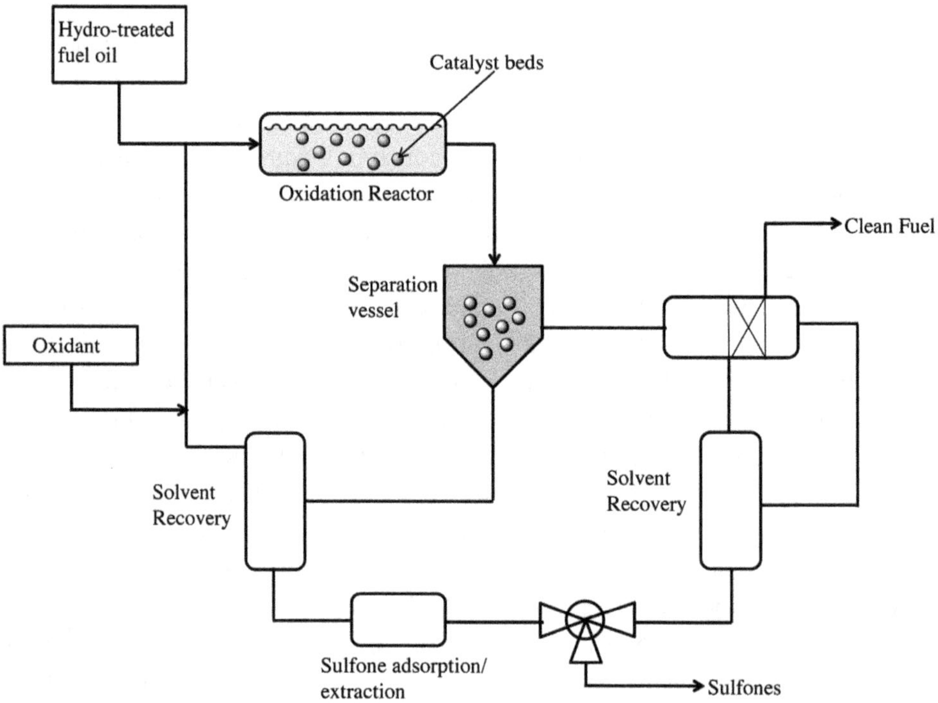

al. 2003; Palomeque *et al.* 2002; Yazu *et al.* 2001; Djangkung *et al.* 2003; Shiraishi *et al.* 2003; Darian *et al.* 1988).

The advantage of the ODS process is that the refractory sulfur-containing compounds can be oxidized and it operates at low reaction temperature and pressure, and in the absence of hydrogen which is expensive. With HDS being able to desulfurize thiophene and acidic sulfur compounds at relatively moderate conditions and the ODS process being capable of desulfurizing refractory sulfur compounds, then the ODS process has great. It is possible to remove sulfur compounds from fuels to meet strict regulatory limits (Tam *et al.* 1984, 1990; Baxendale *et al.* 1946; Venturiello *et al.* 1983; Chica *et al.* 2006; Karas *et al.* 2004, 2008; Kocal *et al.* 2002; Mokhtar *et al.* 2014; Abdullah *et al.* 2015). However, a high oxidant-to-sulfur (O/S) ratio is required to oxidize sulfur compounds to their corresponding sulfoxides and sulfones. This drawback (i.e. the high oxidant constraint), however, can be avoided by use of suitable catalysts.

Figure 9. Simplified illustration of the desulfurization process when applying ionic liquids capable of oxidation. Reprinted with permission from Bhutto et al. (Bhutto et al 2016)

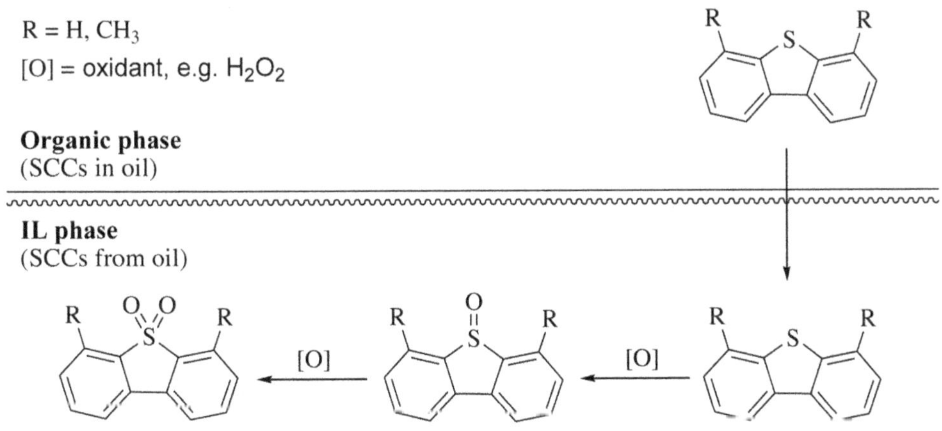

Extractive Desulfurization

Solvent extraction is an avenue that some researchers have explored in quest of solving the issue of achieving ultra-seep desulfurization either coupled with oxidation of the sulfur-containing compounds or on its own. Choice of solvent for the extraction of sulfur in fuel oils is critical. For a solvent to be employed in the extraction process, it must be thermodynamically compatible with the compound to be extracted. Polar solvents with low boiling points such as ethanol, ethanol, and acetonitrile are usually employed for the desulfurization of crude oils, mostly for the extraction of SCCs but these solvents are too volatile for practical application on an industrial scale. (Campos-Martin *et al.* 2010; Mokhtar *et al.* 2014, Abdullahi *et al.* 2015). Polyalkyleneglycol, polyalkylene glycol ether, imidazolidinones, pyrrolidones, pyrimidinones, dimethyl sulfoxide and dimethyl formamide which have much higher boiling points have shown higher extraction levels of SCCs (Campos-Martin *et al.* 2010; Pawelec *et al.* 2011; Petkov *et al.* 2004). However, the boiling points of these solvents are in the same range as some SCCs, and thus may not be easily separated from the extracted compounds, discouraging solvent recovery by distillation. Besides boiling points, poor selectivity still stands as the major deterrent in the application of solvent extraction. The fuel matrix is composed of other polar compounds such as aromatics and heteroatomic compounds that are vulnerable to the extraction process, affecting the octane/cetane numbers of the fuel (Cornelius and Caplan 1952; Ghosh *et al.* 2006; Peralta *et al.* 2012).

Considering the poor selectivity witnessed with the use of organic solvents, researchers redirected their efforts towards the use of ionic liquids (ILs) to try and selectively remove SCCs from fuel. Ionic liquids are described as liquids composed entirely of ions at or below 100^{0}C-more like an ambient temperature molten salt, and they are known to have a broad range of characteristics; they can be toxic or benign, safe or explosive, and non-volatile or Distillable (Ibrahima *et al.* 2017). The cations are normally organic and are a basis upon which much of their physicochemical properties are derived, and the anions can be organic or inorganic. Extraction of SCCs using ILs is based on their solubility that is determined by the chemical interactions involving the acid proton and steric factors in the IL. The use of ionic liquids gained momentum over the past decade with researchers attempting to use various combination of cations and anions, dialkylimidazolium salts being the most commonly used ILs due to their desirable physical properties and since they are easy to synthesize. Although ILs based on $AlCl_3$ and $AlCl_4$ are effective for extraction of SCCs they are sensitive to water and form precipitates. Some researchers have attempted to overcome this challenge by using PF_6^- and BF_4^- anions that are more stable in the presence of water, but selectivity remains a challenge.

Much research is now centred on modifying the chemical structures of ILs to target certain physicochemical properties to obtain what is termed task specific or designer solvents which allow SCCs to be selectively extracted from fuel. In some cases, ILs have been designed to act as both oxidation catalysts and extractants. The ILs catalyse the oxidation of the sulfur compounds, improving their solubility in the ILs which also act as the extractant as described in ODS (Bhutto *et al.* 2016). The performance of the ILs in the presence of oxidant is remarkably higher, although efforts are needed to reduce the amounts of oxidant. Normally polyoxometalates such as vanadium and tungsten are used as the catalytic cations and H_2O_2 as oxidant (Lu *et al.* 2007; Pawelec *et al.* 2011). Researchers have also explored the use of supported ionic liquid phase (SILP) systems where the ionic liquid is dispersed on a support such as silica to increase surface area, reduce mass transfer limitations and for ease of use (Figure 9). Leaching is still a challenge, an area that still requires more attention (Kuhlmann *et al.* 2009; Kohler *et al.* 2010).

Although it is estimated that large-scale production of some ILs will be affordable, some may be too expensive. Still, the ability to regenerate and reuse the ILs makes them applicable on an industrial scale. The use of ILs is also mired by the presence of potent complex-forming agents in the fuel matrix. The presence of water, acids, halide ion, and other impurities are known to significantly affect their structure and ultimately their physicochemical properties. The fuel matrix is a complex mixture that contains such components that may incapacitate the ILs over a number of cycles. Selectivity may also be a challenge in the presence of a wide variety of other aromatics. (Fransisco *et al.* 2010).

Overall, the biggest limit for desulfurization by extraction, in general, is poor selectivity since other aromatic compounds that are important in the fuel are extracted (Pawelec *et al.* 2011). This development has encouraged the use of solid adsorbent materials and synthesis of stimuli-responsive materials that allow the formation of complexes or other types of bonding interactions with SCCs in the presence of other aromatic compounds found in fuels.

Adsorptive Desulfurization

Liquids and gases are retained on unoccupied sites within solid materials due to the presence of unsaturated molecular attractive forces on the material surface. The attractive forces involved are classified as physical (dispersive, polarization or dipole forces) or chemical where the interaction of electrons between adsorbate and adsorbent occurs. Depending on the two forces, either physisorption or chemisorption happens. Chemisorption of sulfur compounds can be considered as reactive adsorption where the sulfur in most cases reacts covalently with certain functional groups in the adsorbent with the aid of hydrogen and elevated temperatures to form sulphide (in most cases) while the sulfur free hydrocarbon is released into the fuel matrix (Tawfik *et al.* 2015). The spent sorbent is then regenerated by oxidizing the trapped sulfur into sulfur oxides (Pawelec *et al.* 2011). Among the physisorption interactions, site specific interaction between sulfur and the metal centre (coordination *via* sulfur) and π-complexation are the most important since they characterize most materials with good selectivity. According to the Pearson acid-base concept (HSAB theory), SCCs prefer to interact with intermediate to soft Lewis acid sites e.g. Cu^{2+}, Co^{2+} and Ni^{2+} (Pearson 1963; Maes *et al.* 2011). In π-complexation, SCCs interact strongly with suitable adsorbents based on electron back-donation. For π-complexation to occur, the metals must have empty s-orbitals and suitable electron density available in the d-orbitals for back donation. Copper(I) with electronic configuration $1s^2 2s^2 2p^6 3s^2 3p^6 3d^{10} 4s^0$ is a very good example. Other cations reported showing good π-complexation are Na^+, Zn^{2+}, Ni^{2+}, Ag^+, Fe^{2+} and Ce^{4+} (Baeza *et al.* 2007). Generally, research around adsorptive desulfurization is receiving more attention because the technique is facile, requires fewer energy inputs, has less consumables such as hydrogen gas or oxidants, is environmentally friendly and can be carried out under mild conditions.

Figure 10. Adsorption of dibenzothiophenes on imprinted polymer nanofibers in an improvised flow set-up. The insert depicts hydrogen bond formation between the oxygen atom of S=O (dibenzothiophene sulfone) and the hydrogen atom of –NH_2 originating from polybenzimidazole nanofibers.

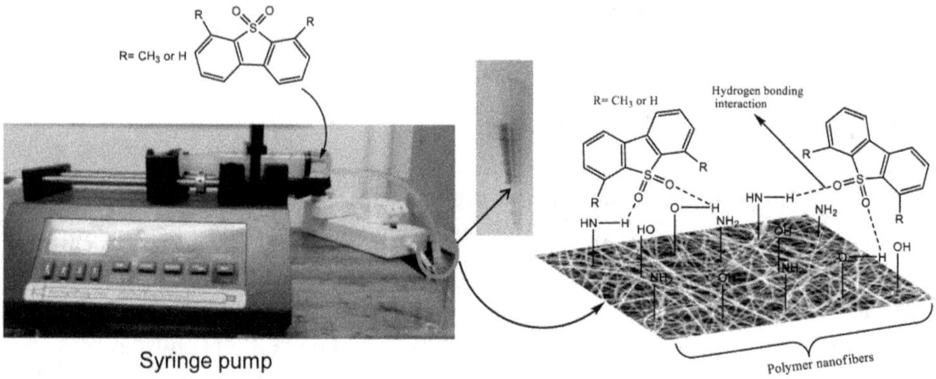

Figure 11. A set-up of the home-made pressurized hot water for desorption of organosulfones from nanofibers

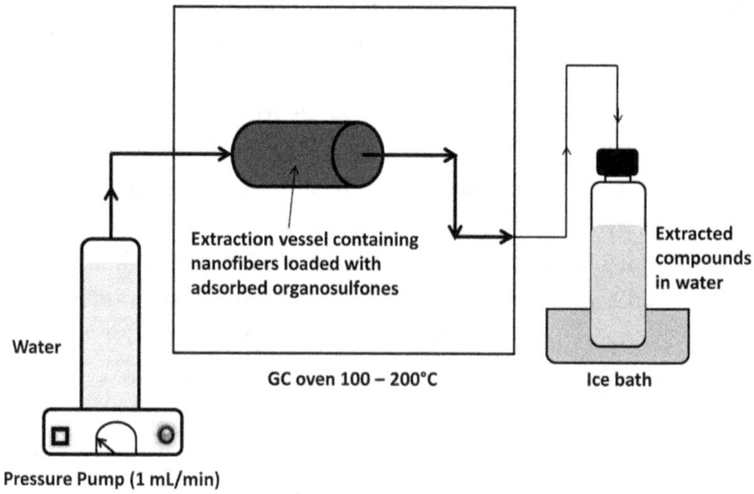

Figure 12. Integrated hydrodesulfurization-oxidative desulfurization (HDS-ODS) processes

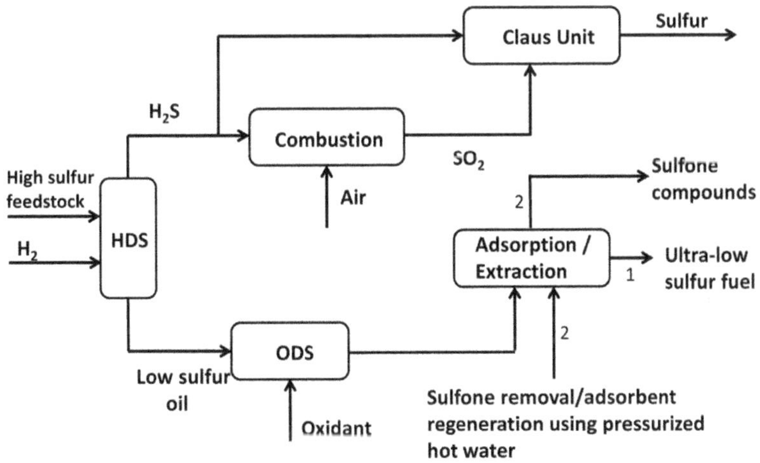

Organic Materials

Several desulfurization attempts have been made with organic materials, usually as organic polymers (Ryu *et al.* 2002; Castro *et al.* 2001; Jorge *et al.* 2004; Ogunlaja *et al.* 2014). To improve selectivity, molecularly imprinted polymer materials have been employed for selective adsorption of oxidized refractory sulfur compounds in hydro-treated fuels (Ryu *et al.* 2002; Castro *et al.* 2001; Jorge *et al.* 2004; Ogunlaja *et al.* 2014). Template imprinting on polymers improves its absorptive properties and also provides selectivity due to polymer flexibility which makes accessing of the recognition site and removal of absorbed template easier. Polymer imprinting takes place through the formation of covalent bonds or through interactions such as hydrogen bonding, π-π interactions, van der Waals forces and electrostatic interactions between the templates (organosulfones) and functional groups of the polymer network.

Most of the studies reported in the literature employed molecularly imprinted polymers in their powdery/monolithic form for the adsorption of refractory sulfur compounds in fuels. A low surface area and porosity are usually observed in the adsorbents, and accessibility of active sites on imprinted polymer sorbents is sometimes compromised. Due to these issues, the polymer adsorption properties can be improved by fabricating them into nanofibers (Ogunlaja *et al.* 2014). The polymer nanofibers can readily be fabricated through a process known as electrospinning. The electrospinning process provides polymer nanofibers with a controlled pore size as well as a high surface-to-volume ratio, and this material decreases adsorption time and mass transfer constraints. Ogunlaja and co-workers employed a continuous flow adsorption process for the elimination of sulfone compounds in oxidized hydro-treated fuel by using sulfone-imprinted polybenzimidazole nanofibers packed to the tip of a Teflon rubber tubing (Figure 10). The imprinted nanofibers show high selectivity in the adsorption of sulfone compounds from oxidized fuel.

Regeneration of spent adsorbent for re-use constitutes an important part of the overall feasibility and cost-effectiveness of the proposed oxidative desulfurization process. The following adsorbent regeneration strategies are possible: (a) use of an organic solvent to wash off the adsorbate under ambient pressure and at elevated temperature, and (b) thermal regeneration at ambient pressure under vacuum. A home-made pressurized hot water system was employed for the regeneration/desorption of sulfones from imprinted nanofibers (Figure 11) to demonstrate the effect of pressure and temperature (Ogunlaja *et al.* 2014). Water was employed as a green solvent for extraction/desorption of sulfone compounds under pressurized subcritical conditions. Water was pressurized to 30 bars at 150°C, and in this subcritical state, the dielectric constant of water is greatly decreased thus imitating the properties of organic solvents.

A combination of the HDS and ODS process is, therefore, proposed as a possible way for the reduction of sulfur in fuels to ultra-low levels as depicted in the schematic of the integrated HDS-ODS process (Figure 12). In this set-up, high sulfur fuel is taken through a hydro-treatment unit where less bulky sulfur compounds such as thiols, thioethers, alkylated disulfides and thiophenes are converted to H_2S as shown in figure 6a. In the Claus unit, H_2S reacts mostly with SO_2 to produce elemental sulfur and water. The resulting fuel (mildly hydro-treated fuel) will be transferred to the oxidative desulfurization chamber where refractory sulfur compounds are oxidized to organosulfone compounds. The removal of the organosulfones from the fuel can be carried out by adsorption to produce an ultra-low sulfur fuel. Molecularly imprinted nanofibers show great promise (Ogunlaja *et al.* 2014), and they can be produced by a process called electrospinning.

Molecular Imprinting

Molecular imprinting is employed for the introduction of recognition sites into matrices (such as a polymer) *via* the formation of an assembly between the imprinting template (molecule: organosulfones) and functional groups within the network (adsorbent). Template removal from the adsorbent creates molecular recognition sites on the spaces vacated within the adsorbent (Holmes *et al.* 1994, Sellergren 2001; Haupt 2003, Mosbach 1994, Wulff 1993, Ryu *et al.* 2002; Castro *et al.* 2001; Jorge *et al.* 2004; Ogunlaja *et al.* 2014).

Figure 13. A schematic representation of the molecular imprinting process through the formation of reversible interactions between the template and polymerizable functionalities with the following possible interactions (A) reversible covalent bond(s), (B) covalently attached polymerizable binding groups that are activated for non-covalent interaction by template cleavage, (C) electrostatic interactions, (D) hydrophobic or van der Waals interactions
(Wulff 1993; Mosbach 1994; Ogunlaja et al. 2014; Sellergren 2001).

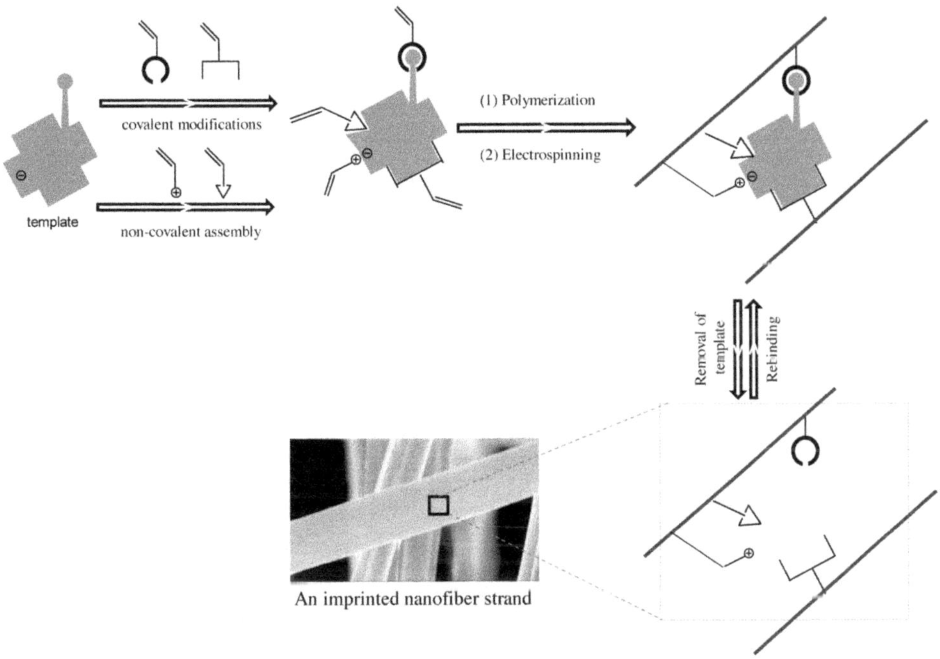

There are three known principal methodologies for assembling recognition site on polymers. These techniques are the covalent binding approach, non-covalent approach, and semi-covalent approach. Covalent bonding between molecules includes of interactions such as σ-bonding, π-bonding or metal ion bonding. In the covalent approach, templates (molecules) which can covalently bind to one or more groups in the polymer matrices are utilized. Compounds such as alcohols, aldehydes, ketones, amines, carboxylic acids, and carboxylic ester linkage can be imprinted with this approach (Wulff 1993). Functional polymer and templates under a non-covalent interaction and this involves interactions such as hydrogen bonding, ion pairing, and dipole-dipole interactions. Non-covalent imprinting approach is important because the methodology is far easier than covalent methods and it produces higher affinity binding sites. This method of imprinting most widely employed due to its simplicity (Mosbach 1994; Ogunlaja *et al.* 2014). Semi-covalent imprinting produces a more

Figure 14. The electrospinning set-up used for the fabrication of polymer nanofibers in an ordinary laboratory

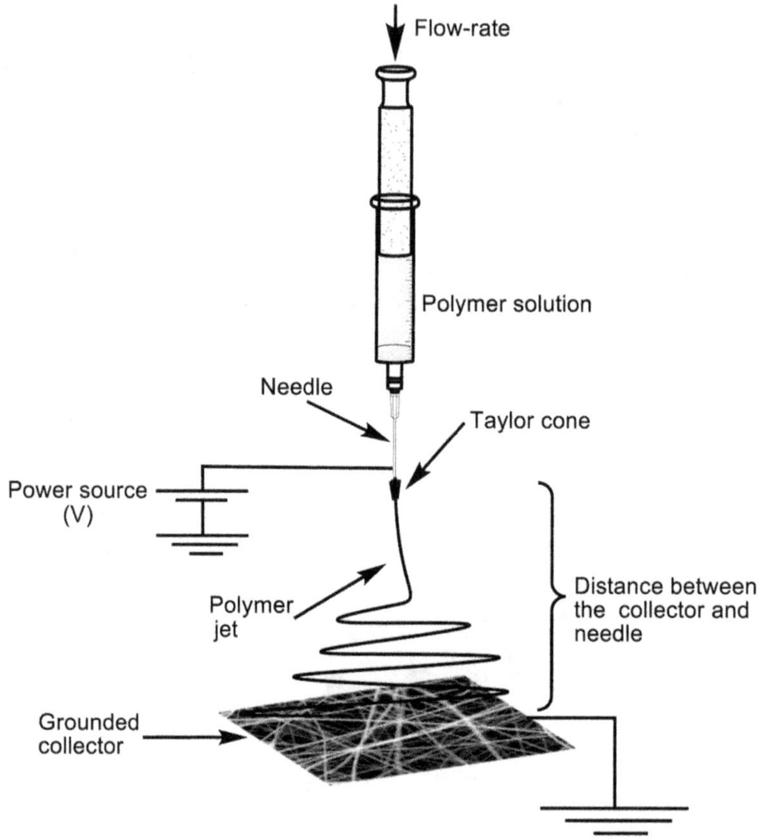

uniform distribution of binding sites on the polymer by combining the technique of covalent and the non-covalent approach. In this approach, templates are covalently bound initially to a polymer group with known functionalities which is recovered after cleavage of the template. Re-binding of the template on the polymer takes place *via* non-covalent interactions without restrictions (Sellergren 2001).

The molecular imprinting process for polymers is shown in Figure 13. However, for molecular imprinting of nanofibers, we use pre-formed polymers and exploit the functionality on the backbone for recognition sites as the polymer strands fold around the template (Ogunlaja *at al.* 2014). This approach is relatively new and unexplored. Figure 13, therefore reflects the typical approach for molecular imprinting but if the polymer is formed first followed by the *in situ* molecular imprinting/electrospinning then that follows the new approach.

Figure 15. Some electrospinning collectors
(http://www.grafen.com.tr/product.php?id=293). Date accessed: 04/05/2019

(A) Rotating drum

(B) Rotating drum

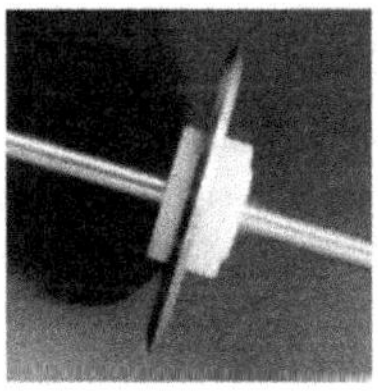

(C) Rotating disk (sharp edge)

(D) Rotating disk (flat edge)

Figure 16. Images of some possible electrospinning spinnerets
(http://www.electro-spinning.com/multi_needle_spinneret.html). Date accessed: 04/05/2019

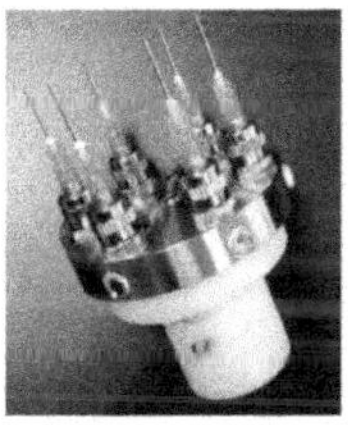

(A) Multiple needle

(B) Rotating drum partly submerged in a thin film of polymer solution

(C) Tri-channel spinneret

(D) Multiple needle

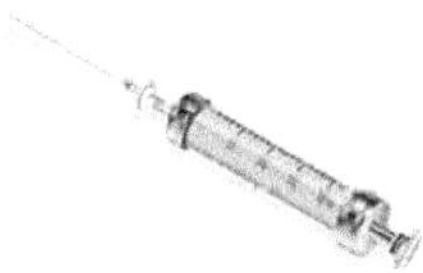

(E) Single needle

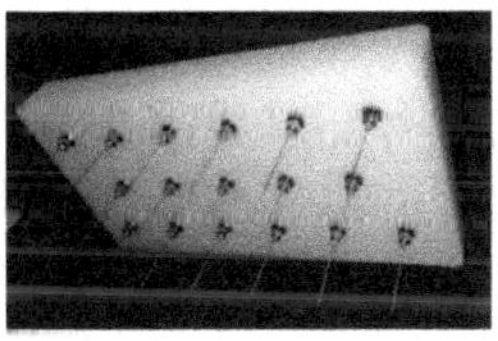

(F) Multiple needle

Electrospinning of Polymer Solutions

Electrospinning is a process whereby high voltage is applied to a polymer solution resulting in an electrified jet that travels through the electric field until it lands on a collector (counter electrode) as nanofibers (Reneker *et al.* 2008). The polymer can be dissolved in a suitable solvent (Reneker *et al.* 2008). A basic laboratory electrospinning set-up consists of a syringe pump, a high voltage and a collector (Figure 14), and it is a relatively simple and inexpensive set-up. During the electrospinning process, a polymer solution in a syringe is pumped to a needle tip attached to the syringe *via* tubing. The solution is held at the needle tip by surface tension, and the application of an electric field using a high-voltage source causes a positive charge to be induced within the polymer resulting in charge repulsion within the solution. This results in an electrostatic force, which opposes and eventually overcomes the surface tension of the polymer solution causing the initiation of a polymer jet. As the jet travels down to the collector that is negatively charged, the polymer solvent evaporates leaving the polymer nanofibers on the collector.

A stable electrospinning jet is known to be comprised of four regions (Reneker *et al.* 1996), and these are: the base, the jet, the splay, and the collection. In the base region, the jet emerges from the needle to form a cone known as the Taylor cone. The shape of the base depends on the surface tension of the liquid and the force of the electric field. Jets can be ejected from surfaces that are essentially flat if the electric field is strong enough, and the solution spreads out evenly and it is collected on a grounded surface. Nanofiber properties such as mechanical properties, degradation rates and functional groups exposure on the surface can be improved by using innovative collectors and spinning techniques (Pham *et al.* 2006). Several electrospinning collectors have been made to improve and control nanofiber orientation. Some of the commercially available collectors employed for electrospinning are presented in Figure 15. Controlled polymer electrospinning conditions can produce nanofibers with uniform morphology and diameters in the nanometer range. The latter is sought after in order to improve the fiber functions by offering a fiber with a high surface-to-volume ratio. Crosslinking of nanofibers enhances their mechanical stability and also prevents the material from dissolving in solvents of similar polarity. It is understood that the crosslinking of imprinted polymers improves their adsorption properties by preserving cavities created by imprinted molecules. However, most crosslinking agents compete with reactive donor atoms on polymer nanofibers with respect to interacting molecules. This occurrence may result in the reduction of polymer active sites available for interaction. Polymer nanofibers sometimes become brittle and surface area compromised upon crosslinking (Ogunlaja *et al.* 2014). Hence, strategies are being sought to mediate this problem.

Figure 17. (A) Elmarco laboratory-scale electrospinning apparatus, A similar Pilot Scale Electrospinning Units Nanospider NS200 & NS500 could be found in; https://www.grafen.com.tr/product.php?id=292. (B) Industrial-scale electrospinning system, Nanospider produced by Elmarco (https://elmarco.com/laboratory-equipment). (C) Industrial-scale electrospinning system, Nanospinner416 produced by Inovenso Ltd. (http://www.inovenso.com/portfolio-view/nanospinner416/)

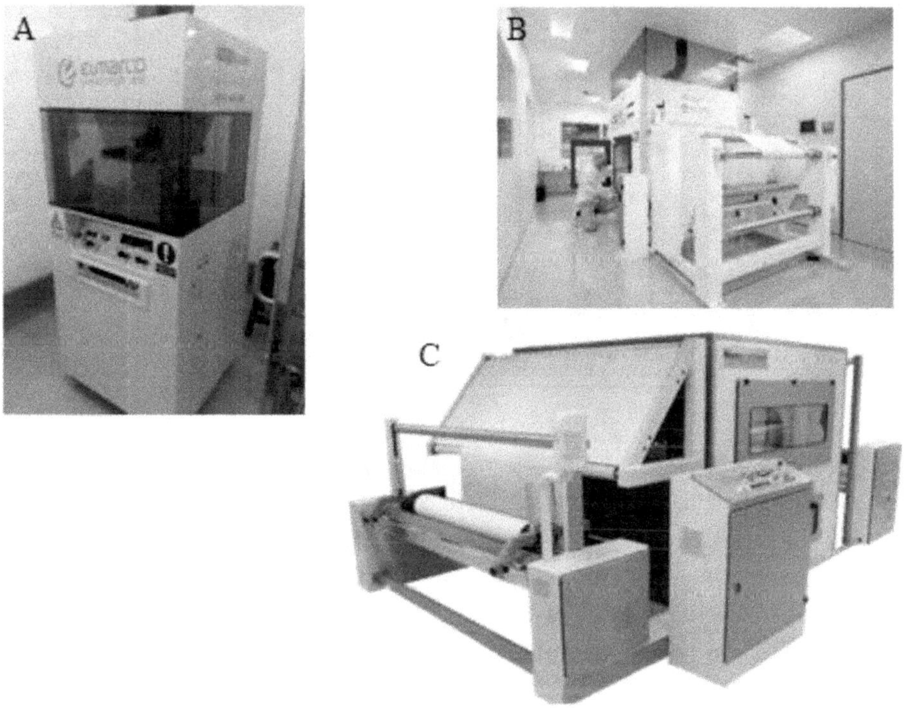

A well-known limitation of the electrospinning process is the level of fiber production. A straightforward method of increasing the productivity of electrospinning in an ordinary laboratory is by increasing the number of spinnerets used in the process. Large-scale fiber production capabilities are important for achieving industrial requirements. Several technological advances have been explored to increase nanofiber production volume. Some reported modifications are as follows: (i) the introduction of multi-spinneret components that allow parallel multiprocessing, and (ii) the development of free surface electrospinning methods (Persano et al. 2013). Electrospinning using multi-spinneret increases fiber mat thickness and quantity (Kidoaki *et al.* 2005; Ding *et al.* 2004; Varesano *et al.* 2009; Theron *et al.* 2005; Teo *et al.* 2005). Figure 16 shows some electrospinning spinnerets which mostly

Figure 18. SEM images of beaded (A) and non-beaded (B) polyvinyl chloride (PVC) nanofibers. Electrospinning conditions: (A) 20 wt% PVC, voltage =18 kV, flow rate = 0.02 mL/h, distance= 18 cm, and (B) 50 wt% PVC, voltage =20 kV, flow rate = 0.05 mL/h, distance= 18 cm

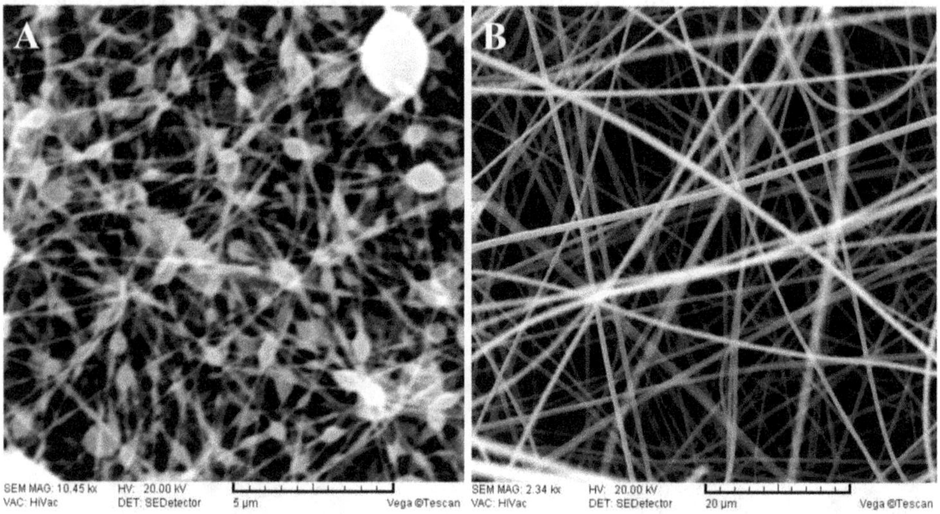

consist of the needle(s). An innovative technology developed by Elmarco (Elmarco's Nanospider) and Stellenbosch Nanofiber Company employs a rotating drum/ball partly submerged in a thin film of the polymer solution to produce nanofibers.

The design and construction of process equipment for controllable, reproducible, continuous and mass electrospinning production have been advanced. The growing interest in electrospinning has already produced increased supply and competition among electrospinning equipment suppliers. Currently, a wide selection of industrial and laboratory-scale electrospinning equipment is available in the market. A list of representative companies that supply electrospinning equipment includes: Elmarco (www.elmarco.com), NaBond (www.electro-spinning.com), Holmarc Opto-Mechatronics (www.holmarc.com), E-Spin Nanotech (www.espinnanotech.com), Linari Engineering (www.linaribiomedical.com), Kato Tech (www.keskato.co.jp), Mecc Co. (www.mecc.co.jp), Toptec (www.toptec.co.kr), Electrospinz (www.electrospinz.co.nz), Electrospunra (www.electrospunra.com), IME Technologies (www.imetechnologies.nl), Yflow (www.yflow.com), and Inovenso (www.inovenso.com) (Persano *et al.* 2013), and Stellenbosch Nanofiber Company (www.sncfibers.com). Some images of industrial and laboratory-scale mass electrospinning apparatus are presented in figure 17.

Figure 19. SEM images of electrospun polybenzimidazole (PBI): (A) In the absence of lithium bromide (LiBr) and (B) in the presence of lithium bromide (LiBr)

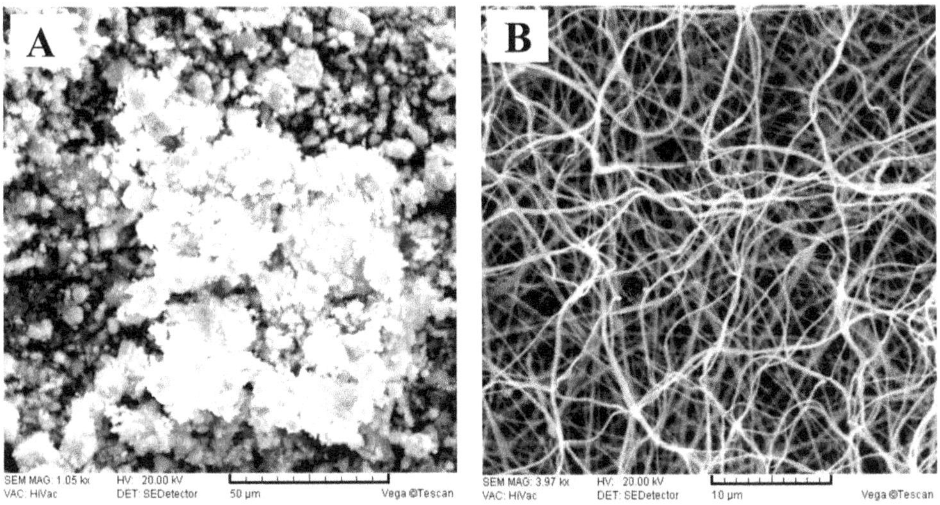

Some representative companies supplying electrospun products for different fields of application as reported by Persano *et al.* 2013 include: Stellenbosch Nanofiber Company (www.sncfibers.com), Donaldson (www.donaldson.com), DuPont (www.dupont.com), Ahlstrom Corporation (www.ahlstrom.com), Espin Technologies (www.espintechnologies.com), Esfil Tehno AS (www.esfiltehno.ee), Finetex Technology (www.finetextech.com), Hemcon Medical Technologies, Inc (www.hemcon.com), Hollingsworth and Vose Company (www.hollingsworth vose. com), Japan Vilene Company (www.vilene.co.jp), Johns Manville (www.jm.com), Kertak Nanotechnology (www.kertaknanotechnology.com), Nanofiber Solutions (www.nanofibersolutions.com), Nano109 (www.nano109.com), NanoSpun (www. nanospuntech.com), Yflow (www.yflow.com), Polynanotec (www.polynanotec.com), SoftMaterials and Technologies s.r.l (www.smtnano.com), and SNS NanoFiber Technology (www.snsnano.com).

Parameters for Electrospinning

Several parameters are known to influence the formation of stable fibers during electrospinning. Doshi *et al.* 1995 classified these parameters as (i) solution properties (viscosity, conductivity, surface tension, polymer molecular weight, dipole moment, and dielectric constant), (ii) controlled variables (distance between the needle tip and collector, voltage and flow-rate) and (iii) ambient parameters (temperature and humidity) (Dzenis 2004; Kumar 2010). Low polymer solution concentrations result

Figure 20. SEM images of (A) imprinted polybenzimidazole and (B) imprinted chitosan nanofibers

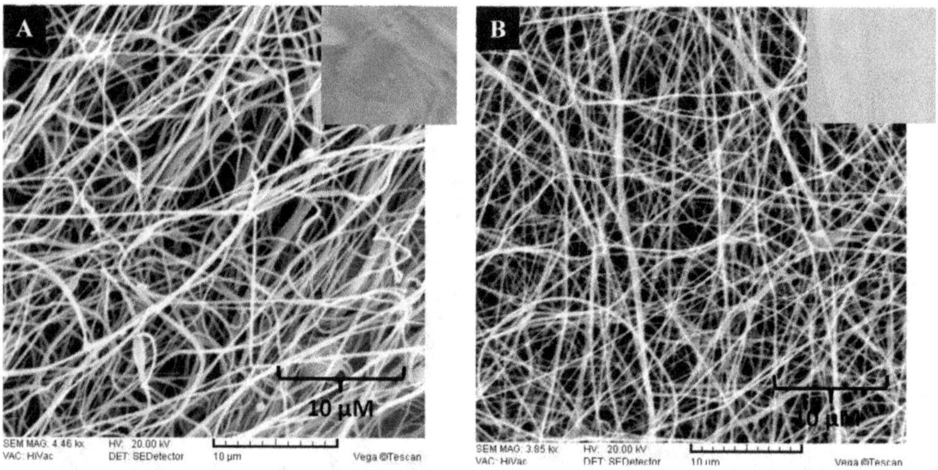

in low viscosity of solutions and tend to form beaded nanofibers when electrospun. Increasing the polymer solution concentration reduces the formation of beaded fibers. Fiber diameters increase with increasing concentration/viscosity (Ki *et al.* 2005; Jiang *et al.* 2004; Huang *et al.* 2001; Son *et al.* 2004; Fong *et al.* 1999; Koski *et al.* 2004; Zhang *et al.* 2005; Lee *et al.* 2004; Ding *et al.* 2002; Teo *et al.* 2006; Demir *et al.* 2002; Mit-Uppatham *et al.* 2004, Chen *et al.* 2004). Varying the concentration of polymer solution affects polymer fiber morphology. At very high polymer solution concentrations, charged solutions dry out at the tip of the needle hence preventing electrospining (Mit-Uppatham *et al.* 2004). SEM images of beaded and non-beaded polyvinyl chloride nanofibers are shown in figure 18 below. Polymer weight also determines the morphology of fibers, most polymers with low molecular weight result in beaded fibers (Fong *et al.* 1999; Koski *et al.* 2004).

Increasing polymer solution conductivity can produce more uniform nanofibers with fewer beads. A simple approach to increasing polymer solution conductivity is through the use of alcohols as dissolution solvents and also through the addition of a salt (Reneker *et al.* 1995; Mit-Uppatham *et al.* 2004, Chen *et al.* 2004; Geng *et*

Figure 21. Chemical structure of polybenzimidazole (PBI)

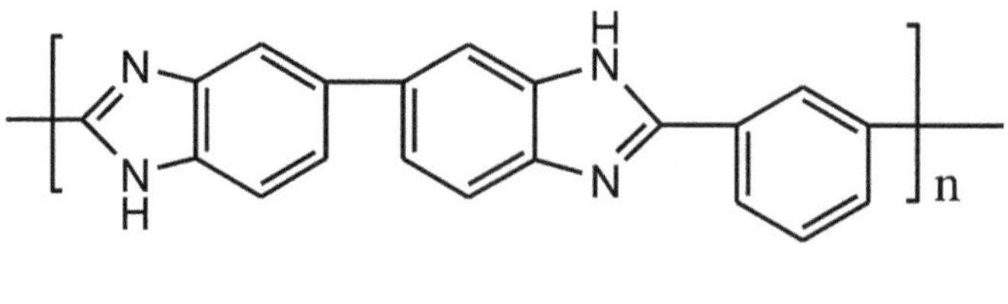

al. 2005; Kim *et al.* 2005; Lin *et al.* 2004; Liu *et al.* 2005; Zuo *et al.* 2005). The use of alcohols as solvents in the electrospinning of poly(hydroxybutyrate-co-valerate) (PHBV) produced bead-free nanofibers (Zuo *et al.* 2005). In the electrospining of polybenzimidazole (PBI), lithium bromide salt was added in order to achieve polymer nanofibers (Figure 19). The salt allowed for the formation of uniform nanofibers rather than particles (Ogunlaja et al. 2014; Kim *et al.* 2005).

Varying polymer electrospinning flow rate also affects the fiber size and morphology. Low flow-rates yielded nanofibers with smaller diameters, and this may be due to the longer time it takes for the travelling charged polymer to reach the collector. The long jet travelling time allows polymer solvent to evaporate, thus producing dry fibers. High flow-rates produced beaded nanofibers which do not dry upon reaching the collector due to short polymer jet traveling time (Zhang *et al.* 2005; Zuo *et al.* 2005; Gupta *et al.* 2005; Duan *et al.* 2004; Jarusuwannapoom *et al.* 2005; Wannatong *et al.* 2004; Zong *et al.* 2002; Yuan *et al.* 2004). Varying the distance between the needle tip and grounded collector has resulted in nanofibers having different diameters and morphology. The minimum distance required in obtaining smooth fibers is said to be a function of the fiber drying time and the volatility of the solvent. The shape and design of a collector affect fiber morphology as porous fiber structures are obtained when a porous collector is employed. Also aligned fibers and yarns were obtained using a conductive frame, rotating drum, or a wheel-like bobbin collector (Walmsley *et al.* 2012; Huang *et al.* 2003; Fong *et al.* 1999; Teo *et al.* 2006). Voltage variation has been the most controlled parameter during electrospinning. A proper voltage or field strength produces a Taylor cone which gives rise to bead-free fibers. When a Taylor cone is formed, the force of the electric field has overcome the surface tension of the solution (Lee *et al.* 2004; Ding *et al.* 2002; Teo *et al.* 2006; Demir *et al.* 2002; Mit-Uppatham *et al.* 2004, Chen *et al.* 2004; Kim *et al.* 2005). Ambient conditions such as a change in temperature and humidity during the polymer electrospinning process affect nanofibers morphology and diameter (Mit-Uppatham *et al.* 2004). Reneker *et al.* 1996 electrospun polymer solutions under vacuum and the fibers produced under this condition gave rise to large diameters.

Adsorptive Desulfurization Using Molecularly Imprinted Nanofibers

Adsorptive separation work we have carried out on oxidized hydrotreated diesel using molecularly imprinted polybenzimidazole and chitosan nanofibers is presented. Scanning Electron Micrograph (SEM) images of the imprinted polybenzimidazole and chitosan nanofibers are shown in figure 20 (Ogunlaja *at al.* 2014).

Figure 22. GC–AED chromatograms of carbon (A) at 179 nm, and sulfur (B) at 181 nm in hydrotreated diesel after oxidation, and chromatograms of carbon (C) at 179 nm and sulfur (D) at 181 nm in oxidized hydrotreated diesel after using organosulfones-imprinted polybenzimidazole nanofibers as adsorbent

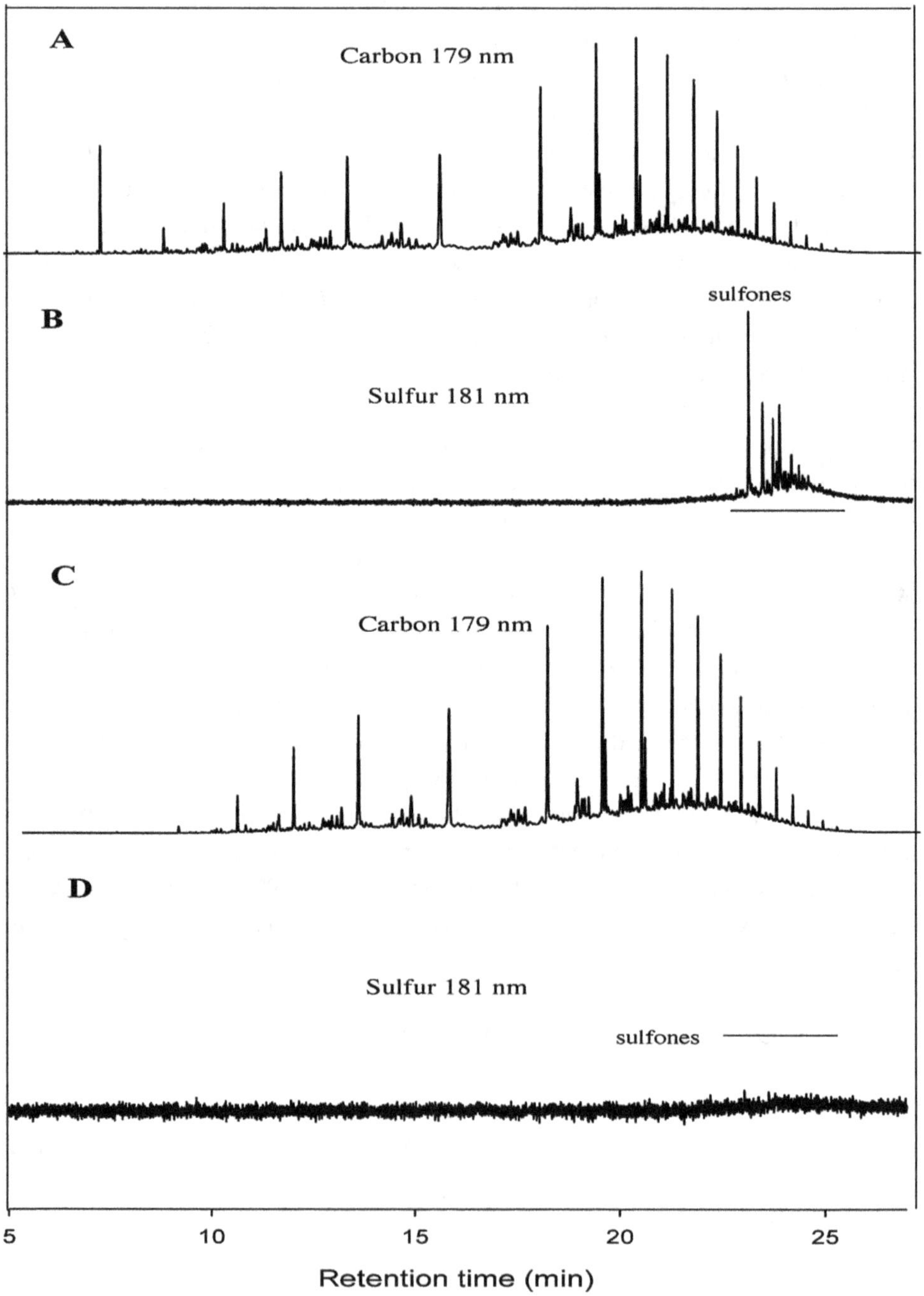

Figure 23. HOMO and LUMO locations of PBI-dibenzothiophene (PBI-DBTO$_2$) adduct: (A) HOMO and (B) LUMO. Red, yellow and grey colours represent oxygen, sulfur and carbon atoms respectively

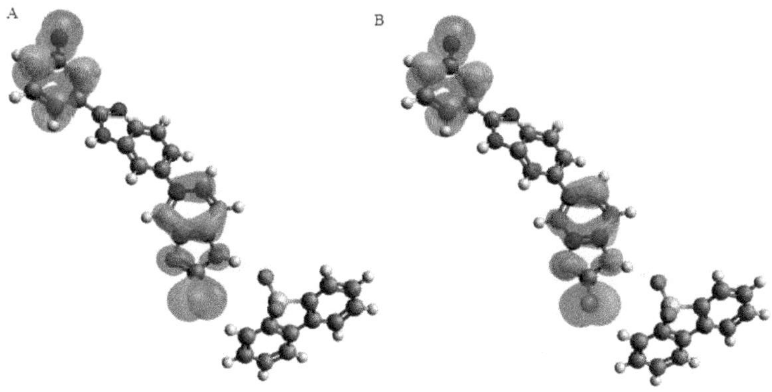

Figure 24. HOMO and LUMO locations of PBI-4,6-dimethyldibenzothiophene (PBI-4,6-DMDBTO$_2$) adduct: (A) HOMO and (B) LUMO. Red, blue, yellow and grey colours represent oxygen, nitrogen, sulfur, and carbon atoms respectively

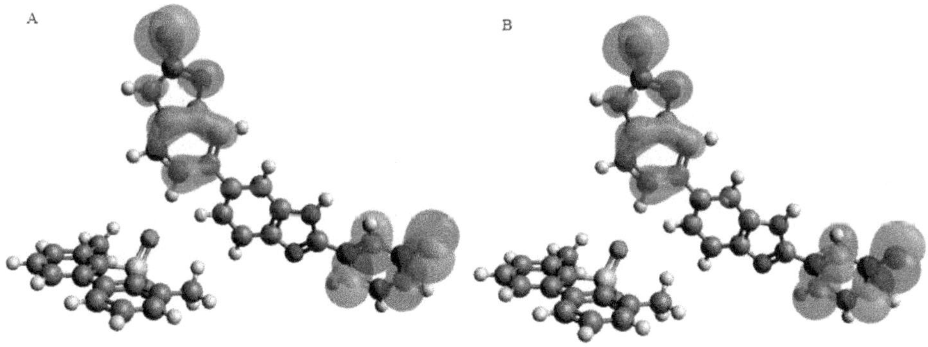

Figure 25. Chemical structure of chitosan, an amino polysaccharide with free amine and hydroxyl groups which allow for chemical modifications

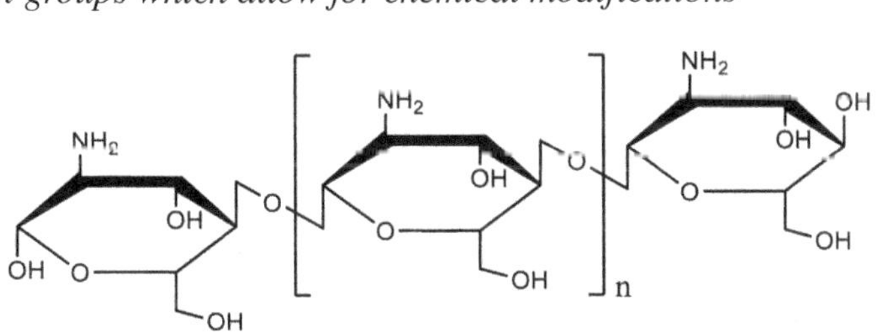

Figure 26. GC–AED chromatograms of carbon (A) at 179 nm, and sulfur (B) at 181 nm in hydrotreated diesel after oxidation, and chromatograms of carbon (C) at 179 nm and sulfur (D) at 181 nm in oxidized hydrotreated diesel after using organosulfones-imprinted chitosan nanofibers as adsorbent

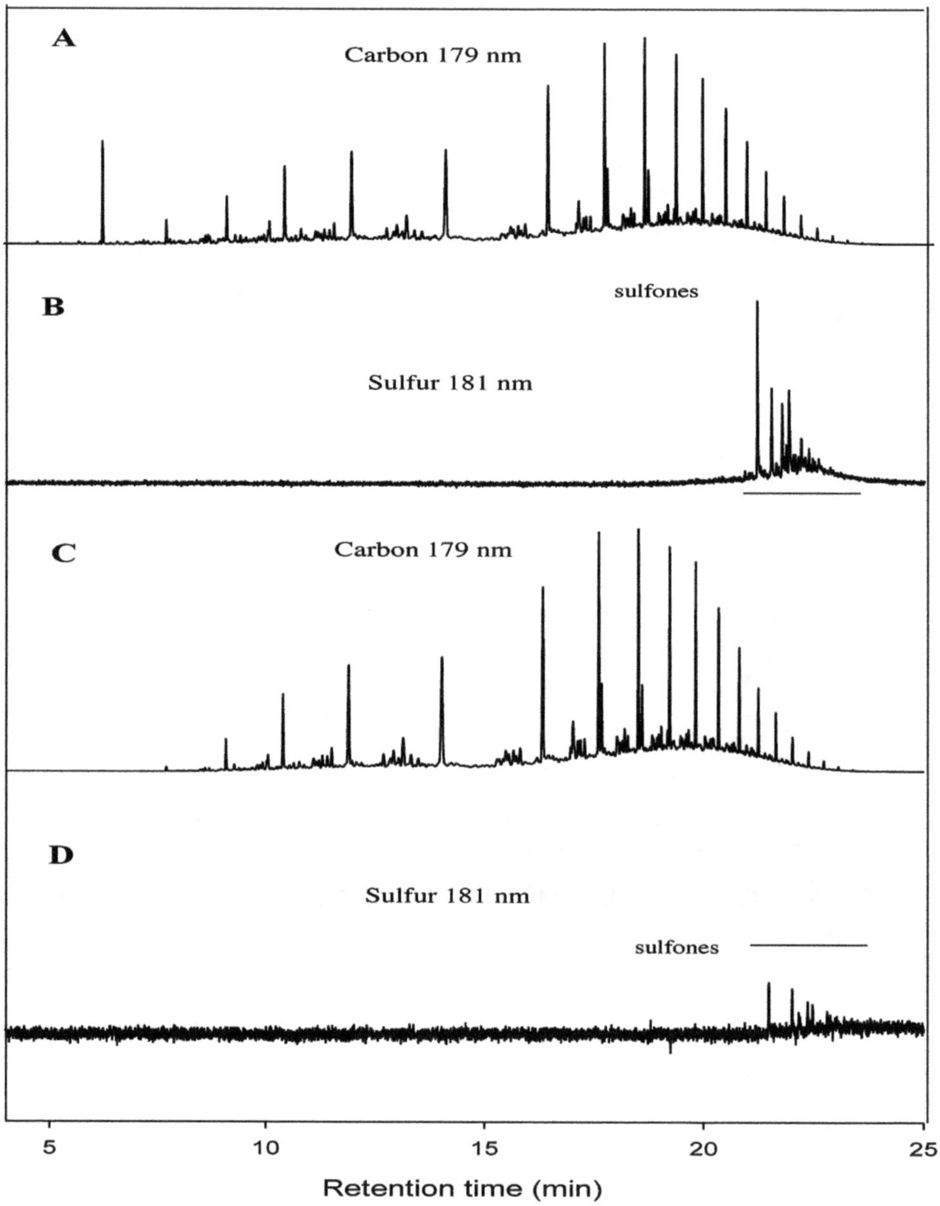

Figure 27. HOMO and LUMO locations of chitosan-dibenzothiophene (chitosan-$DBTO_2$) adduct: (A) HOMO and (B) LUMO. Red, yellow and grey colours represent oxygen, sulfur and carbon atoms respectively

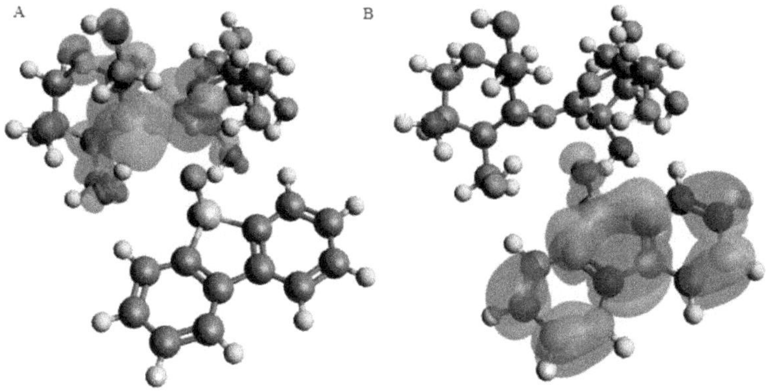

Figure 28. HOMO and LUMO locations of chitosan-4,6-dimethyldibenzothiophene (chitosan-4,6-$DMDBTO_2$) adduct: (A) HOMO and (B) LUMO. Red, blue, yellow and grey colours represent oxygen, nitrogen, sulfur, and carbon atoms respectively

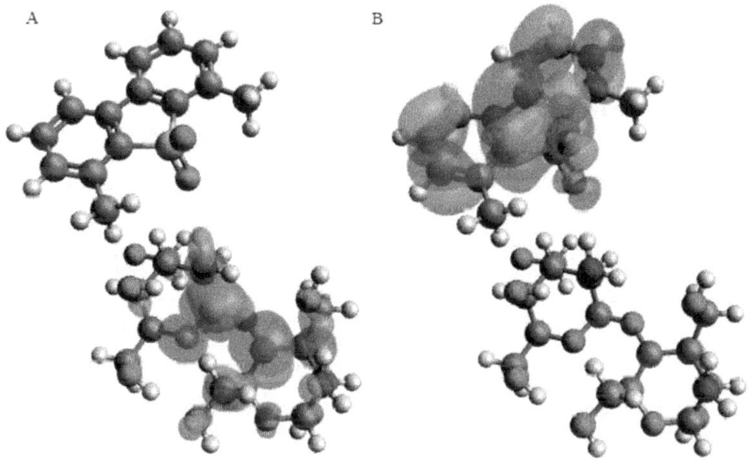

1. Polybenzimidazole (PBI)

Polybenzimidazole (PBI) is a heterocyclic polymer which comprises of repeating units of benzimidazole (Figure 20). The -NH groups of the benzimidazole allows for hydrogen-bonding interactions with polar compounds such as sulfones produced by the oxidation of organosulfur compounds in fuels (Ogunlaja *et al.* 2014). PBI provides excellent chemical and thermal stability properties (Walmsley *et al.* 2012). Polybenzimidazole (PBI) nanofibers were obtained by electrospinning dissolved PBI powder in *N,N*-dimethylacetamide (Ogunlaja *et al.* 2014).

Organosulfones-imprinted polybenzimidazole nanofibers were employed for the adsorption of organosulfone compounds in oxidized hydro-treated diesel, containing 394 ± 4.2 ppm S. A continuous flow process was employed in the adsorption process shown in figure 10. Prior to adsorption, a 150 mg quantity of the respective organosulfone-imprinted PBI nanofibers was conditioned with a suitable solvent. The oxidized fuel was pumped through the nanofibers at a flow-rate of 1 $mL.h^{-1}$. The sulfur left within the hydro-treated fuel after the adsorption process was found to be below the limit of detection of the GC-AED (Figure 22). The saturated organosulfone adsorption capacity of the molecularly imprinted adsorbents from diesel fuel was 5.3 ± 0.4 mg/g. Polybenzimidazole showed excellent adsorption capacities, and this was attributed to the imprinting effect as well as other factors, mostly molecular interactions which were confirmed from the molecular modelling interactions between known organosulfone compounds and polybenzimidazole.

PBI-$DBTO_2$ adduct indicated that the HOMO and LUMO center around PBI, and this resulted in the formation of π-π interaction (π-π stacking) between polybenzimidazole (PBI) and dibenzothiophene sulfone ($DBTO_2$), hydrogen bonding interaction between -S=O of $DBTO_2$ and –NH of PBI was also visible (Figure 23).

PBI-4,6-$DMDBTO_2$ adduct indicated that the HOMO and LUMO centre on PBI, and this resulted in the formation of π-π stacking between polybenzimidazole (PBI) and 4,6-dimethyldibenzothiophene sulfone (4,6-$DMDBTO_2$), in a similar manner to the interaction observed with $DBTO_2$. Hydrogen bonding formation between -S=O of 4,6-$DMDBTO_2$ and –NH of PBI was however not observed, and this might be due to the steric hindrance caused by the methyl groups in 4,6-$DMDBTO_2$ (Figure 24).

2. Chitosan (CHI)

Chitosan, an amino polysaccharide usually obtained by the deacetylation of chitin (Figure 25), has free amine and hydroxyl groups which allow for chemical modifications through the formation of hydrogen bonding interactions and has been investigated for the development of sorption systems (Ryu *et al.* 2002; Ruiz *et al.* 2000; Ngah *et al.* 1999; Xiaoliang *et al.* 2002. Chitosan also has the ability to swell

Figure 29. The basic principle of the S-Zorb sulfur removal technology, a process developed by Phillips petroleum for sulfur removal from liquid fuel at elevated temperatures under low H_2 pressure

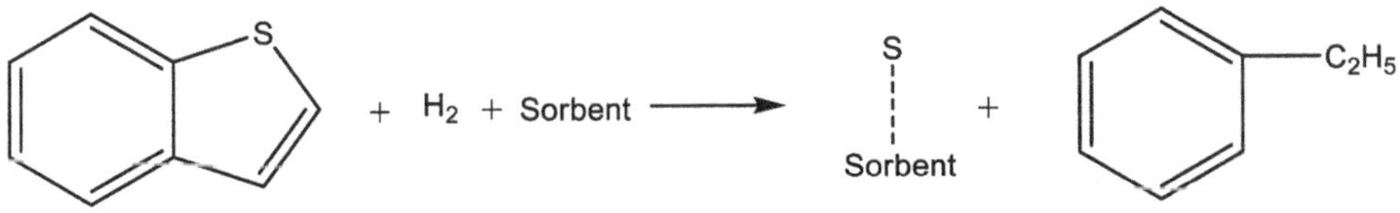

when in contact with a thermodynamically compatible solvent which is another favourable property (Fariba *et al.* 2010; Monteiro *et al.* 1999).

A similar adsorption process as reported for polybenzimidazole was also carried out by employing 300 mg of conditioned organosulfones-imprinted chitosan nanofibers. The adsorption process was also allowed to proceed at a flow-rate of 1 mL/h under room temperature. Sulfur analysis after the adsorption process indicated that a total of 62 ± 3.2 ppm S was left in the diesel (Figure 26). The saturated sulfur adsorption capacity of the molecularly imprinted adsorbents was 2.2 ± 0.2 mg/g. Molecular modelling was also employed in discussing the possible mode of binding between the organosulfone compounds and chitosan.

Chitosan-$DBTO_2$ adducts, on the other hand, showed that the HOMO originates from chitosan while the LUMO comes from $DBTO_2$, this indicated that interactions took place *via* electron donation between HOMO and LUMO centers. Hydrogen bond interaction is also viable due to the presence of free $–NH_2$ and –OH groups on chitosan (Figure 27).

Chitosan-4,6-$DMDBTO_2$ adducts also showed that the HOMO originates from chitosan while the LUMO comes from 4,6-$DMDBTO_2$, and this is similar to the results obtained with Chitosan-$DBTO_2$ adduct confirming that interaction took place *via* electron donation between chitosan and 4,6-$DMDBTO_2$ (Figure 28).

Inorganic Materials

Various porous inorganic materials have always been popular in the adsorption of small organic molecules ranging from helium to triethyl amine based on their critical diameters (0.20 – 0.84 nm). These are commonly identified as molecular sieves, for example aluminosilicates (zeolites) and metalosilicoaluminophosphates (MAPSO) (Jiménez-Cruz, F. and Laredo, G.C. 2004; www.sigmaaldrich.com). Several researchers have extensively attempted to use these general adsorbents to selectively remove SCCs. Inorganic materials ranging from molecular sieves, reduced metals, metal sulfides, metal oxides to activated carbon have been extensively investigated for the adsorption of SCCs (Ma *et al.* 2002; Hernandez *et al.* 2004; Zaki et al. 2005;

Figure 30. Illustrations of 1, 2 and 3-dimensional structures of coordination polymers

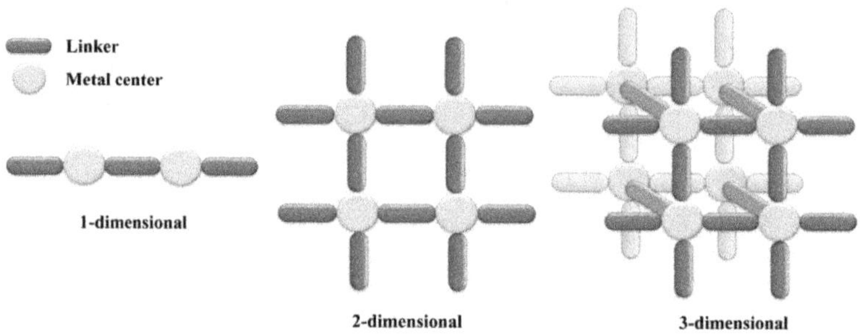

Bhandari *et al.* 2006; Kim *et al.* 2006; Liu *et al.* 2007; Wang *et al.* 2007; Jeon *et al.* 2009; Blanco *et al.* 2011; Tawfik *et al.* 2015).

Most these studies and several others in literature have been fixated on trapping the SCCs into the cavities of the adsorbents based on their sizes (critical diameters, volumes) or interactive (dispersion) forces between functional groups in the adsorbents and the compounds and coordination of the sulfur to unsaturated coordination sites in the materials. Although most of these studies yielded wonderful results in terms of loading capacities and recoverability of the adsorbents when tested in model fuel samples, the major challenge with applying them on real fuel samples is selectivity where other important compounds with matching critical dimensions and dispersion properties are also adsorbed.

Several attempts have been made over the years to improve the simple inorganic adsorbents to target SCCs and have better adsorption capacities and modified versions of these simple adsorbents have propped up with most researchers focusing on coordination of the sulfur compounds to the unsaturated coordination sites in the materials. Many researchers have attempted to improve selectivity by loading of metals (especially those capable of π complexation) into porous inorganic materials such as alumina and silica, zeolites (Yang *et al.* 2003; Sarda *et al.* 2012; Bhandari *et al.* 2006) Transition metals such as Pd, Cu, Ag, Zn, Fe, Ni, Mn have been impregnated into inorganic adsorbents (Patil *et al.* 2014; Hernandez *et al.* 2004, 2005; Yang *et al.* 2003, 2006; Zhang *et al.* 2008; King *et al.* 2006; Shan *et al.* 2008; Ania *et al.* 2006; Dasgupta *et al.* 2013; Velu *et al.* 2003). Other researchers have also taken the route of the blending of metal oxides. In 2001, Phillips petroleum licensed a process called "S-Zorb" for the selective removal of sulfur. The adsorbent is based on zinc oxide, silica, alumina, and nickel (Sughrue *et al.* 2001). The S-Zorb process is carried out in the presence of hydrogen and modified zinc oxide. Sulfur compounds are attracted to hydrogen sulfide, and the sulfur is bound by zinc oxide as zinc sulfide by the process known as chemisorption (Figure 29).

Table 2. Summary of results obtained by Cychosz et al. when they investigated a group of MOFs for their potential in removing thiophenic compounds from fuel

MOF	Metal Center	Linker	Solvent System	Max Uptake of Sulfur Compounds (wt%S)		
				BT	DBT	DMDBT
MOF-5	$Zn_4O(COO)_6$	1,4-benzene dicarboxylate	I	1.6	3.3	1.4
			I/T	-	1.0	0.7
MOF-177	$Zn_4O(COO)_6$	1,3,5-benzene tribenzoate	I	0.8.	1.8	0.8
HKUST1	$Cu_2(CO_2)_4$	1,3,5-benzene tricarboxylate	I	2.5	4.5	1.6
			I/T	-	1.0	0.8
MOF-505	$Cu_2(CO_2)_4$	3,3′,5,5-′biphenyl tetracarboxylate	I	5.1	3.9	2.7
			I/T	-	3.3	0.9
UMCM150	$Cu_3(CO_2)_6$	biphenyl 3,4′,5-tricarboxylate	I	4.0	8.3	4.1
			I/T	-	2.5	0.7

(Cychosz et al. 2008).

Benzothiophene (BT), dibenzothiophene (DBT), dimethyl dibenzothiophene (DMDBT), isooctane (I), toluene (T) and isooctane/toluene (85:15-v:v) (I/T).

Particle size also matters where smaller particles tend to show better adsorption capacity for SCCs as demonstrated by Zhang *et al.* using ZnO (Zhang *et al.* 2012). Mansouri *et al.* presented a nanometric adsorbent made out of nickel and copper impregnated on $\gamma\text{-}Al_2O_3$ which showed ultra-deep desulfurization on a model fuel sample although they didn't carry out selectivity studies (Mansouri *et al.* 2014). The regeneration and reuse of inorganic materials are quite easy. A number of authors have reported several techniques for regenerating spend inorganic adsorbents, REFS. Outstanding is a joint technique by Li *et al.* where they integrated biodesulfurization for regenerating spend inorganic adsorbents. (Li *et al.* 2009). Recently, coordination polymers show promise as adsorbents for adsorptive ultra-deep desulfurization.

Coordination Polymers (CPs)

It is obvious that combining both properties of inorganic and organic sorbents into one material produces a more effective sorbent where these desirable properties work synergistically to achieve good selectivity and loading capacity. Coordination polymers (CPs) are such hybrid structures, consisting of repeating coordination entities between transition-metal clusters or ions and multidentate organic linkers extending in 1, 2 or 3 dimensions as illustrated in Figure 30. Unlike inorganic materials which are rigid and organic materials which are non-porous, they possess the ideal selectivity features owed to organic materials which can be easily be fine-

Table 3. Summary of results obtained in experiments conducted by Maes et al. where they determined maximum uptake of sulfur compounds by the different MOFs in different solvent systems

MOF	Metal Center	Linker	Solvent	TPH	BT	DBT
MIL-100(Fe)	Fe	1,3,5-benzene tricarboxylate	H/T	4	6	7
			T/H	1	<1	<1
MIL-100(Cr)	Cr	1,3,5-benzene tricarboxylate	H/T	2	7	7
			T/H	<1	<1	1
MIL-100(Al)	Al	1,3,5-benzene tricarboxylate	H/T	2	6	6
			T/H	<1	1	<1
MIL-101(Cr)	Cr	1,3,5-benzene tricarboxylate	H/T	3	5	8
			T/H	1	<1	1
HKUST1	Cu	1,3,5-benzene tricarboxylate	H/T	10	10	12
			T/H	5	2	2
CPO-27(Ni)	Ni	2,5-dihydroxy-1,4-benzenedicarboxylate	T/H	12	19	20
CPO-27(Co)	Co	2,5-dihydroxy-1,4-benzenedicarboxylate	T/H	13	20	20
MIL-47/MIL-53	V/Al	1,4-benzene dicarboxylate	H/T	<1	-	<1
			T/H	<1	-	<1

(Maes, *et al.* 2011).

Indole (IND), 2-methylindole (2MI), 1,2-dimethylindole (1,2DMI), carbazole (CBZ), N-methyl carbazole (NMC), thiophene (TPH), benzothiophene (BT), dibenzothiophene (DBT), heptane/toluene (H/T, ratio 80:20 v/v) and toluene/ heptane (T/H, 80:20 v/v).

tuned, as well as the coordinative potential, rigidity, pores and high surface areas found in inorganic materials. MOFs offer exceptional features such as selective and hysteretic adsorption, molecular recognition, gate opening and magnetic bistability (Nagarkar *et al.* 2008) Due to these combined desirable hybrid features, CPs are gathering a lot of interest in various applications including adsorptive separation. Some structures of CPs are dynamic and respond to external stimuli as a result of their flexible ligands and/or coordination sites. (Amo-Ochoa *et al.* 2015). One-dimensional structure CPs (1D-CPs) can be applied in adsorptive separation where the coordination equilibrium between 2 stable switchable states can be exploited, i.e. a stable unsaturated state which is capable of coordination to molecules of interest to give a stable saturated state which can easily release the adsorbed molecules under certain conditions to reproduce the initial unsaturated state (Osborne *et al.* 2015; Xiao *et al.* 2015). In their 3-dimensional form, CPs are crystalline and are known as metal-organic frameworks (MOFs). MOFs have exceptional adsorption capabilities due to the additional features, i.e. high surface area, high total pore volume and uniformly sized pores (Lee *et al.* 2013). MOFs have more potential in the removal of SCCs from fuel, with better selectivity due to the synergy between cavity sizes and dispersion forces in one material (Van de Voorde *et al.* 2014). The use of 1D-CPs and MOFs in adsorptive separation is discussed below as well as design strategies to enable such materials to be applied in adsorptive desulfurization.

1. Dimensional Coordinative Polymers (1D-CPs)

The adsorption site is more imperative when considering the use of 1D-CPs in adsorptive separation. As such, attention will be around the ability of suitable unsaturated metal centers in CPs to coordinate with the SCCs under certain conditions and to release them when the conditions are changed while adopting the initial stable unsaturated geometry. Limited study has been reported for the use of 1D polymer in the selective adsorption of SCCs due to the lack of cavities. Recently, Tella *et al*, 2018 demonstrated that the distorted octahedral zig-zag 1D coordination polymer, copper(II) $[Cu(4\text{-mba})_2(H_2O)_3]$, shows promise for the uptake of sulfur compounds *via* pi-pi stacking.

Metal Organic Frameworks (MOFs)

While very little research has been done around the use of 1D-CPs in adsorptive separation considering that they do not have pores and have relatively smaller surface areas, 3D-CPs (MOFs) have been extensively investigated as potential adsorbents for various compounds, including SCCs. The beauty of metal-organic frameworks (MOFs) lies in how easy it is to engineer them for particular applications due to simple design strategies used to obtain particular pore sizes and modify their pore properties (Schneemann *et al.* 2014).The design strategies include control of their composition, structure, scale and the bulk properties (Falcaro *et al.* 2014; Kokado, 2017). Cychosz *et al.* pioneered the use of MOFs in adsorptive desulfurization where they investigated removal of thiophenic compounds [benzothiophene (BT), dibenzothiophene (DBT) and dimethyl dibenzothiophene (DMDBT)] from an isooctane/toluene hydrocarbon stream using 5 MOFs; MOF5, HKUST1, MOF177, MOF505 and UMCM150 (Table 2) (Cychosz *et al.* 2008). Their major finding was that the MOFs showed much better adsorption than zeolite and that the performance of copper MOFs was superior to that of zinc MOFs. Highest performance observed for MOF505 and UMCM150 was attributed to structural open metal sites (unsaturated coordination sites) in copper.

Figure 31. Schematic representation of a MOF containing an adsorbed molecule illustrating the key design strategies used to improve selectivity

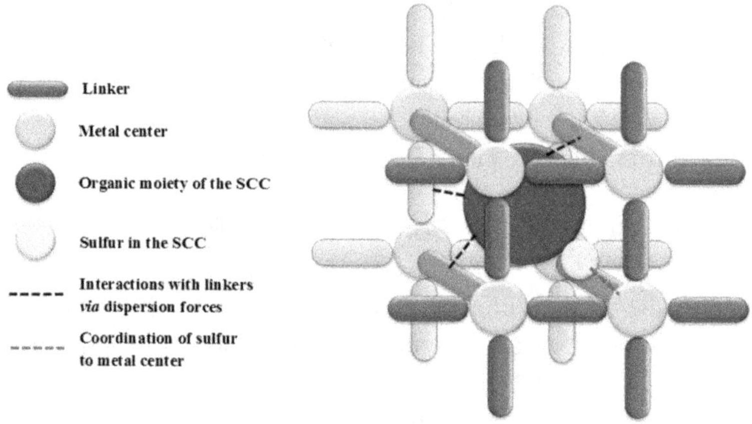

Work done by Maes *et al.* also confirmed the crucial role played by unsaturated metal sites in MOFs for significant adsorption to occur in mixtures containing competing aromatic toluene molecules. MOFs lacking such sites, e.g. MIL-47 and MIL-53 had poor adsorption (Table 3) (Maes, *et al.* 2011).

Xin *et al.* also compared two MOFs based on the ligand biphenyl-3,4',5-tricarboxylate (H_3BHTC), *i.e.* UMCM-150 [$Cu_3(BHTC)_2$] and its heterobimetallic analogue $Co_1Cu_2(BHTC)_2$, and concluded that Cu^{2+} interacts with sulfur better than Co^{2+} compared to Cu^{2+} (both d^1-d^9 transition metal cations) since Cu^{2+} has more d electrons. After characterizing and noting that UMCM-150 and its heterobimetallic analogue has similar topology and porosity they attributed any differences in adsorption of SCCs to replacement of the trinuclear copper cluster with a cobalt atom. (Xin and Wang 2018). Liu *et al.* also investigated the use of rare earth metal-organic frameworks in adsorptive desulfurization and although they showed good potential copper would be more favorable since they are expensive. (Liu *et al.* 2014). Since better performance in copper MOFs over several other likely transition metal candidates was also attributed to the easily obtainable unsaturated coordination sites, these can also be obtained in other MOFs through post-synthesis removal of coordinated waters as well as changing oxidation states of the metal centers to create unsaturated coordination sites in MOFs is also possible. (Pilloni *et al.* 2015).

While it has become common knowledge that doping MOFs with metals capable of π-complexation with sulfur drastically increases the adsorption capacity of the MOFs and in some cases selectivity, leaching has always been a big challenge (Chang *et al.* 2014). Huang *et al.* managed to apply the MOF [(Cr)-MIL-101-SO_3Ag] which contains Ag(I) immobilized *via* sulfonic groups in the adsorption of SCCs and the

Figure 32. An illustration of the solvothermal MOF synthesis approach.
Reprinted with permission from Springer Nature Customer Service Centre GmbH: Springer Nature Korean J. Chem. Eng. Synthesis of metal-organic frameworks: A mini review. Lee et al., Copyright 2019.

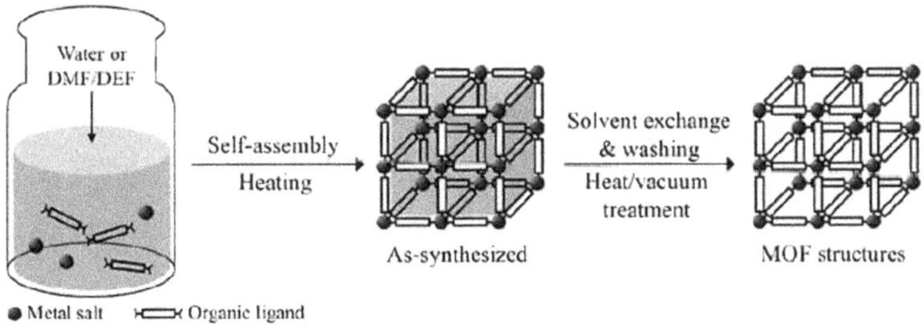

MOF retained its activity after 5 cycles (Huang *et al.* 2015). Stability of the MOFs under the conditions and in environments where they will be applied must not be overlooked. Stability in the presence of moisture, acids and other chemicals is crucial for MOFs to be used in adsorptive desulfurization. MIL-101 is reported to show outstanding stability to moisture and other chemicals.(Lee *et al.* 2013).

Other researchers have attempted to use composite materials where they incorporate other materials such as activated carbon into MOFs to improve their performance. One such example is work done by Zhu *et al.* where they dispersed MOF-5 crystallites onto 200 meshes activated carbon. The composite material showed a much higher performance than the parent materials. Expansion in the pore volume of MOF-5, an increase in average pore size and better exposed MOF channels were determined, aspects that were attributed to improved performance. (Zhu *et al.* 2018). Peng *et al.*, through a composite they developed for ODS by encapsulating Keggin-type phosphotungstic acid ($H_3PW_{12}O_{40}$) in size-matched UiO-67 (i.e. a Zr MOF), also demonstrated that composite materials show better collective performance. In their case, the UiO-67 was meant to attract SCCs and prevent in order to avoid leaching of the catalytic phosphotungstic acid. (Peng *et al.* 2018).

The role of the organic linkers applied in the MOFs should not be underestimated. To accentuate this, from the studies conducted by Cychosz *et al.*, it can be clearly noted that the uptake of SCCs decreased drastically in the presence of toluene (Cychosz *et al.* 2008). Secondly, comparing the maximum uptake of sulfur compounds in HKUST1, MOF-505 and UMCM150, which contain 1,3,5-benzene tricarboxylate; 3,3′,5,5-′biphenyl tetracarboxylate and biphenyl-3,4′,5-tricarboxylate as linkers, respectively, it can be noted that the differences in their adsorption performances were inspired by their different linkers and they were also affected to different extends by the presence of toluene in the system (Table 2). The MOFs in table 3 also shows a drastic decrease in the maximum uptake of the SCCs when the 80:20 toluene/heptane solvent system is used compared to the 80:20 heptane/toluene

Figure 33. An illustration of synthesis of MOFs via the mechanochemical route.
Reprinted by permission from Springer Nature Customer Service Centre GmbH: Springer Nature Korean J. Chem. Eng. Synthesis of metal-organic frameworks: A mini review. Lee et al., Copyright 2019.

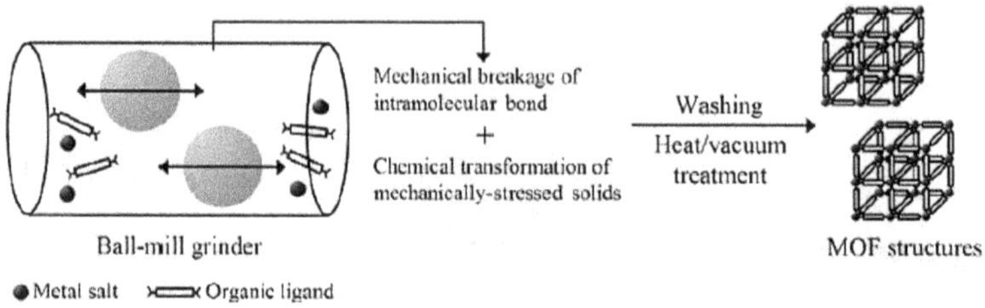

solvent system. This basically proves that dispersion forces play a major role in the uptake of SCCs by MOFs. The electronic characteristics of the adsorbate and the surface nature of the adsorbent play a major role, as was previously outlined under organic adsorbates. Lu *et al.* presented an in-depth review on designing linker to achieve certain properties such as unparalleled surface area, pore aperture, molecular recognition and stability (Lu *et al.* 2014).

From the majority of discussions around MOFS presented so far, it can be noted that most research on developing MOFs for removing SCCs from fuel has been based on testing and screening random MOFs. This method can be daunting, and in most cases, selectivity is not achieved. The biggest shortfall in most of the MOFs is the lack of synergy among the features applied when carrying out adsorption, especially the metal center and linker properties. It is assumed that the entity can either be adsorbed based on its interaction with the linker or through interaction with the metal center, but not effectively with both at the same time. That could also explain the lack of coherence in the results reported by authors so far as they change linkers, target molecules and when toluene was added. No resolute trend can be interpreted from the data. However, from this random testing of general MOFs conducted by the researchers, pinnacles upon which MOFs can be specifically tailored to effectively target the SCCs in fuel can be derived. As such the following areas will be key in designing such MOFs:

1. Making sure that the sizes of the cavities match the critical dimensions of the SCCs.
2. Using suitable metal centers which allow for coordinative interaction with the SCCs.
3. Using organic moieties with suitable functionality as linkers to enhance specific interactions (dispersion forces) with the SCCs.

Figure 34. General schemes of the covalent (top), dative (middle) and deprotection (bottom) post-synthetic modifications as described by Cohen.
Reprinted with permission from Cohen, S.M. (2012) Postsynthetic Methods for the Functionalization of Metal-Organic Frameworks, Chem. Rev., 112, 970–1000. Copyright 2019 American Chemical Society.

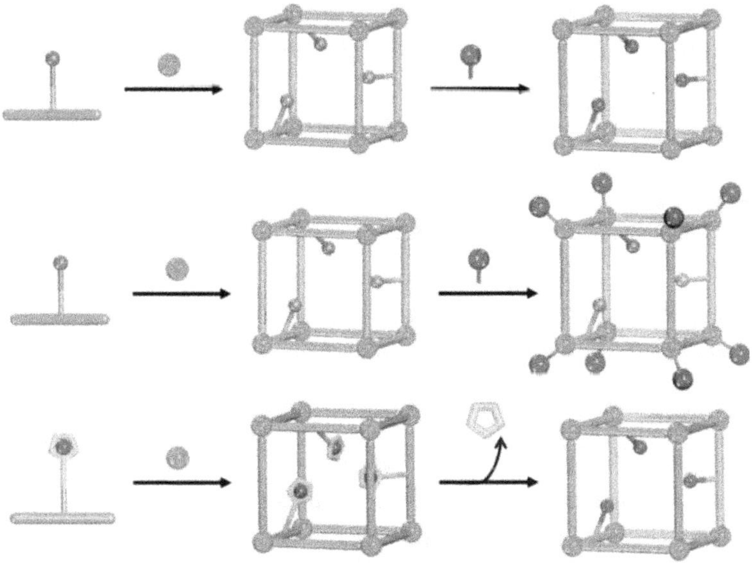

These features are summarized pictorially in Figure 31.

Khan *et al.* also raised structural flexibility as an aspect that plays a crucial role in the efficient adsorption of SCCs by MOFs. Structural flexibility entails dynamic reaction of the crystalline MOFs when an external stimulus is applied. They demonstrated that flexible MOFs possessing μ_2-OH groups, for example, hydroxo terephthalate/MIL-53 type MOFs, showed improved adsorption of SCCs in the presence of oxygen containing compounds (OCCs), an unusual occurrence, unlike rigid MOFs such as MIL-68 and MIL-100. The display of such physical properties is based on the 3 pinnacle strategies already outlined. (Khan *et al.* 2016). Research into the use of MOFs for adsorptive desulfurization is now moving towards incorporating those 3 keys strategies when designing the MOFs. It is also important to stress how incorporating the power of computational modeling to design, predict and screen MOFs possessing the most desirable properties that allow efficient and selective adsorption of SCCs can simplify the work (François-Xavier and Fuchs 2015).

Synthesis of Metal Organic Frameworks

Several techniques have been applied in the synthesis of MOFs. Some of the commonly used ones are hydrothermal, mechanochemical, ionothermal, microwave, sonochemical, electrochemical, microfluidics and dry gel conversion. Solvothermal and mechanochemical techniques are the widely applied techniques in the synthesis of MOFs for adsorptive desulfurization. The solvothermal technique involves reacting stoichiometric quantities of reagents (metal centre and linkers) dissolved in a suitable solvent and sealed in an enclosed vessel (usually a parr teflon-lined stainless-steel reactor) and heated over a period (Figure 32). The majority of MOFs have discovered through solvothermal synthesis and the technique is also mostly used to optimize synthesis protocols. However, most of the solvothermal syntheses conducted in laboratories required long synthesis periods under harsh conditions such as high temperature and pressure and large amounts of organic solvents. In some cases scaling up solvothermal synthesis can present challenges in terms of output and cost depending on the MOF to be synthesized. Solvothermal syntheses can be attractive where water is the solvent.

MOF synthesis *via* the mechanochemical route is driven by mechanical breakage of intermolecular bonds and successive formation of new ones (Figure 33). Syntheses are normally carried out solvent-free or with minimal amounts of solvent (solvent assisted) at room temperature with high yields over a short time. Metal oxides are preferred over metal salts as a starting material and the only side product is water. Although quite an attractive option, the mechanochemical technique is limited to selected types of MOF and scaling up is difficult. A liquid assisted mechanochemical synthesized Fe(III) trimesate, a MOF isostructural with MIL-100(Fe), had double the maximum adsorption capacity compared to the solvothermal synthesized one. (Pilloni *et al.* 2015).

Several post-synthesis modifications (PSM) are also possible with MOFs, a feature that has also made them attractive to research development (Deria *et al.* 2014; Cohen 2012). These modifications are mostly carried out to add, activate or enhance certain properties in the MOFs, usually by adding chemical functionality onto the linkers or the metal centers of the as-synthesized forms. One of the most primary post-synthetic modifications is desolvation of synthesis solvents to open up the cavities. Other non-covalent PMS techniques include drying with supercritical CO_2 and ion exchange of guest molecules, and these are reported to radically alter the properties of some MOFs. PSMs can also be covalent, dative (coordinate covalent) and deprotection depending on the bond formed or broken (Cohen 2012). In covalent PSM a reagent is used to modify a component of the MOF with the formation of a covalent bond. With dative PSM a reagent that forms a dative (metal-ligand) bond with a component of the MOF is added (Figure 34). With deprotection, a reaction occurs that results in the cleavage of a chemical bond within an intact framework of the MOF. Multiple PSMs can be performed to achieve a certain desired outcome in the MOF, especially when it comes to improving selectivity. It is important to make sure that there is no degradation or collapse of the framework when carrying out PSMs.

The capital required in the synthesis MOFs is huge, a discouragement when bearing in mind their application on an industrial scale (Blanco *et al.* 2013). However, considering that MOFs have shown remarkable recoverability and the fact their use has minimal consumables and running costs, their ascendency in the future of desulfurization is not questionable. However, more considerations must be made on the separation of MOFs from fuel and regeneration after use since they determine applicability in the fuel refinery industry as well as contribution to costs.

Jin *et al.* designed and fabricated an iron MOF [Fe_3O_4@PAA@MOF-199] which they magnetized for easy magnetic separation from the fuel matrix. (Jin *et al.* 2014). Magnetic bistability, where desolvation and resolvation result in reversible changes of the magnetic (solvatomagnetic effect), has also been reported. (Ferrando-Soria *et al.* 2012) Carrying adsorption under flow conditions would also ease the challenges associated with separating adsorbent. It is to the greatest of our expectations that if the 3-key design strategies are adopted in the development MOFs they will go a long way in fostering the complete removal of SCCs from the fuel in an economical way.

PERSPECTIVES

The heightened concerns for cleaner fuel and the increasingly strict regulations on sulfur contents in fuels have made the desulfurization process progressively more important. The sulfur problem is becoming more serious as most of the oil reserves are sourer, *i.e.* they possess a high sulfur content. Researchers are continually exploring possible ways of removing organosulfur compounds in fuels without compromising the quality of the fuel. Sulfur compounds in fuel are mostly removed *via* hydro-treatment but this is not sufficient to achieve the ultra-low levels without compromising the fuel properties. In recent studies, refractory organosulfur compounds in fuels were oxidized by employing a heterogeneous catalyst followed by separation of the oxidized sulfur compounds using imprinted polymers, metal-organic frameworks (MOFs) or solvents to remove the organosulfur and organosulfones. This process is termed oxidative desulfurization, and it shows great promise for the future of the fuel industry. Progress has been made in the area of ODS by Petroleum Technologies Corp. (www.altpetrol.com), which piloted an oxidative desulfurization technology, known as Sulfex, that can remove sulfur from liquid hydrocarbons at near-ambient pressures and temperatures (Sulfex™ 2019). New approaches, for example, those applying nanomaterials for selective adsorptive removal of organosulfur compounds in fuels, are necessary for the production of affordable ultraclean transportation fuels. In recent studies, refractory organosulfur compounds in fuels are separated from fuel by using metal-organic frameworks (MOFs). Porous materials and MOFs offer a large surface area and high porosity as well as the mechanically robust and thermally stable adsorbent, and in this way moieties such as organosulfur compounds with functionalities for π-complexation, hydrogen bonding, van der Waals interactions, and other interactions are available without compromising its chemical stability. Research in adsorptive desulfurization of fuel using MIPs, CPs, and MOFs is still at its infancy and requires more extensive investigations due to selectivity, mechanical and thermal stability challenges. A systematic approach in improving material's selectivity, mechanical and thermal stabilities during adsorptive removal of refractory organosulfur compounds from the fuel is required. This short chapter, therefore, demonstrated the use of solid sorbents MOFs, CPs and MIPs as potential adsorbents for the selective removal of refractory organosulfur compounds in fuel oils in order to achieve the mandated sulfur levels of less than 10 ppmS. It is desirable that the fuel industry can find possibilities for this technology, and it will be the role of academia to present attractive solutions in this regard.

REFERENCES

Abdullah, W., Nazwanie, W., Bakar, W. A., Azelee, W., Rusmidah, A., & Zaidi, E. (2015). Oxidative desulfurization of commercial diesel catalyzed by *tert*-butyl hydroperoxide polymolybdate on alumina: Optimization by Box-Behnken design. *Clean Technologies and Environmental Policy*, *17*(2), 433–441. doi:10.100710098-014-0797-5

Aida, T., & Yamamoto, D. (1994). Annual book of ASTM standards, 2005. Standard test method for determination of sulfur compounds in natural gas and gaseous fuels by gas chromatography and chemiluminescence. *Preprints of Papers- American Chemical Society. Division of Fuel Chemistry*, *39*(5), 623.

Amo-Ochoa, P., Hassanein, K., Gómez-García, C. J., Benmansour, S., Perles, J., Castillo, O., ... Zamora, F. (2015). Reversible stimulus-responsive Cu(I) iodide pyridine coordination polymer. *Chemical Communications (Cambridge)*, *51*(76), 14306–14309. doi:10.1039/C5CC04746C PMID:26264525

Andersson, J. T., Panda, S. K., & Schrader, W. (2007). Analysis of large sulfur aromatic compounds in fossil fuels by FT-ICR MS. In O. R. Koseoglu & S. Kasztelan (Eds.), *Hydrotreating/Hydrocracking Processing*. Washington, DC: American Chemical Society.

Andersson, J. T., & Schade, T. (2006). Speciation of alkylated dibenzothiophenes in a deeply desulfurized diesel fuel. *Energy & Fuels*, *20*(4), 1614–1620. doi:10.1021/ef0502507

Ania, C. O., & Bandosz, T. J. (2006). Metal-loaded polystyrene-based activated carbons as dibenzothiophene removal media via reactive adsorption. *Carbon*, *44*(12), 2404–2412. doi:10.1016/j.carbon.2006.05.016

Anisimov, A. V., Fedorova, E. V., Lesnugin, A. Z., Senyavin, V. M., Aslanov, L. A., Rybakov, V. B., & Tarakanova, A. V. (2003). Vanadium peroxocomplexes as oxidation catalysts of sulfur organic compounds by hydrogen peroxide in bi-phase systems. *Catalysis Today*, *78*(1-4), 319–325. doi:10.1016/S0920-5861(02)00311-5

Babich, V. I., & Moulijn, J. A. (2003). Science and technology of novel processes for deep desulfurization of oil refinery streams: A review. *Fuel*, *82*(6), 607–631. doi:10.1016/S0016-2361(02)00324-1

Bacsa, J., Eve, D., & Dunbar, K. R. (2005). Catena-Poly[[diaquacobalt(II)]-μ-Oxalato]. *Acta Crystallographica. Section C, Crystal Structure Communications*, *61*(1), m58–m60. doi:10.1107/S0108270104030409 PMID:15640580

Baeza, P., Aguila, G., Gracia, F., & Araya, P. (2007). Desulfurization by adsorption with copper supported on zirconia. *Catalysis Communications*, *9*. doi:10.1016/j.catcom

Baran, E. J. (2014). Review: Natural oxalates and their analogous synthetic complexes. *Journal of Coordination Chemistry*, *67*(23-24), 3734–3768. doi:10.1080/00958972.2014.937340

Barbara, P., Rufino, M. N., Campos-Martin, J. M., & Fierro, J. L. G. (2011). Towards near zero-sulfur liquid fuels: A perspective review. *Catalysis Science & Technology*, *1*(1), 23–42. doi:10.1039/c0cy00049c

Baxendale, J. H., Evans, M. G., & Park, G. S. (1946). The mechanism and kinetics of the initiation of polymerisation by systems containing hydrogen peroxide. *Transactions of the Faraday Society*, *42*, 155–169. doi:10.1039/tf9464200155

Bhandari, V. M., Ko, C. H., Park, J. G., Han, S., Cho, S., & Kim, J. (2006). Desulfurization of diesel using ion-exchanged zeolites. *Chemical Engineering Science*, *61*(8), 2599–2608. doi:10.1016/j.ces.2005.11.015

Bhutto, A. W., Abro, R., Gao, S., Abbas, T., Chen, X., & Yu, G. (2016). Oxidative desulfurization of fuels using ionic liquids: A review. *Journal of the Taiwan Institute of Chemical Engineers*, *62*, 84–97. doi:10.1016/j.jtice.2016.01.014

Blanco-Brieva, G., Campos-Martin, J. M., Al-Zahrani, S. M., & Fierro, J. L. G. (2011). Effectiveness of metal–organic frameworks for removal of refractory organo-sulfur compound present in liquid fuels. *Fuel*, *90*(1), 190–197. doi:10.1016/j.fuel.2010.08.008

Blanco-Brieva, G., Campos-Martin, J. M., Al-Zahrani, S. M., & Fierro, J. L. G. (2013). Thermal regeneration of the metal organic frameworks used in the adsorption of refractory organosulfur compounds from liquid fuels. *Fuel*, *105*, 459–465. doi:10.1016/j.fuel.2012.08.003

Booth, B., Buglass, J. G., & Unsworth, J. F. (2001). Tailoring fuels for the new millennium. *Topics in Catalysis*, *16-17*(1/4), 39–46. doi:10.1023/A:1016670528451

Campestrini, S., Conte, V., Di Furia, F., Modena, G., & Bortolini, O. (1988). Metal catalysis in oxidation by peroxides. 30. Electrophilic oxygen transfer from anionic, coordinatively saturated molybdenum peroxo complexes. *The Journal of Organic Chemistry*, *53*(24), 5721–5724. doi:10.1021/jo00259a021

Campos-Martin, J. M., Capel-Sanchez, M. C., Perez-Presas, P., & Fierro, J. L. G. (2010). Oxidative processes of desulfurization of liquid fuels. *Journal of Chemical Technology and Biotechnology (Oxford, Oxfordshire)*, *85*(7), 879–890. doi:10.1002/jctb.2371

Castro, B., Whitcombe, M. J., Vulfson, E. N., Vazquez-Duhalt, R., & Barzana, E. (2001). Molecular imprinting for the selective adsorption of organosulfur compounds present in fuels. *Analytica Chimica Acta*, *435*(1), 83–90. doi:10.1016/S0003-2670(01)00799-1

Chang, G., Bao, Z., Ren, Q., Deng, S., Zhang, Z., Su, B., ... Yang, Y. (2014). Fabrication of cuprous nanoparticles in MIL-101: An efficient adsorbent for the separation of olefin–paraffin mixtures. *RSC Advances*, *4*(39), 20230–20233. doi:10.1039/C4RA02125H

Chang, J. H., Chang, Y. K., Cho, K. S., & Chang, H. N. (2000). Desulfurization of model and diesel oils by resting cells of Gordona sp. *Biotechnology Letters*, *22*(3), 193–196. doi:10.1023/A:1005637013879

Chang, J. H., Rhee, S. K., Chang, Y. K., & Chang, H. N. (1998). Desulfurization of diesel oils by a newly isolated dibenzothiophene-degrading Norcadia sp. strain CYKS2. *Biotechnology Progress*, *14*(6), 851–855. doi:10.1021/bp9800788 PMID:9841646

Chen, H., & Hsieh, Y. L. (2004). Ultrafine hydrogel fibers with dual temperature- and pH-responsive swelling behaviors. *Journal of Polymer Science. Part A, Polymer Chemistry*, *42*(24), 6331–6339. doi:10.1002/pola.20461

Chica, A., Corma, A., & Dómine, M. E. (2006). Catalytic oxidative desulfurization (ODS) of diesel fuel on a continuous fixed-bed reactor. *Journal of Catalysis*, *242*(2), 299–308. doi:10.1016/j.jcat.2006.06.013

Chica, A., Gatti, G., Moden, B., Marchese, L., & Iglesia, E. (2006). Selective catalytic oxidation of organosulfur compounds with *tert*-butylhydroperoxide. *Chemistry (Weinheim an der Bergstrasse, Germany)*, *12*(7), 1960–1967. doi:10.1002/chem.200500858 PMID:16385590

Chong, E. J., Phan, T. T., Lim, I. J., Zhang, Y. Z., Bay, B. H., Ramakrishna, S., & Lim, C. T. (2007). Evaluation of electrospun PCL/gelatin nanofibrous scaffold for wound healing and layered dermal reconstitution. *Acta Biomaterialia*, *3*(3), 321–330. doi:10.1016/j.actbio.2007.01.002 PMID:17321811

Clean Fuel, I. I. (2019). Retrieved from http://www.sapia.org.za/key-issues/cleaner-fuels-ii

Cohen, S. M. (2012). Postsynthetic Methods for the Functionalization of Metal-Organic Frameworks. *Chemical Reviews*, *112*(2), 970–1000. doi:10.1021/cr200179u PMID:21916418

Collins, F. M., Lucy, A. R., & Sharp, C. (1997). Oxidative desulfurization of oils via hydrogen peroxide and heteropolyanion catalysis. *Journal of Molecular Catalysis A Chemical*, *117*(1-3), 397–403. doi:10.1016/S1381-1169(96)00251-8

Crude oil reserves and spatial distribution. (n.d.). Retrieved September 13, 2013, from http://web2.mendelu.cz/af_291_projekty2/vseo/stranka.php?kod=312

Cychosz, K. A., Wong-Foy, A. G., & Matzger, A. J. (2008). Liquid phase adsorption by microporous coordination polymers. *Journal of the American Chemical Society*, *130*(22), 6938–6943. doi:10.1021/ja802121u PMID:18465854

Darian, S. T., & Arabshshi, S. H. (1988). *Process for upgrading diesel oils*. US Patent 4746420.

Dasgupta, S., Gupta, P., Anshu, A., Nanoti, A. N., Goswami, M. O., Garg, E. T., ... Stöcker, M. (2013). Adsorptive desulfurization of diesel by regenerable nickel based adsorbents. *Fuel*, *108*, 184–189. doi:10.1016/j.fuel.2012.12.060

Davis, F. J. (2004). Polymer chemistry-A practical approach. Oxford University Press.

Dekker, M., Yocum, R.H., & Nyquist, E.B. (1973). *Functional monomers; their preparation, polymerization, and application*. Academic Press.

Demir, M. M., Yilgor, I., Yilgor, E., & Erman, B. (2002). Electrospinning of polyurethane fibers. *Polymer*, *43*(11), 3303–3309. doi:10.1016/S0032-3861(02)00136-2

Deria, P., Mondloch, J. E., Karagiaridi, O., Bury, W., Hupp, J. T., & Farha, O. K. (2014). Beyond post-synthesis modification: Evolution of metal–organic frameworks *via* building block replacement. *Chemical Society Reviews*, *43*(16), 5896–5912. doi:10.1039/C4CS00067F PMID:24723093

Di-Giuseppe, A., Crucianelli, M., De Angelis, F., Crestini, C., & Saladino, R. (2009). Efficient oxidation of thiophene derivatives with homogeneous and heterogeneous MTO/H_2O_2 systems: A novel approach for, oxidative desulfurization (ODS) of diesel fuel. *Applied Catalysis B: Environmental*, *89*(1-2), 239–245. doi:10.1016/j.apcatb.2009.02.009

Ding, B., Kim, H. Y., Lee, S. C., Shao, C. L., Lee, D. R., Park, S. J., ... Choi, K. J. (2002). Preparation and characterization of a nanoscale poly(vinyl alcohol) fiber aggregate produced by an electrospinning method. *Journal of Polymer Science. Part B, Polymer Physics*, *40*(13), 1261–1268. doi:10.1002/polb.10191

Ding, B., Kimura, E., Sato, T., Fujita, S., & Shiratori, S. (2004). Fabrication of blend biodegradable nanofibrous nonwoven mats via multi-jet electrospinning. *Polymer*, *45*(6), 1895–1902. doi:10.1016/j.polymer.2004.01.026

Djangkung, S., Murti, S., Yang, H., Choi, K., Kora, Y., & Mochida, I. (2003). Influences of nitrogen species on the hydrodesulfurization reactivity of a gas oil over sulfide catalysts of variable activity. *Applied Catalysis A, General*, *252*(2), 331–346. doi:10.1016/S0926-860X(03)00468-X

Doshi, J., & Reneker, D. H. (1995). Electrospinning process and applications of electrospun fibers. *Journal of Electrostatics*, *35*(2-3), 151–160. doi:10.1016/0304-3886(95)00041-8

Dreizler, R. M., & Gross, E. K. U. (1990). *Density functional theory*. Berlin: Springer. doi:10.1007/978-3-642-86105-5

Duan, B., Dong, C. H., Yuan, X. Y., & Yao, K. D. (2004). Electrospinning of chitosan solutions in acetic acid with poly(ethylene oxide). *Journal of Biomaterials Science. Polymer Edition*, *15*(6), 797–811. doi:10.1163/156856204774196171 PMID:15255527

Dzenis, Y. (2004). Spinning continuous fibers for nanotechnology. *Science*, *304*, 1917–1919.

El-Newehy, M.H., Al, D.S.S., Kenawy, E.R., & Abdel, M.A. (2011). Nanospider technology for the production of nylon 6-nanofibers for biomedical applications. *Journal of Nanomaterial,* 1-8.

EPA sulfur distribution. (n.d.). Retrieved January 28, 2013, from http://www.eia.gov/todayinenergy/detail.cfm?id=7110

Falcaro, P., Ricco, R., Doherty, C. M., Liang, K., Hill, A. J., & Styles, M. J. (2014). MOF positioning technology and device fabrication. *Chemical Society Reviews*, *43*(16), 5513–5560. doi:10.1039/C4CS00089G PMID:24802634

Fariba, G., Vasheghani-Farahani, S., & Vasheghani-Farahani, E. (2010). Theoretical description of hydrogel swelling: A review. *Iranian Polymer Journal*, *19*, 375–398.

Ferrando-Soria, J., Ruiz-Garcia, R., Cano, J., Stiriba, S., Vallejo, J., Castro, I., ... Pardo, E. (2012). Reversible Solvatomagnetic Switching in a Spongelike Manganese(II)-Copper(II) 3D Open Framework with a Pillared Square/Octagonal Layer Architecture. *Chemistry (Weinheim an der Bergstrasse, Germany)*, *18*(6), 1608–1617. doi:10.1002/chem.201103308 PMID:22262643

Fong, H., Chun, I., & Reneker, D. H. (1999). Beaded nanofibers formed during electrospinning. *Polymer*, *40*(16), 4585–4592. doi:10.1016/S0032-3861(99)00068-3

Francisco, M., Arce, A., & Soto, A. (2010). Ionic liquids on desulfurization of fuel oils. *Fluid Phase Equilibria*, *294*(1-2), 39–48. doi:10.1016/j.fluid.2009.12.020

François-Xavier, C., & Fuchs, A. H. (2015). Computational characterization and prediction of metal–organic framework properties. *Coordination Chemistry Reviews*, *307*, 211–236.

Geng, X. Y., Kwon, O. H., & Jang, J. H. (2005). Electrospinning of chitosan dissolved in concentrated acetic acid solution. *Biomaterials*, *26*(27), 5427–5432. doi:10.1016/j.biomaterials.2005.01.066 PMID:15860199

Ghosh, P., Hickey, K. J., & Jaffe, S. B. (2006). Development of a Detailed Gasoline Composition-Based Octane Model. *Industrial & Engineering Chemistry Research*, *45*(1), 337–345. doi:10.1021/ie050811h

Gupta, P., Elkins, C., Long, T. E., & Wilkes, G. L. (2005). Electrospinning of linear homopolymers of poly(methyl methacrylate): Exploring relationships between fiber formation, viscosity, molecular weight and concentration in a good solvent. *Polymer*, *46*(13), 4799–4810. doi:10.1016/j.polymer.2005.04.021

Hallensleben, M. L. (2000). *Polyvinyl compounds*, others. In *Ullmann's encyclopedia of industrial chemistry*. Weinheim: Wiley-VCH; doi:10.1002/14356007.a21_743

Haupt, K. (2003). Molecularly imprinted polymers: The next generation. *Analytical Chemistry*, *75*(17), 376A–383A. doi:10.1021/ac031385h PMID:14632031

Hernandez-Maldonado, A. J., Qi, G., & Yang, R. T. (2005). Desulfurization of commercial fuels by p-complexation monolayer CuCl/g-Al_2O_3. *Applied Catalysis B: Environmental*, *61*(3-4), 212–218. doi:10.1016/j.apcatb.2005.05.003

Hernańdez-Maldonado, A. J., Stamatis, S. D., Yang, R. T., He, A. Z., & Cannella, W. (2004). New sorbents for desulfurization of diesel fuels via p-complexation: Layered beds and regeneration. *Industrial & Engineering Chemistry Research*, *43*(3), 769–776. doi:10.1021/ie034108+

Hernańdez-Maldonado, A. J., & Yang, R. T. (2004). Desulfurization of diesel fuels via π-complexation with nickel(II)-exchanged X- and Y-zeolites. *Industrial & Engineering Chemistry Research*, *43*(4), 1081–1089. doi:10.1021/ie034206v

Hernandez-Maldonado, A. J., & Yang, R. T. (2004). Desulfurization of diesel fuels by adsorption *via* π-complexation with vapor-phase exchanged Cu(I)-Y zeolites. *Journal of the American Chemical Society*, *126*(4), 992–993. doi:10.1021/ja039304m PMID:14746446

Heunis, T., Bshena, O., Klumperman, B., & Dicks, L. (2011). Release of bacteriocins from nanofibers prepared with combinations of poly(D,L-lactide) (PDLLA) and poly(ethylene oxide) (PEO). *International Journal of Molecular Sciences*, *12*(4), 2158–2173. doi:10.3390/ijms12042158 PMID:21731433

Heunis, T. D. J., Botes, M., & Dicks, L. M. T. (2010). Encapsulation of *Lactobacillus plantarum* 423 and its bacteriocin in nanofibers. *Probiotics and Antimicrobial Proteins*, *2*(1), 46–51. doi:10.100712602-009-9024-9 PMID:26780900

Heunis, T. D. J., & Dicks, L. M. T. (2010). Nanofibers offer alternative ways to the treatment of skin infections. *Journal of Biomedicine & Biotechnology*, *2010*, 1–10. doi:10.1155/2010/510682 PMID:20798871

Hizaddin, H. F., Hashim, M. A., & Anantharaj, R. (2013). Evaluation of molecular interaction in binary mixture of ionic liquids+ heterocyclic nitrogen compounds: Ab Initio method and COSMORS model. *Industrial & Engineering Chemistry Research*, *52*(50), 18043–18058. doi:10.1021/ie403032t

Holmes, A. B., & Stephenson, G. R. (1994). Molecular imprinting. *Chemistry & Industry*, *18*, 304.

Huang, L., Nagapudi, K., Apkarian, R. P., & Chaikof, E. L. (2001). Engineered collagen-PEO nanofibers and fabrics. *Journal of Biomaterials Science. Polymer Edition*, *12*, 979–993. doi:10.1163/156856201753252516 PMID:11787524

Huang, M., Chang, G., Su, Ye., Xing, H., Su, B., Zhang, Z., ... Chen, B. (2015). A metal-organic framework with immobilized Ag(I) for highly efficient desulfurization of liquid fuels. *Chemical Communications*, *51*(61), 12205–12207. doi:10.1039/C5CC03648H PMID:26136210

Huang, Z. M., Zhang, Y. Z., Kotaki, M., & Ramakrishna, S. (2003). A review on polymer nanofibers by electrospinning and their applications in nanocomposites. *Composites Science and Technology*, *63*(15), 2223–2253. doi:10.1016/S0266-3538(03)00178-7

Hulea, V., Fajula, F., & Bousquet, J. (2001). Mild oxidation with H_2O_2 over Ti-containing molecular sieves-a very efficient method for removing aromatic sulfur compounds from fuels. *Journal of Catalysis*, *198*(2), 179–186. doi:10.1006/jcat.2000.3149

Ibrahima, M. H., Hayyan, M., Hashima, M. A., & Hayyan, A. (2017). Article. *Renewable & Sustainable Energy Reviews*, *76*, 1534–1549. doi:10.1016/j.rser.2016.11.194

Isarankura-Na-Ayudhya, C., Nantasenamat, C., Buraparuangsang, P., Piacham, T., Ye, L., Bülow, L., & Prachayasittikul, V. (2008). Computational insights on sulfonamide imprinted polymers. *Molecules (Basel, Switzerland)*, *13*(12), 3077–3091. doi:10.3390/molecules13123077 PMID:19078850

Ishihara, A., Wang, D. H., Dumeignil, F., Amano, H., Qian, E. W. H., & Kabe, T. (2005). Oxidative desulfurization and denitrogenation of a light gas oil using an oxidation/adsorption continuous flow process. *Applied Catalysis A, General*, *279*(1-2), 279–287. doi:10.1016/j.apcata.2004.10.037

Jarusuwannapoom, T., Hongroijanawiwat, W., Jitjaicham, S., Wannatong, L., Nithitanakul, M., Pattamaprom, C., ... Supaphol, P. (2005). Effect of solvents on electro-spinnability of polystyrene solutions and morphological appearance of resulting electrospun polystyrene fibers. *European Polymer Journal*, *41*(3), 409–421. doi:10.1016/j.eurpolymj.2004.10.010

Jeon, H., Ko, C. H., Kim, S. H., & Kim, J. (2009). Removal of refractory sulphur compounds in diesel using activated carbon with controlled porosity. *Energy & Fuels*, *23*(5), 2537–2543. doi:10.1021/ef801050k

Jiang, H. L., Fang, D. F., Hsiao, B. S., Chu, B., & Chen, W. L. (2004). Optimization and characterization of dextran membranes prepared by electrospinning. *Biomacromolecules*, *5*(2), 326–333. doi:10.1021/bm034345w PMID:15002991

Jiang, Z., Lu, H., Zhang, Y., & Li, C. (2011). Oxidative desulfurization of fuel oils. *Chinese Journal of Catalysis*, *32*(5), 707–715. doi:10.1016/S1872-2067(10)60246-X

Jiménez-Cruz, F., & Laredo, G. C. (2004). Molecular size evaluation of linear and branched paraffins from the gasoline pool by DFT quantum chemical calculations. *Fuel*, *83*(16), 2183–2188. doi:10.1016/j.fuel.2004.06.010

Jin, T., Yang, Q., Meng, C., Xu, J., Liu, H., Hu, J., & Ling, H. (2014). Promoting desulfurization capacity and separation efficiency simultaneously by the novel magnetic Fe_3O_4@PAA@MOF-199. *RSC Advances*, *4*(79), 41902–41909. doi:10.1039/C4RA06515H

Jorge, A., & Le Borgne, S. (2004). Selective adsorption of dibenzothiophene sulfone by an imprinted and stimuli-responsive chitosan hydrogel. *Macromolecules*, *37*(8), 2938–2943. doi:10.1021/ma049961e

Karas, L. J., Grey, R. A., & Lynch, M. W. (2008). *Desulfurization process*. US Patent 2008047875.

Karas, L. J., Han, Y. Z., & Leyshon, D. W. (2004). *Organosulfur oxidation process*. US Patent 2004178122.

Kenta, K. (2017). Network polymers derived from the integration of flexible organic polymers and rigid metal–organic frameworks. *Polymer Journal*, *49*(4), 345–353. doi:10.1038/pj.2016.122

Khan, N. A., Yoon, J. W., Chang, J., & Jhung, S. (2016). Enhanced adsorptive desulfurization with flexible metal–organic frameworks in the presence of diethyl ether and water. *Chemical Communications (Cambridge)*, *52*(56), 8667–8670. doi:10.1039/C6CC03976F PMID:27284597

Ki, C. S., Baek, D. H., Gang, K. D., Lee, K. H., Um, I. C., & Park, Y. H. (2005). Characterization of gelatin nanofiber prepared from gelatin-formic acid solution. *Polymer*, *46*(14), 5094–5102. doi:10.1016/j.polymer.2005.04.040

Kidoaki, S., Kwon, I. K., & Matsuda, T. (2005). Mesoscopic spatial designs of nano- and micron-fiber meshes for tissue-engineering matrix and scaffold based on newly devised multilayering and mixing electrospinning techniques. *Biomaterials*, *26*(1), 37–46. doi:10.1016/j.biomaterials.2004.01.063 PMID:15193879

Kim, B., Park, H., Lee, S. H., & Sigmund, W. M. (2005). Poly(acrylic acid) nanofibers by electrospinning. *Materials Letters*, *59*(7), 829–832. doi:10.1016/j.matlet.2004.11.032

Kim, H. S., Kim, K., Jin, H. J., & Chin, I. J. (2005). Morphological characterization of electrospun nano-fibrous membranes of biodegradable poly(L-lactide) and poly(lactideco-glycolide). *Macromolecular Symposia*, *224*(1), 145–154. doi:10.1002/masy.200550613

Kim, J. H., Ma, X., Zhou, A., & Song, C. (2006). Ultra-deep desulfurization and denitrogenation of diesel fuel by selective adsorption over three different adsorbents: A study on adsorptive selectivity and mechanism. *Catalysis Today*, *111*(1-2), 74–83. doi:10.1016/j.cattod.2005.10.017

King, D. L., & Li, L. (2006). Removal of sulfur components from low sulfur gasoline using copper exchanged zeolite Y at ambient temperature. *Catalysis Today*, *116*(4), 526–529. doi:10.1016/j.cattod.2006.06.026

Klobes, P., Meyer, K., & Munro, R. G. (2006). Porosity and specific surface area measurements for solid materials. *National Institute of Standards and Technology Special Publication*, *89*, 960–917.

Kocal, J. A., & Branvold, T. A. (2002). *Removal of sulfur-containing compounds from liquid hydrocarbon streams*. US Patent 6368495.

Kodama, K., Umehara, K., Shimizu, K., Nakatani, S., Minoda, Y., & Yamada, K. (1973). Identification of microbial products from dibenzothiophene and its proposed oxidation pathway. *Agricultural and Biological Chemistry*, *37*(1), 45–50. doi:10.1080/00021369.1973.10860640

Kohler, F., Roth, D., Kuhlmann, E., Wasserscheid, P., & Haumann, M. (2010). Continuous gas-phase desulfurisation using supported ionic liquid phase (SILP) materials. *Green Chemistry*, *12*(6), 979–984. doi:10.1039/c004883f

Koski, A., Yim, K., & Shivkumar, S. (2004). Effect of molecular weight on fibrous PVA produced by electrospinning. *Materials Letters*, *58*(3-4), 493–497. doi:10.1016/S0167-577X(03)00532-9

Kuhlmann, E., Haumann, M., Jess, A., Seeberger, A., & Wasserscheid, P. (2009). Ionic Liquids in Refinery Desulfurization: Comparison between Biphasic and Supported Ionic Liquid Phase Suspension Processes. *ChemSusChem*, *2*(10), 969–977. doi:10.1002/cssc.200900142 PMID:19798713

Kumar, A. (2010). *Nanofibers*. Vukovar: Intech. doi:10.5772/45660

Kuznetsova, L., Detusheva, L., Kuznetsov, N., Duplyakin, V., & Likholobov, V. (2008). Liquid-phase oxidation of benzothiophene and dibenzothiophene by cumylhydroperoxide in the presence of catalysts based on supported metal oxides. *Kinetics and Catalysis*, *49*(5), 644–652. doi:10.1134/S002315840805008X

Lee, J. S., Choi, K. H., Do Ghim, H., Kim, S. S., Chun, D. H., Kim, H. Y., & Lyoo, W. S. (2004). Role of molecular weight of a tactic poly(vinyl alcohol) (PVA) in the structure and properties of PVA nanofabric prepared by electrospinning. *Journal of Applied Polymer Science*, *94*(4), 1638–1646. doi:10.1002/app.20602

Lee, Y., Kim, J., & Ahn, W. (2013). Synthesis of metal-organic frameworks: A mini review. *Korean Journal of Chemical Engineering*, *30*(9), 1667–1680. doi:10.100711814-013-0140-6

Li, W., Tang, H., Liu, Q., Xing, J., Li, Q., Wang, D., ... Liu, H. (2009). Deep desulfurization of diesel by integrating adsorption and microbial method. *Biochemical Engineering Journal*, *44*(2-3), 297–301. doi:10.1016/j.bej.2008.12.016

Lin, T., Wang, H. X., Wang, H. M., & Wang, X. G. (2004). The charge effect of cationic surfactants on the elimination of fibre microspheres in the electrospinning of polystyrene. *Nanotechnology*, *15*(9), 1375–1381. doi:10.1088/0957-4484/15/9/044

Liu, B. S., Xu, D. F., Chu, J. X., Liu, W., & Au, C. T. (2007). Deep desulfurization by the adsorption process of fluidized catalytic cracking (FCC) diesel over mesoporous Al-MCM- 41 materials. *Energy & Fuels*, *21*(1), 250–255. doi:10.1021/ef060249n

Liu, J., & Kumar, S. (2005). Microscopic polymer cups by electrospinning. *Polymer*, *46*(10), 3211–3214. doi:10.1016/j.polymer.2004.11.116

Liu, W., Wang, J., Li, Q., Sai, J., Zhang, T., & Ji, S. (2014). Synthesis of rare earth metal-organic frameworks (Ln-MOFs) and their properties of adsorption desulfurization. *Journal of Rare Earths*, *32*(2), 189–194. doi:10.1016/S1002-0721(14)60050-8

Lu, W., Wei, Z., Gu, Z., Liu, T., Park, J., Park, J., ... Zhou, H. (2014). Tuning the structure and function of metal–organic frameworks *via* linker design. *Chemical Society Reviews*, *43*(16), 5561–5593. doi:10.1039/C4CS00003J PMID:24604071

Lu, L., Cheng, S.F., Gao, J.B., Gao, G.H., & He, M.Y. (2007). Title. *Energy and Fuels, 21*, 383-384.

Ma, X., Lu, S., Yin, Z., & Song, C. (2001). New approaches to deep desulfurization of diesel fuel, jet fuel, and gasoline by adsorption for ultra-clean fuels and for fuel cell applications. *American Chemical Society. Division of Fuel Chemistry*, *46*(2), 648–649.

Ma, X., Sakanishi, K., Isoda, T., & Mochida, I. (1995). Quantum chemical calculation on the desulfurization reactivities of heterocyclic sulfur compounds. *Energy & Fuels*, *9*(1), 33–37. doi:10.1021/ef00049a005

Ma, X., Sakanishi, K., & Mochida, I. (1994). Hydrodesulfurization reactivities of various sulfur compounds in diesel fuel. *Industrial & Engineering Chemistry Research*, *33*(2), 218–222. doi:10.1021/ie00026a007

Ma, X., Sakanishi, K., & Mochida, I. (1996). Hydrodesulfurization reactivities of various sulfur compounds in vacuum gas oil. *Industrial & Engineering Chemistry Research*, *35*(8), 2487–2494. doi:10.1021/ie960137r

Ma, X. L., Sprague, M., Sun, L., & Song, C. S. (2002). Deep desulfurization of gasoline by SARS process using adsorbent for fuel cells. Prep. Am. Chem. Soc. DiV. *Fuel Chem.*, *47*, 452–453.

Maes, M., Trekels, M., Boulhout, M., Schouteden, S., Vermoortele, F., Alaerts, L., ... De Vos, D. E. (2011). Selective Removal of N-Heterocyclic Aromatic Contaminants from Fuels by Lewis Acidic Metal-Organic Frameworks. *Angewandte Chemie International Edition*, *50*(18), 4210–4214. doi:10.1002/anie.201100050 PMID:21472941

Mansouri, A., Khodadadi, A. A., & Mortazavi, Y. (2014). Ultra-deep adsorptive desulfurization of a model diesel fuel on regenerable Ni–Cu/γ-Al2O3 at low temperatures in absence of hydrogen. *Journal of Hazardous Materials*, *271*, 120–130. doi:10.1016/j.jhazmat.2014.02.006 PMID:24632365

Meyer, K., Klobes, P., & Röhl-Kuhn, B. (1997). Certification of reference material with special emphasis on porous solids. *Crystal Research and Technology*, *32*(1), 173–183. doi:10.1002/crat.2170320118

Mit-Uppatham, C., Nithitanakul, M., & Supaphol, P. (2004). Ultrafine electrospun polyamide-6 fibers: Effect of solution conditions on morphology and average fiber diameter. *Macromolecular Chemistry and Physics*, *205*(17), 2327–2338. doi:10.1002/macp.200400225

Miyazaki, T., Takeda, Y., Akane, S., Itou, T., Hoshiko, A., & En, K. (2010). Role of boric acid for a poly (vinyl alcohol) film as a cross-linking agent: Melting behaviors of the films with boric acid. *Polymer*, *51*(23), 5539–5549. doi:10.1016/j.polymer.2010.09.048

Mokhtar, W. N. A. W., Bakar, W. A. W. A., Ali, R., & Abdul-Kadir, A. A. (2014). Deep desulfurization of model diesel by extraction with *N,N*-dimethylformamide: Optimization by Box–Behnken design. *Journal of the Taiwan Institute of Chemical Engineers*, *45*(4), 1542–1548. doi:10.1016/j.jtice.2014.03.017

Molecular sieves-Technical Information Bulletin: Sigma-Aldrich. (n.d.). Retrieved from www.sigmaaldrich.com/chemistry/chemical-synthesis/learning-center/technical-bulletins/al 1430/molecular-sieves.html

Monteiro, O. A. C. Jr, & Airoldi, C. (1999). Some studies of crosslinking chitosan–glutaraldehyde interaction in a homogeneous system. *International Journal of Biological Macromolecules*, *26*(2-3), 119–128. doi:10.1016/S0141-8130(99)00068-9 PMID:10517518

Mosbach, K. (1994). Molecular imprinting. *Trends in Biochemical Sciences*, *19*(1), 9–14. doi:10.1016/0968-0004(94)90166-X PMID:8140624

Nagarkar, S. S., Das, R., Poddar, P., & Ghosh, S. K. (2008). Bistable dynamic coordination polymer showing reversible structural and functional transformations. *Inorganic Chemistry*, *51*(15), 8317–8321. doi:10.1021/ic301481p PMID:22827720

Nantasenamat, C., Isarankura-Na-Ayudhya, C., Naenna, T., & Prachayasittikul, V. (2007). Quantitative structure-imprinting factor relationship of molecularly imprinted polymers. *Biosensors & Bioelectronics*, *22*(12), 3309–3317. doi:10.1016/j.bios.2007.01.017 PMID:17317143

Nantasenamat, C., Naenna, T., Isarankura Na Ayudhya, C., & Prachayasittikul, V. (2005). Quantitative prediction of imprinting factor of molecularly imprinted polymers by artificial neural network. *Journal of Computer-Aided Molecular Design*, *19*(7), 509–524. doi:10.100710822-005-9004-4 PMID:16244792

Nantasenamat, C., Tantimongcolwat, T., Naenna, T., Isarankura-Na-Ayudhya, C., & Prachayasittikul, V. (2006). Prediction of selectivity index of pentachlorophenol-imprinted polymers. *EXCLI Journal*, *5*, 150–163.

Ngah, W. S. W., & Liang, K. H. (1999). Adsorption of gold(III) ions onto chitosan and *N*-carboxymethyl chitosan: Equilibrium studies. *Industrial & Engineering Chemistry Research*, *38*(4), 1411–1414. doi:10.1021/ie9803164

Nicholls, I. A., Adbo, K., Andersson, H. S., Andersson, P. O., Ankarloo, J., Hedin-Dahlström, J., ... Wikman, S. (2001). Can we rationally design molecularly imprinted polymers? *Analytica Chimica Acta*, *435*(1), 9–18. doi:10.1016/S0003-2670(01)00932-1

Nolte, T., Posch, T. N., Huhn, C., & Andersson, J. T. (2013). Desulfurized fuels from Athabasca bitumen and their polycyclic aromatic sulfur heterocycles. Analysis based on capillary electrophoresis coupled with TOF MS. *Energy & Fuels*, *27*(1), 97–107. doi:10.1021/ef301424d

Ogunlaja, A. S., Coombes, M. J., Torto, N., & Tshentu, Z. R. (2014). The adsorptive extraction of oxidised sulfur-containing compounds from fuels by using molecularly imprinted chitosan materials. *Reactive & Functional Polymers*, *81*, 61–76. doi:10.1016/j.reactfunctpolym.2014.04.006

Ogunlaja, A. S., du Sautoy, C. W., Torto, N., & Tshentu, Z. R. (2014). Design, fabrication and evaluation of intelligent sulfone-selective polybenzimidazole nanofibers. *Talanta*, *126*, 61–72. doi:10.1016/j.talanta.2014.03.035 PMID:24881535

Ogunlaja, A. S., Khene, S., Antunes, E., Nyokong, T., Torto, N., & Tshentu, Z. R. (2013). The development of catalytic oxovanadium(IV)-containing microspheres for the oxidation of various organosulfur compounds. *Applied Catalysis A, General*, *462-463*, 157–167. doi:10.1016/j.apcata.2013.05.004

Ogunlaja, A. S., Walmsley, R. S., du Sautoy, C. W., Torto, N., & Tshentu, Z. R. (2013). Development of a continuous flow system for the oxidative desulfurization of refractory organosulfur compounds in hydrotreated diesel. *Energy & Fuels*, *27*(12), 7714–7723. doi:10.1021/ef401708f

Otsuki, S., Nonaka, T., Qian, W., Ishihara, A., & Kabe, T. (1999). Oxidative desulfurization of middle distillate using ozone. *Journal of the Japan Petroleum Institute*, *42*(5), 315–320. doi:10.1627/jpi1958.42.315

Otsuki, S., Nonaka, T., Takashima, N., Qian, W., Ishihara, A., Imai, T., & Kabe, T. (2000). Oxidative desulfurization of light gas oil and vacuum gas oil by oxidation and solvent extraction. *Energy & Fuels*, *14*(6), 1232–1239. doi:10.1021/ef000096i

Palomeque, J., Clacens, J. M., & Figueras, F. (2002). Oxidation of dibenzothiophene by hydrogen peroxide catalysed by solid bases. *Journal of Catalysis*, *211*(1), 103–108. doi:10.1016/S0021-9517(02)93721-8

Parr, R. G., & Yang, W. (1989). *Density functional theory of atoms and molecules*. Oxford, UK: Oxford University Press.

Patil, S. V., Sorokhaibam, L. G., Bhandari, V. M., Killedar, D. J., & Ranade, V. V. (2014). Investigating role of sulphur specific carbon adsorbents in deep desulphurization. *Journal of Environmental Chemical Engineering*, *2*(3), 1495–1505. doi:10.1016/j.jece.2014.07.009

Pawelec, B., Navarro, R.M., Campos-Martin, J.M., & Fierro, J.L.G. (2011). *Towards near zero-sulfur liquid fuels: a perspective review*. Academic Press.

Paybarah, A., Bone, R. L., & Corcoran, W. H. (1982). Selective oxidation of dibenzothiophene by peroxybenzoic acid formed in situ. *Industrial & Engineering Chemistry Product Research and Development*, *21*, 426–431.

Pearson, R. G. (1963). Hard and Soft Acids and Bases. *Journal of the American Chemical Society*, *85*(22), 3533–3539. doi:10.1021/ja00905a001

Pearson, R. G. (1986). Absolute electronegativity and hardness correlated with molecular orbital theory. *Proceedings of the National Academy of Sciences of the United States of America*, *83*(22), 8440–8441. doi:10.1073/pnas.83.22.8440 PMID:16578791

Peng, Y., Liu, J., Zhang, H., Luo, D., & Li, D. (2018). A size-matched POM@ MOF composite catalyst for highly efficient and recyclable ultra-deep oxidative fuel desulfurization. *Inorganic Chemistry Frontiers*, *5*(7), 1563–1569. doi:10.1039/C8QI00295A

Peralta, D., Chaplais, G., Simon-Masseron, A., Barthelet, K., & Pirngruber, G. D. (2012). Separation of C6 Paraffins Using Zeolitic Imidazolate Frameworks: Comparison with Zeolite 5A. *Energy & Fuels*, *26*, 4953–4960. doi:10.1021/ef300762z

Persano, L., Camposeo, A., Tekmen, C., & Pisignano, D. (2013). Industrial upscaling of electrospinning and applications of polymer nanofibers: A review. *Macromolecular Materials and Engineering*, *298*(5), 504–520. doi:10.1002/mame.201200290

Pham, Q. P., Sharma, U., & Mikos, A. G. (2006). Electrospinning of polymeric nanofibers for tissue engineering applications: A review. *Tissue Engineering*, *12*(5), 1197–1211. doi:10.1089/ten.2006.12.1197 PMID:16771634

Pilloni, M., Padella, F., Ennas, G., Lai, S., Bellusci, M., Rombi, E., ... Ferino, I. (2015). Liquid-assisted mechanochemical synthesis of an iron carboxylate metal organic framework and its evaluation in diesel fuel desulfurization. *Microporous and Mesoporous Materials*, *213*, 14–21. doi:10.1016/j.micromeso.2015.04.005

Prasad, V. V. D. N., Jeong, K. E., Chae, H. J., Kim, C. U., & Jeong, S. Y. (2008). Oxidative desulfurization of 4,6-dimethyl dibenzothiophene and light cycle oil over supported molybdenum oxide catalysts. *Catalysis Communications*, *9*(10), 1966–1969. doi:10.1016/j.catcom.2008.03.021

Reneker, D. H., & Chun, I. (1996). Nanometre diameter fibres of polymer, produced by electrospinning. *Nanotechnology*, *7*(3), 216–223. doi:10.1088/0957-4484/7/3/009

Reneker, D. H., & Yarin, A. L. (2008). Electrospinning jets and polymer nanofibers. *Polymer*, *49*(10), 2837–2425. doi:10.1016/j.polymer.2008.02.002

Ritter, H. L., & Drake, L. C. (1945). Pore size distribution in porous materials. *Industrial & Engineering Chemistry Fundamentals*, *17*, 782–787.

Ruiz, M., Sastre, A. M., & Guibal, E. (2000). Comparative adsorption of Cu(II), Zn(II), and Pb(II) ions in aqueous solution on the crosslinked chitosan with epichlorohydrin. *Reactive & Functional Polymers*, *45*, 155–173. doi:10.1016/S1381-5148(00)00019-5

Ryu, K., Heo, J., & Yoo, I. K. (2002). Removal of dibenzothiophene and its oxidized product in anhydrous water-immiscible organic solvent by immobilized cytochrome c. *Biotechnology Letters*, *24*(2), 143–146. doi:10.1023/A:1013806830105

Sami, M., & Hatch, L. F. (1994). Chemistry of petrochemical processes, Gulf publishing company, Houston, Texas, p11-17. Monticello, D.J. (2000). Biodesulfurization and the upgrading of petroleum distillates. *Current Opinion in Biotechnology*, *11*, 540–546.

Sarda, K. K., Bhandari, A., Pant, K. K., & Jain, S. (2012). Deep desulfurization of diesel fuel by selective adsorption over Ni/Al_2O_3 and Ni/ZSM-5 extrudates. *Fuel*, *93*, 86–91. doi:10.1016/j.fuel.2011.10.020

Schneemann, A., Bon, V., Schwedler, I., Senkovska, I., Kaskel, S., & Fischer, R. A. (2014). Flexible metal–organic frameworks. *Chemical Society Reviews*, *43*(16), 6062–6096. doi:10.1039/C4CS00101J PMID:24875583

Schulz, H., Böhringer, W., Ousmanov, F., & Waller, P. (1999). Refractory sulfur compounds in gas oils. *Fuel Processing Technology*, *61*(1-2), 5–41. doi:10.1016/S0378-3820(99)00028-4

Schulz, H., Böhringer, W., Waller, P., & Ousmanov, F. (1999). Gas oil deep hydrodesulfurization: Refractory compounds and retarded kinetics. *Catalysis Today*, *49*(1-3), 87–97. doi:10.1016/S0920-5861(98)00412-X

Sellergren, B. (2001). *In molecularly Imiprinted polymers: man-made mimics of antibodies and their applications in analytical chemistry.* In B. Sellergren (Ed.), *Techniques and instrumentation in analytical chemistry* (Vol. 23, pp. 113–184). Amsterdam: Elsevier.

Shan, J. H., Liu, X. Q., Sun, L. B., & Cui, R. (2008). Cu-Ce Bimetal ion-exchanged Y-Zeolites for selective adsorption of thiophenic sulfur. *Energy & Fuels, 22*(6), 3955–3959. doi:10.1021/ef800296n

Si, X., Cheng, S., Lu, Y., Gao, G., & He, M. Y. (2008). Oxidative desulfurization of model oil over Au/Ti-MWW. *Catalysis Letters, 122*(3-4), 321–324. doi:10.100710562-007-9380-6

Sing, K. S. W., Everett, D. H., Haul, R. A., Moscou, L., Pierotti, R. A., Rouquerol, J., & Siemieniewska, T. (1985). Reporting physisorption data for gas/solid systems with special reference to the determination of surface area and porosity. *Pure and Applied Chemistry, 57*(4), 603–619. doi:10.1351/pac198557040603

Soleimani, M., Bassi, A., & Margaritis, A. (2007). Biodesulfurization of refractory organic sulfur compounds in fossil fuels. *Biotechnology Advances, 25*(6), 570–596. doi:10.1016/j.biotechadv.2007.07.003 PMID:17716849

Son, W. K., Youk, J. H., Lee, T. S., & Park, W. H. (2004). The effects of solution properties and polyelectrolyte on electrospinning of ultrafine poly(ethylene oxide) fibers. *Polymer, 45*(9), 2959–2966. doi:10.1016/j.polymer.2004.03.006

Song, C. (1999). Designing sulfur-resistant, noble-metal hydrotreating catalysts. *Chemtech, 29*(3), 26–30.

Song, C. (1999). Sulfur-resistant noble metal catalysts based on shape-selective exclusion and hydrogen spillover. In *Shape-selective catalysis, chemicals synthesis and hydrocarbon processing* (ACS Symposium Series 738). American Chemical Society.

Song, C., & Ma, X. L. (2003). New design approaches to ultra-clean diesel fuels by deep desulfurization and deep dearomatization. *Applied Catalysis B: Environmental, 41*(1-2), 207–238. doi:10.1016/S0926-3373(02)00212-6

Song, C., & Reddy, K. M. (1999). Mesoporous molecular sieve MCM-41 supported Co-Mo catalyst for hydrodesulfurization of dibenzothiophene in distillate fuels. *Applied Catalysis A, General, 176*(1), 1–10. doi:10.1016/S0926-860X(98)00230-0

Song, C., Turaga, U. T., & Ma, X. (2006). Desulfurization. Encyclopedia of Chemical Processing, 651-661. Doi:10.1081/E-ECHP-120007732

Song, C. S. (2003). An overview of new approaches to deep desulfurization for ultra-clean gasoline, diesel fuel and jet fuel. *Catalysis Today*, *86*(1-4), 211–263. doi:10.1016/S0920-5861(03)00412-7

Song, C. S., & Ma, X. L. (2004). Ultra-deep desulfurization of liquid hydrocarbon fuels: Chemistry and process. *International Journal of Green Energy*, *1*(2), 167–191. doi:10.1081/GE-120038751

Stellenbosch nanofiber company (SNC). (2015). Retrieved from http://www.sncfibers.com/#!products/c1jm9

Sughrue, E. L., Khare, G. P., Bertus, B. J., & Johnson, M. M. (2001). *US Patent 6.254.766*. US Patent Office.

Sulfex™ – Desulfurization simplified. (2019). Retrieved from https://www.altpetrol.com/sulfex/

Tam, P. S., & Kittrell, J. R. (1984). *Process for purifying hydrocarbonaceous oils*. U.S. Patent 4, 485, 007.

Tam, P. S., Kittrell, J. R., & Eldridge, J. W. (1990). Desulfurization of fuel oil by oxidation and extraction. 1. Enhancement of extraction oil yield. *Industrial & Engineering Chemistry Research*, *29*(3), 321–324. doi:10.1021/ie00099a002

Tam, P. S., Kittrell, J. R., & Eldridge, J. W. (1990). Desulfurization of fuel oil by oxidation and extraction. 2. Kinetic modeling of oxidation reaction. *Industrial & Engineering Chemistry Research*, *29*(3), 324–329. doi:10.1021/ie00099a003

Tawfik, A. S., Gaddafi, I. D., & Taye, D. S. (2015). Nanocomposites and Hybrid Materials for Adsorptive Desulfurization. Advances in Chemical and Materials Engineering, 129-153.

Tella, A. C., Owalude, S. O., Olatunji, S. J., Oloyede, S. O., Ogunlaja, A. S., & Bourne, S. A. (2018). Synthesis, crystal structure and desulfurization properties of zig-zag 1D coordination polymer of copper(II) containing 4-methoxybenzoic acid ligand. *Journal of Sulfur Chemistry*, *39*(6), 588–606. doi:10.1080/17415993.2018.1489808

Teo, W. E., & Ramakrishna, S. (2005). Electrospun fibre bundle made of aligned nanofibres over two fixed points. *Nanotechnology*, *16*(9), 1878–1884. doi:10.1088/0957-4484/16/9/077

Teo, W. E., & Ramakrishna, S. (2006). A review on electrospinning design and nanofibre assemblies. *Nanotechnology*, *17*(14), 89–106. doi:10.1088/0957-4484/17/14/R01 PMID:19661572

Theron, S. A., Yarin, A. L., Zussman, E., & Kroll, E. (2005). Multiple jets in electrospinning: Experiment and modelling. *Polymer*, *46*(9), 2889–2899. doi:10.1016/j.polymer.2005.01.054

Topsoe, H., Clausen, B. S., & Massoth, F. E. (1996). Hydrotreating catalysis. In *Science and Technology* (p. 310). Berlin: Springer-Verlag.

Torres-García, E., Galano, A., & Rodriguez-Gattorno, G. (2011). Oxidative desulfurization (ODS) of organosulfur compounds catalysed by peroxo-metallate complexes of WOx–ZrO_2: Thermochemical, structural, and reactivity indexes analyses. *Journal of Catalysis*, *282*(1), 201–208. doi:10.1016/j.jcat.2011.06.010

Van de Voorde, B., Bueken, B., Denayer, J., & De Vos, D. (2014). Adsorptive separation on metal-organic frameworks in the liquid phase. *Chemical Society Reviews*, *43*(16), 5766–5788. doi:10.1039/C4CS00006D PMID:24647892

Varesano, A., Carletto, R. A., & Mazzuchetti, G. (2009). Experimental investigations on the electrospinning process. *Journal of Materials Processing Technology*, *209*(11), 5178–5185. doi:10.1016/j.jmatprotec.2009.03.003

Velu, S., Ma, X., & Song, C. (2003). Selective adsorption for removing sulfur from jet fuel over zeolite-based adsorbents. *Industrial & Engineering Chemistry Research*, *42*(21), 5293–5304. doi:10.1021/ie020995p

Venturiello, C., Alneri, E., & Ricci, M. (1983). A new effective catalytic system for epoxidation of olefins by hydrogen peroxide under phase-transfer conditions. *The Journal of Organic Chemistry*, *48*(21), 3831–3833. doi:10.1021/jo00169a052

Walmsley, R. S., Hlangothi, P., Litwinski, C., Torto, N., Nyokong, T., & Tshentu, Z. R. (2012). Catalytic oxidation of thioanisole using oxovanadium(IV)-functionalized electrospun polybenzimidazole nanofibers. *Journal of Applied Polymer Science*, *127*(6), 4719–4725. doi:10.1002/app.38067

Walmsley, R. S., Ogunlaja, A. S., Coombes, M. J., Chidawanyika, W., Litwinski, C., Torto, N., ... Tshentu, Z. R. (2012). Imidazole-functionalized polymer microspheres and fibers – useful materials for immobilization of oxovanadium(IV) catalysts. *Journal of Materials Chemistry*, *22*(12), 5792–5800. doi:10.1039/c2jm15485d

Wang, D. H., Qian, E. W. H., Amano, H., Okata, K., Ishihara, A., & Kabe, T. (2003). Oxidative desulfurization of fuel oil - Part I. Oxidation of dibenzothiophenes using *tert*-butylhydroperoxide. *Applied Catalysis A, General*, *253*(1), 91–99. doi:10.1016/S0926-860X(03)00528-3

Wang, Y., & Yang, R. T. (2007). Desulfurization of liquid fuels by adsorption on carbon-based sorbents and ultrasound-assisted sorbent regeneration. *Langmuir*, *23*(7), 3825–3831. doi:10.1021/la063364z PMID:17315903

Wannatong, L., Sirivat, A., & Supaphol, P. (2004). Effects of solvents on electrospun polymeric fibers: Preliminary study on polystyrene. *Polymer International*, *53*(11), 1851–1859. doi:10.1002/pi.1599

Washburn, E. W. (1921). The dynamics of capillary flow. *Physical Review*, *17*(3), 273–283. doi:10.1103/PhysRev.17.273

Whitehurst, D. D., Isoda, T., & Mochida, I. (1998). Present state of the art and future challenges in the hydrodesulfurization of polyaromatic sulfur compounds. *Advances in Catalysis*, *42*, 345–471. doi:10.1016/S0360-0564(08)60631-8

Wulff, G. (1993). The role of binding-site interactions in the molecular imprinting of polymers. *Trends in Biotechnology*, *11*(3), 85–87. doi:10.1016/0167-7799(93)90056-F PMID:7763512

Xiaoliang, M., Lu, S., & Song, C. (2002). A new approach to deep desulfurization of gasoline, diesel fuel and jet fuel by selective adsorption for ultra-clean fuels and for fuel cell applications. *Catalysis Today*, *77*(1-2), 107–116. doi:10.1016/S0920-5861(02)00237-7

Xin, C., & Wang, S. (2018). Desulfurization by MOFs as Sorbents for Thiophene Sulfides. *Earth and Environmental Science*, *108*, 042035.

Yang, R. T., Hernandez-Maldonado, A. J., & Yang, F. H. (2003). Desulfurization of transportation fuels with zeolites under ambient conditions. *Science*, *301*(5629), 79–81. doi:10.1126cience.1085088 PMID:12843388

Yang, X., Erickson, L. E., Hohn, K. L., Jeevanandam, P., & Klabunde, K. K. (2006). Sol-gel Cu-Al_2O_3 adsorbents for selective adsorption of thiophene out of hydrocarbon. *Industrial & Engineering Chemistry Research*, *45*(18), 6169–6174. doi:10.1021/ie060559t

Yazu, K., Furuya, T., Miki, K., & Ukegawa, K. (2003). Tungstophosphoric acid-catalysed oxidative desulfurization of light oil with hydrogen peroxide in a light oil/acetic acid biphasic system. *Chemistry Letters*, *32*(10), 920–921. doi:10.1246/cl.2003.920

Yazu, K., Yamamoto, Y., Furuya, T., Miki, K., & Ukegawa, K. (2001). Oxidation of dibenzothiophenes in an organic biphasic system and its application to oxidative desulfurization of light oil. *Energy & Fuels*, *15*(6), 1535–1536. doi:10.1021/ef0101412

Yuan, X. Y., Zhang, Y. Y., Dong, C. H., & Sheng, J. (2004). Morphology of ultrafine polysulfone fibers prepared by electrospinning. *Polymer International*, *53*(11), 1704–1710. doi:10.1002/pi.1538

Zahedi, P., Rezaeian, I., Ranaei-Siadat, S. O., Jafari, S. H., & Supaphol, P. (2010). A review on wound dressings with an emphasis on electrospun nanofibrous polymeric bandages. *Polymers for Advanced Technologies*, *21*, 77–95.

Zaki, T., Riad, M., Saad, L., & Mikhail, S. (2005). Selected oxide materials for sulfur removal. *Chemical Engineering Journal*, *113*(1), 41–46. doi:10.1016/j.cej.2005.08.004

Zannikos, F., Lois, E., & Stournas, S. (1995). Desulfurization of petroleum fractions by oxidation and solvent extraction. *Fuel Processing Technology*, *42*(1), 35–45. doi:10.1016/0378-3820(94)00104-2

Zhang, C. X., Yuan, X. Y., Wu, L. L., Han, Y., & Sheng, J. (2005). Study on morphology of electrospun poly(vinyl alcohol) mats. *European Polymer Journal*, *41*(3), 423–432. doi:10.1016/j.eurpolymj.2004.10.027

Zhang, Y., Yang, Y., Han, H., Yang, M., Wang, L., Zhang, Y., Jiang, Z., & Li, C. (2012). Ultra-deep desulfurization via reactive adsorption on Ni/ZnO: The effect of ZnO particle size on the adsorption performance. *Applied Catalysis B: Environmental*, *119– 120*, 13– 19.

Zhang, Z. Y., Shi, T. B., Jia, C. Z., Ji, W. J., Chen, Y., & He, M. Y. (2008). Adsorptive removal of aromatic organosulfur compounds over the modified Na-Y zeolites. *Applied Catalysis B: Environmental*, *82*(1-2), 1–10. doi:10.1016/j.apcatb.2008.01.006

Zhou, X. R., Zhao, C. X., Yang, J. Z., & Zhang, S. F. (2007). Catalytic oxidation of dibenzothiophene using cyclohexanone peroxide. *Energy & Fuels*, *21*(1), 7–10. doi:10.1021/ef060441p

Zhou, Y., Yang, D., Chen, X., Xu, Q., Lu, F., & Nie, J. (2008). Electrospun water-soluble carboxyethyl chitosan/poly(vinyl alcohol) nanofibrous membrane as potential wound dressing for skin regeneration. *Biomacromolecules*, *9*(1), 349–354. doi:10.1021/bm7009015 PMID:18067266

Zhu, L., Jia, X., Bian, H., Huo, T., Duan, Z., Xiang, Y., & Xia, D. (2018). Structure and adsorptive desulfurization performance of the composite material MOF-5@AC. *New Journal of Chemistry*, *42*(5), 3840–3850. doi:10.1039/C7NJ04192F

Zong, X. H., Kim, K., Fang, D. F., Ran, S. F., Hsiao, B. S., & Chu, B. (2002). Structure and process relationship of electrospun bioabsorbable nanofiber membranes. *Polymer*, *43*(16), 4403–4412. doi:10.1016/S0032-3861(02)00275-6

Zuo, W. W., Zhu, M. F., Yang, W., Yu, H., Chen, Y. M., & Zhang, Y. (2005). Experimental study on relationship between jet instability and formation of beaded fibers during electrospinning. *Polymer Engineering and Science*, *45*(5), 704–709. doi:10.1002/pen.20304

Chapter 7
Advances in the Reduction of the Costs Inherent to Fossil Fuel Biodesulfurization Towards Its Potential Industrial Applications

Susana M. Paixão
https://orcid.org/0000-0003-0955-4467
Bioenergy Unit, National Laboratory of Energy and Geology (LNEG), Lisbon, Portugal

Tiago P. Silva
Bioenergy Unit, National Laboratory of Energy and Geology (LNEG), Lisbon, Portugal

Bruno F. Arez
Bioenergy Unit, National Laboratory of Energy and Geology (LNEG), Lisbon, Portugal

Luís Alves
Bioenergy Unit, National Laboratory of Energy and Geology (LNEG), Lisbon, Portugal

ABSTRACT

The biodesulfurization (BDS) process consists of the use of microorganisms for the removal of sulfur from fossil fuels. Through BDS it is possible to treat most of the organosulfur compounds recalcitrant to the conventional hydrodesulfurization (HDS), the petroleum industry's solution, at mild operating conditions, without the need for molecular hydrogen or metal catalysts. This technique results in lower emissions, smaller residue production, and less energy consumption, which makes BDS an eco-friendly process that can complement HDS making it more efficient. BDS has been extensively studied and much is already known about the process.

DOI: 10.4018/978-1-7998-2146-5.ch007

Clearly, BDS presents advantages as a complementary technique to HDS; however, its commercial use has been delayed by several limitations both upstream and downstream the process. This study will comprehensively review and discuss key issues, like reduction of the BDS costs, advances, and/or challenges for a competitive BDS towards its potential industrial application aiming ultra-low sulfur fuels.

INTRODUCTION

The combustion of fossil fuel generates emissions of sulfur as sulfur dioxide (SO_2), which is corrosive and toxic, and as fine particulate matter of metal sulfates. These emissions are responsible for damage in many different areas. Gaseous chemical compounds of sulfur constitute a major health hazard when present in the air: the large-ring thiophenes, such as dibenzothiophene, abundant in crude oil, are toxic to mammals (Murphy, Amin, Coletta, & Hoffman, 1992); SO_2 gas at high levels can cause bronchial irritation and trigger asthma attacks in susceptible individuals and long-term exposure to combustion-related one particulate air pollution is an important risk factor for cardio-pulmonary and lung cancer mortality (Pope et al., 2002; Mohebali & Ball, 2008). In addition, incomplete burning of liquid fossil fuels causes emissions of aromatic sulfur compounds to the air (Ho & Li, 2002), and the oxidation of sulfur compounds in the atmosphere eventually leads to an aerosol of sulphuric acid. This aerosol causes acid rains, which are responsible for the corrosion of many infrastructures and monuments, and even affect several living organisms including agricultural crops, thus causing direct damage to the economy (Bender & Weigel, 2011). The aerosol is also harmful to the stratospheric ozone contributing to the hole on the Earth's protective ozone layer (Denis, 2010). Lastly, sulfur compounds even prevent the functioning of all major pollution control technologies such as automobile catalytic converters (Maricq, Chase, Xu, & Laing, 2002), making it more difficult to fight against pollution.

Since gasoline, diesel and non-transportation fuels account for 75 to 80% of the total refinery products (Babich & Moulijn, 2003), it is only natural that countries find the reductions of sulfur concentration in fuels as the most effective way to decrease the amount of SO_2 emitted in to the air and limit its prejudicial effects (Mohebali, Ball, Kaytash, & Rasekh, 2008).

Therefore, in response to the increasing concerns with environmental and health effects of the SO_x molecules, several countries have started to impose strict limits on the levels of sulfur present in fossil fuels. This forced the petroleum industry to develop techniques which remove the sulfur from the fuels, such as hydrodesulfurization (HDS), a process that combines high temperatures and pressures with molecular hydrogen in the presence of complex metal catalysts. However, this process is not very effective at removing heterocyclic sulfur compounds, which can account

Figure 1. Dibenzothiophene and two of its alkyl derivatives, 4-methyl dibenzothiophene, and 4,6-dimethyl dibenzothiophene

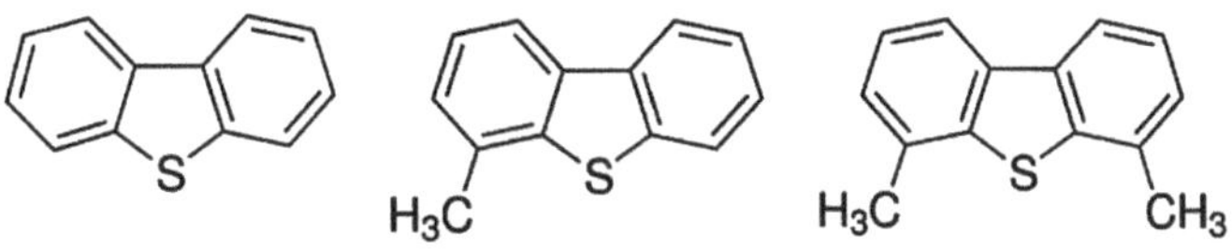

to up to 70% of the sulfur in petroleum (Borgne & Quintero, 2003), requiring harsher conditions to meet the strict EU sulfur regulations (10 ppm). This deeper desulfurization, with even higher temperatures and pressures, results in an increase of pollution (with > CO_2 release), a rise of production costs (resulting from higher energy consumption) and, sometimes, a loss of octane value (Khedkar & Shanker, 2014; Mohebali & Ball, 2008; Srivastava, 2012).

For this reason, oil desulfurization has become an increasingly studied area, and there have been many different approaches to solve these problems, such as the development of new combinations of metal catalysts for HDS, the study of techniques such as catalytic oxidation, desulfurization by adsorption, chemical desulfurization, physical desulfurization, photochemical desulfurization, photocatalytic desulfurization, and biodesulfurization (Srivastava, 2012).

The biodesulfurization (BDS) process consists of the use of microorganisms for the removal of sulfur from fossil fuels. Through BDS it is possible to treat most of the compounds recalcitrant to HDS at mild operating conditions, without the need for molecular hydrogen, or metal catalysts (Mohebali & Ball, 2008). If successfully implemented, it will result in lower environmental costs, with a reduction of 70–80% of CO_2 emissions, smaller residue production, and reduced energy consumption, which in turn translates in reduced capital (two thirds of HDS) and operational costs (10-15% lower) (Kaufman et al., 1997; McFarland et al., 1998; Tuli & Kumar, 2008; Vazquez-Duhalt, Torres, Brenda, & Le Borgne, 2002). This makes BDS an eco-friendly process that can complement the traditional methods of desulfurization making them more efficient (Kaufman, Harkins, & Borole, 1998; Linguist & Pacheco, 1999; Pacheco et al., 1999).

BDS has been extensively studied in the last decades, and much is already known about the process. Its efficiency is traditionally determined by following the degradation of model compounds, either on aqueous medium or on model oils composed of organic solvents. When using complex substrates such as real oil derivatives, BDS efficiency can also be determined by measuring the reduction in total sulfur concentration.

The most commonly used molecule for BDS studies is dibenzothiophene (DBT). This model compound consists of two benzene rings fused with a central thiophene ring and has a great variety of alkyl derivatives which are very abundant in crude oil (Figure 1). Among all the organosulfur compounds present in petroleum and its refined products, DBT and its derivatives present the greatest challenge to the desulfurization process (Rivera, Navarro-Santos, Guerra-Gonzalez, & Lima, 2014), because their structure makes it very hard to remove the sulfur atom contained in the thiophene ring without damaging the benzene structure, making them ideal to study.

Many microorganisms are able to use DBT as a sulfur source, and there are several pathways described for DBT desulfurization. However, most of the known BDS metabolic pathways involve the mineralization of the sulfur atom and complete or partial destruction of the benzene ring which makes them unappealing for the biodesulfurization process (Aggarwal, Karimi, Kilbane, & Lee, 2012; Gupta, Roychoudhury, & Deb, 2005; Lee, Senius, & Grossman, 1995).

The most studied and widely used pathway in BDS is the "4S pathway" (Kilbane, 2006). Through it, microorganisms are able to remove the sulfur atom under aerobic conditions while maintaining both benzene rings intact (Figure 2). This results in little loss of calorific potential making it an industrially interesting process (Mcfarland, 1999; Mohebali & Ball, 2008).

The "4S pathway" is composed of four enzymes (Ohshiro & Izumi, 1999). Two are monooxygenases: a DBT-monooxygenase (DBT-MO or DszC, encoded by the *dszC* gene) and a DBT-sulfone monooxygenase (DBTO2-MO or DszA, encoded by the *dszA* gene), which are flavin-dependent and require a third enzyme, the flavin reductase (DszD, encoded by the *dszD* gene) for activity. The last enzyme, hydroxyphenyl benzene sulfonate-desulfinase (DszB, encoded by the *dszB* gene), completes the reaction sequence, which results in the production of 2-hydroxybiphenyl (2-HBP), released into the culture medium, and SO_3^{2-} that is metabolized by the cell (Boniek, Figueiredo, dos Santos, & Stoianoff, 2014; Gray, Mrachko, & Squires, 2003)

This metabolic pathway is widespread throughout several genera of microorganisms, such as *Arthrobacter*, *Agrobacterium*, *Brevibacterium*, *Klebsiella*, *Mycobacterium*, *Nocardia*, *Paenibacillus*, *Pseudomonas*, *Xanthomonas*, *Gordonia* and *Rhodococcus*, with these two last ones being the most promising.

Using biocatalysts, researchers have successfully been able to remove sulfur from such different substrates as gasoline, kerosene, diesel, HDS diesel, and light and heavy crude oil. The desulfurization values described range from ~30 to ~100%, using different microorganisms and on different distillates, being achieved ultra-low sulfur fuels (Bhatia & Sharma, 2012; Guchhait, Biswas, Bhattacharya & Chowdhury, 2005a; Kilbane, 2006; Nuhu, 2013). As can be observed from Table 1, which summarizes the best desulfurizing strains studied with the respective BDS activity achieved, the highest desulfurization was obtained with a genetically modified strain

Figure 2. Scheme of the 4S pathway for DBT microbial desulfurization. The sulfur atom is removed from DBT keeping C-C structure

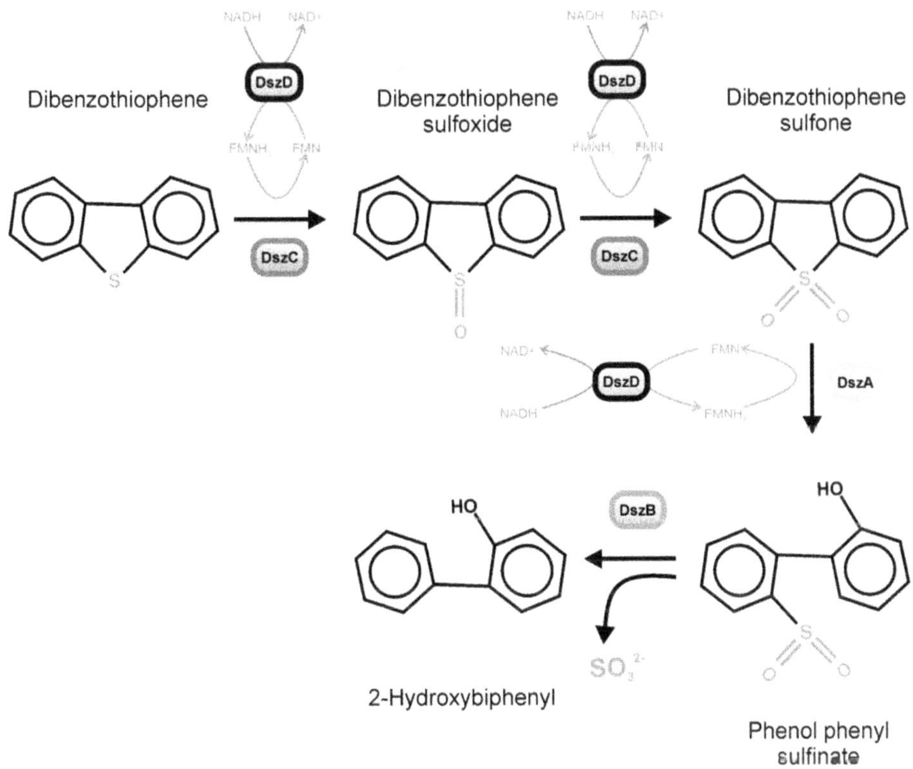

Table 1. Cultures with the highest desulfurization activities

Microbial Strain	Specific Activity (μmol g^{-1} (DCW) h^{-1})	References
Rhodococcus erythropolis KA 2-5-1, wild type	74	Kobayashi et al., 2000
Rhodococcus erythropolis KA 2-5-1, cloned dsz genes	280	Hirasawa, Ishii, Kobayashi, Koizumi, & Maruhashi, 2001
Rhodococcus erythropolis KA 2-5-1, cloned *dsz* genes	250	M. Konishi et al., 2005
Rhodococcus erythropolis IGTS8	72	Kilbane & LeBorgne, 2004
Mycobacterium sp. G3, wild type	178	Takada et al., 2005
Mycobacterium sp. G3, hsp60 promoter	211	

(adapted from Kilbane, 2006)

of *R. erythropolis* KA 2-5-1. This strain has achieved the maximal biodesulfurization activity observed until nowadays (280 μmol g^{-1} h^{-1} on DBT). However, this value is still much lower than the estimated for a successful commercial process, which would require a biocatalyst with desulfurization activity on petroleum of 1.2 mmol g^{-1} h^{-1} to 3 mmol g^{-1} h^{-1} (Kilbane, 2006; Kirimura et al., 2004; Monticello, 2000; Okada et al., 2003).

Within a biological desulfurization process, the microorganism must first be cultivated in order to produce large amounts of highly active biomass. This biomass is then used in a bioreactor to desulfurize an organic phase. At the end of the process, biomass, water, and organic phase must be separated. So, even if the levels of biodesulfurization activity are brought up to industrial levels there are other bottlenecks to the implementation of a cost appealing BDS process.

BDS COST REDUCTION

To make the BDS process economically competitive with the deep hydrodesulfurization process that is currently used by the petroleum industry, it is necessary to improve several factors, including the cost of producing the biocatalyst, and its biocatalytic activity (Mohebali & Ball, 2008).

There is no economically suitable method for large-scale production of biocatalysts (Ma et al., 2006), and a crucial limitation, common within many other biotechnological processes, is the cost of the culture medium, which represents 30 to 40% of the total production costs (Silva, Silva, Kamida, Goes-Neto, & Koblitz, 2014). So, an important line of work to turn BDS scale-up less costly is the optimization of the culture medium aiming to reduce expenses with maximal biocatalyst production.

Cost-Effective Culture Media

In the last years, culture media for the BDS process have been greatly optimized, targeting a reduction in overall components for minimum costs, and the elimination of all possible inhibitors. Aiming for a cost-effective culture medium for biocatalyst production within the BDS process, three main aspects must be considered: sulfur source, BDS promoters, and carbon source.

Sulfur Source

It is a well-known fact that the nature and concentration of the sulfur source used for the growth of the biocatalyst greatly influence the biodesulfurization activity (M. Z. Li, Squires, Monticello, & Childs, 1996; Oldfield, Pogrebinsky, Simmonds,

Olson, & Kulpa, 1997). The promoter and regulatory regions of the *dsz* operon are strongly repressed in the presence of readily bioavailable sulfur, *i.e.*, sulfate, sulfide, methionine, and cysteine, (M. Z. Li et al., 1996), because desulfurization enzymes are sulfate-starvation-induced proteins (Tanaka, Yoshikawa, Maruhashi, & Kurane, 2002). Also, in the presence of sulfate other enzymes such as sulfite oxiredutase have their activities leading to a preferential consumption of sulfate instead of DBT as the source of sulfur to support cell growth (Aggarwal et al., 2012).

In fact, even small concentrations of sulfate or other easy-access sulfur sources produce an inhibitory effect on BDS. Mohebali & Ball (2008) describe that for *G. alkanivorans* strain RIPI90A 14.4 mg l^{-1} sulfate repressed the Dsz phenotype. Silva, Paixão, Teixeira, Roseiro, & Alves (2013) reported 50% inhibition at 13.6 ± 0.6 mg l^{-1}sulfate and complete repression at 60 mg l^{-1} on growing cells. Moreover, bacteria have very low sulfur requirements. In *Rhodococcus sp.*, the cells were found to require 0.1 mM of sulfur for normal growth (Reichmuth, Hittle, Blanch, & Keasling, 2000). Only 1% of bacterial dry weight is sulfur (Stoner et al., 1990), which implies a very low need for this element. Therefore, it is important to take into account the nature of the sulfur source used when preparing the medium for the biocatalyst.

In order to avoid BDS inhibition, most studies are performed using DBT as the sulfur source for biocatalysts production (Mohebali & Ball, 2008). This however becomes an impractical method for large-scale production of biocatalysts because of DBT high price, low water solubility, and the growth inhibition caused by the 2-HBP, the end product of DBT desulfurization (Guobin et al., 2006; Honda, Sugiyama, Saito, & Kobayashi, 1998; Kilbane & Le Borgne, 2004; Ohshiro, Suzuki, & Izumi, 1996; Yoshikawa et al., 2002). So, to achieve a commercially efficient method of avoiding repression it becomes important to explore alternatives.

Mixed Sources

One alternative that could slightly decrease the cost of the sulfur source is the use of small concentrations of sulfate mixed with DBT. If the concentration of sulfate is limiting, the biocatalyst will consume sulfate until it is spent and it will then turn towards the DBT. Silva et al. (2013) demonstrated that *G. alkanivorans* strain 1B cells cultivated with 0.063 mM of sulfate and 0.4 mM of DBT were able to express desulfurization activity, despite a reduction of 22.4% when compared with the DBT control. Also, Ma et al. (2006) were able to obtain increased desulfurization activity of *Rhodococcus* sp. cells cultured with mixed sulfur sources. Maximum activity of 39 μmol g^{-1} h^{-1} was achieved when the concentrations of DBT and sulfate were 0.16 mM and 0.04 mM, respectively, which is very similar to what was observed with 0.20 mM DBT. This method allows a small reduction in DBT concentration while maintaining some BDS activity.

Induction

Another alternative is to use sulfate, which is cheap, as the sulfur source to produce large amounts of high-density bacterial culture with no desulfurization abilities, then wash the biocatalyst to remove the sulfate, and finally proceed with the induction of the expression of the *dsz* operon, by exposing the cells to DBT. This translates in a reduction of the amount of DBT necessary to produce an active catalyst, leading to a less costly biocatalyst production. Ma et al. (2006) were able to obtain *Rhodococcus* sp. biomass with a maximum value of the desulfurization activity of about 15.6 μmol g^{-1} h^{-1} after about 8 h of induction. Longer induction lowered the desulfurization activity, probably due to the accumulation of toxic metabolites in the medium (Chang et al., 2001; Honda et al., 1998; Monticello & Finnerty, 1985).

Alternative Sulfur Sources

The best option towards a cheaper biocatalyst production would probably be to use alternative sulfur sources, which do not inhibit the Dsz phenotype expression, unlike sulfate, and are not as expensive as DBT. In this context, several sulfur sources have been tested. In 2002, Yoshikawa et al. tested different water-soluble sulfur compounds such as dimethyl-sulfoxide (DMSO), 2-aminoethanesulfonic acid, methanesulfonic acid, cysteic acid, p-aminobenzene sulfonic acid, or sodium sulfate as the sole sulfur source for the cultivation of *R. erythropolis* KA 2-5-1. It was found that 2-aminoethanesulfonic acid was the most effective to produce a high density of cells, even when compared with DBT. Its use resulted in a BDS activity of 111 μmol (2-HBP) g^{-1} h^{-1}, with a high growth rate (μ_{max} = 0.37 h^{-1}) and a final cell concentration of 20 g dry cells weight (DCW) after 89 h.

Many researchers have reported that DMSO can be used as an alternative sulfur substrate that does not repress desulfurization activity. (Abbad-Andaloussi, Lagnel, Warzywoda, & Monot, 2003; M. Z. Li et al., 1996; Luo et al., 2003; Ma et al., 2006; Mohebali, Ball, Kaytash, & Rasekh, 2007, 2008; Oldfield et al., 1997; Olmo, Santos, Alcon, & Garcia-Ochoa, 2005; Tang & Hong, 2014). This happens because DMSO is not a substrate for the DBT desulfurizing enzymes (M. Z. Li et al., 1996). So, in the presence of DMSO there is a de-repression of the *dsz* operon in the absence of more readily bioavailable sulfur sources (M. Z. Li et al., 1996; Oldfield et al., 1997). Ma et al. (2006) used 1 mM of DMSO as the single sulfur source to grow *Rhodococcus* sp. cells and obtained a cell density of 4.0 g l^{-1} and desulfurization activity of 51.6 μmol g^{-1} h^{-1}. Based on the cell density and desulfurization activity and comparing with results using DBT and Na_2SO_4, DMSO proved to be the cheapest sulfur source for preparing the biocatalyst for oil desulfurization.

Mohebali et al. (2008) showed that the Dsz phenotype was expressed through the 4S pathway in the presence of DMSO as the sulfur source for the growth of *G. alkanivorans* strain RIPI90A. When using DMSO the growth rate was higher than with DBT, however, concentrations higher than 200 μg ml^{-1} induced inhibition. This study also concluded that DMSO should be used as the sulfur source for mass production of *G. alkanivorans* RIPI90A, and there should be an induction step with DBT to increase resting cell activity. Prior to induction, the specific activity detected was 1.4 μmol (2-HBP) g^{-1} h^{-1}, and following incubation (5 h) the highest specific activity observed was 5.11 μmol (2-HBP) g^{-1} h^{-1}.

Another option is the use of hydrodesulfurized oil or waste engine oil as the sulfur source. This method employed by Mukhopadhyaya, Chowdhury, & Bhattacharya (2005), consists on the supplementation of the culture medium with an oil or derivative rich in sulfur, as a way of stimulating desulfurization activity of the biocatalyst, while promoting pre-adaptation of the culture to the BDS conditions. These sulfur sources are cheap, and/or easily available in a refinery making them very interesting for the BDS process. However, these oils are very complex and could have inhibitors that limit high-density microbial growth. Moreover, the emulsification resulting from the interaction of water, bacteria, and oil should difficult the separation of the biomass making it impractical for some processes.

Biodesulfurization Promoters

Nitrogen Source

Glutamic acid can be both a carbon and nitrogen source, so Martin, Alcon, Santos, & Garcia-ochoa (2004) tested the importance of the presence of an inorganic nitrogen source for the growth and desulfurization by *Pseudomonas putida* CECT5279. They determined that the presence or absence of NH_4Cl in the culture medium with glutamic acid did not influence the growth of the microorganism. But, when both nitrogen sources were present the desulfurization rate was 10 to 20% higher. So, the optimal conditions for desulfurization were determined to be an initial concentration of 20 g l^{-1} glutamic acid supplemented with 670 ppm of NH_4Cl.

NADH Promoting Substances

The 4S pathway is NADH dependent since this molecule is necessary to regenerate the DszA and DszC enzymes. Moreover, the regeneration of NADH also implies the use of other key cofactors, such as ATP, the basic source of energy for most metabolic reactions. ATP may be even more important since it seems to be implicated in the uptake of DBT (Z. Wang et al., 2011), and in the formation and consumption of sulfite and sulfate for biomass production during DBT metabolism (Aggarwal et al., 2012).

The industrial BDS process will depend on the reuse of the biocatalyst, for economic efficiency. However, in long-term use of the biocatalyst there is a loss of BDS activity that is associated with a loss in redox potential and energy. So, in order to allow the greater reuse of the bacterial cells, it became important to study ways to regenerate molecules such as NADH, FMNH, and ATP. In this context, Martinez, Santos, Alcon, & Garcia-Ochoa (2014) studied some of the compounds related to the Krebs cycle dehydrogenases able of producing NADH from NAD^+ as possible co-substrates to enhance the BDS process. Citric and succinic acids were used in order to study both extremes of the Krebs cycle, and acetic acid was selected due to the metabolic importance of its active derivative, acetyl-CoA. The results obtained from this work showed that the addition of these cofactors resulted in an increase of the yield and rate of biodesulfurization by *P. putida* CECT5279 cells. Amongst the three co-substrates tested, acetic acid proved to be the most influential, at a concentration of 1.5% the yield of the process increased up to 140% when using single-aged cells in a batch process and to 122% using an optimized mixture of cells in a fed-batch process.

Besides NADH, FMNH is also a fundamental cofactor for the regeneration of the 4S pathway. So, some works on how to maintain the levels of intercellular cofactors of the biocatalyst during the BDS process have been carried out. Nicotinamide and riboflavin are precursors of $NADH_2$ and $FMNH_2$, respectively. As a consequence, to reduce the loss of BDS efficiency due to immobilization and excessive reuse of the biocatalyst, Derikvand, Etemadifar, & Biria (2014) applied the two co-factor precursors and recorded a reduced loss in cell viability and desulfurization activity after each step, enhancing the BDS efficiency by more than 30% after four reuse steps. Therefore, by providing essential co-factors in 4S pathway, it is possible to increase the efficiency of the biocatalyst, allowing for more cycles of BDS for each batch of cells produced, thus effectively lowering the cost of the process.

Carbon Sources

Alongside the sulfur source, carbon is probably the most important component of a culture medium. Depending on the chosen carbon source the biocatalyst may present great changes on BDS efficiency and growth rate. Traditionally, most microbial biotechnological studies were performed with glucose as the single carbon source to demonstrate the biocatalysts activity. Since most microorganisms are glucophilic, it is common sense that this will result in a better expression of the strains metabolic properties (Alves & Paixão, 2014b; Alves et al., 2005; Mohebali et al., 2008; Nuhu, 2013).

However, in recent years, there have been some BDS studies which have revealed that other carbon sources might be better suited for this process. In this context, some theoretical studies were performed based on the sequenced genomes of *R. erithropolis and G. alkanivorans*. These studies analyzed the effects on growth rate and desulfurization efficiency of several carbon sources, such as acetate, citrate, ethanol, formate, fructose, fumarate, gluconate, glucose, glutamate, glycerol, lactate, malate, oxaloacetate, oxoglutarate, pyruvate, and succinate.

According to results presented by Aggarwal, Karimi, & Ivan (2013) and Aggarwal, Karimi, & Lee (2011), ethanol appears to be the best for both growth and desulfurization on both genera. For *G. alkanivorans* the carbon sources as compared to ethanol have the 100% basis rank as follows: fumarate (80%) > oxoglutarate (78.79%) > pyruvate (78.43%) > glutamate (78.24%) > succinate (78%) > acetate > fructose=glucose=lactate (76.86%) > glycerol (75%) > citrate (71.88%) > oxaloacetate (69.70%) > malate (69.11%) > formate (50%). Aggarwal et al. (2013) proceeded to explain that NADH production and usage could justify why ethanol is the best carbon source. Since the desulfurization activity requires high levels of NADH (4 moles for each 1 of DBT), and the cell still requires NADH to grow, priority should be given to a carbon source that provides more NADH during its metabolism. And this was the case for ethanol, since it led to the production of two additional moles of NADH for each one of its own mol. This was the highest among all tested substrates, and thus it seems to be the best substrate for both growth and desulfurization.

However, Alves & Paixão (2014b) reported a novel behavior of a strain isolated in 2005. *G. alkanivorans* strain 1B has now been described as fructophilic bacterium. This means that, contrary to what was predicted, this strain prefers fructose to glucose for both growth and desulfurization. In fact, the lowest values for the growth rate (0.025 h^{-1}), 2-HBP production rate (1.80 μM h^{-1}) and specific production rate ($q_{2\text{-HBP}}$ = 1.22 μmol g^{-1} h^{-1}) were obtained in glucose grown cultures. When cultivated with an equal mix of glucose and fructose as a carbon source, the value of $q_{2\text{-HBP}}$ increased to 1.90 μmol g^{-1} h^{-1}. The highest values for cell growth (μ = 0.091 h^{-1}), 2-HPB production (9.29 μM h^{-1}) and specific production ($q_{2\text{-HBP}}$ = 2.12 μmol g^{-1} h^{-1}) were obtained when strain 1B was desulfurizing DBT, in the presence of fructose as the only carbon source. These results translate into an increase of over 5-fold in the overall 2-HBP production rate and of about 74% in $q_{2\text{-HBP}}$, which clearly demonstrate the strain fructophilic properties.

The metabolic characteristics observed in *G. alkanivorans* strain 1B highlight that there are still to many unknown factors associated to each strain behavior, and thus theoretical models, even if well performed, can only be applied to each strain individually and not extrapolated to an entire species.

Alternative Carbon Sources

The high cost of the carbon source necessary to cultivate the biocatalyst is one of the biggest barriers for the applicability of BDS at the industrial level. Thus, in order to reduce the costs associated with this process, it is important to search for cheaper alternative carbon sources that can contribute to producing high quantities of microbial biomass with the reducing equivalents required for BDS reactions. In the last decade several examples of alternative feedstocks have been explored, many of which based on the use of agroindustrial residues because of their low cost, and high availability. However, the use of these complex carbon sources brought upon several additional problems, such as the need to perform prior hydrolysis, and/or the need to remove excess sulfate (Silva et al., 2013; Werther, Saenger, Hartge, Ogada, & Siagi, 2000).

Glycerol

Glycerol is a byproduct of the biodiesel production, so with the increasing production of biodiesel as an alternative fuel, there will be an increase in the availability of this low-cost carbon source. Abo-state, El-Gendy, El-Temtamy, Mahdy, & Nassar (2014) and Tang & Hong, (2014) studied the use of glycerol to produce biocatalysts with desulfurization ability. However, only Tang & Hong (2014) reported that glycerol is the most suitable carbon source for the growth and desulfurization activity of their biocatalyst (strain HT1). They determined that with 5 g l^{-1} glycerol, 102 mg l^{-1} DBT and 2 g l^{-1} NH_4NO_3, at pH 7 and 30°C for four days cultivation time, the maximum desulfurization was obtained by strain HT1.

Recycled Paper Sludge

The first report on the utilization of alternative raw materials as carbon sources in biodesulfurization studies was performed by Alves, Marques, Matos, Tenreiro, & Gírio (2008). They proposed the use of recycled paper sludge (RPS) as a carbon source since RPS is a very abundant type of waste resulting from the pulp and paper industry. The sludge generated by the water treatment facilities is still a major disposal problem which needs to be dealt with (Oral et al., 2005). However, after neutralization, RPS is very rich in cellulose (35%), xylan (10%) and lignin (20%), % dry weight, which can be hydrolyzed into glucose, xylose, and cellobiose making RPS an interesting feedstock.

In their work, Alves, Marques, et al. (2008) produced several different hydrolyzates by enzymatic saccharification of RPS and tested them as nutrient source for low-cost DBT desulfurization by *G. alkanivorans* strain 1B at a concentration of 10 g l^{-1} of glucose. Under these conditions strain, 1B was able to grow, but it wasn't able to desulfurize. This fact was due to the concentration of sulfate present in the RPS hydrolyzate. So, further work was carried accounting the prior sulfate removal. The sulfate was removed by simple dialysis, and after that strain, 1B was able to consume 250 μM DBT in 96 h displaying a maximum $q_{2\text{-HBP}}$ of 1.1 μmol g^{-1} h^{-1}. The maximum specific growth rates, μ_{max}, for growth with non-dialyzed and dialyzed hydrolyzate were 0.051 h^{-1} and 0.035 h^{-1}, respectively. Both these values are above to what is described for glucose μ_{max} = 0.025 h^{-1} (Alves & Paixão, 2014b). Further investigation on the formulation led to the development of an ideal medium of RPS hydrolyzate supplemented with phosphates, ammonia, magnesium, and zinc. Zinc was found to have a great influence on both growth and BDS activity by *G. alkanivorans* strain 1B (Alves, Matos, Tenreiro, & Gírio, 2008).

Therefore the results obtained by Alves, Marques, et al. (2008) clearly indicate that RPS hydrolyzates can be used as a cheaper alternative carbon source for DBT desulfurization by *G. alkanivorans* strain 1B. Since RPS wastes are harmful to the environment and must be disposed of, their potential as carbon source makes them very interesting for biotechnological processes.

Carob Pulp

The use of byproducts from the carob processing industry as an alternative carbon source is already common for several bioprocesses (Carvalheiro, Moniz, Duarte, Esteves, & Gírio, 2011; Mendes et al., 2007; Sánchez-Segado et al., 2012). Carob industrial wastes are small carob kibbles with a high content of soluble sugars, mainly sucrose, glucose, and fructose, which are easily extractable by water producing sugar-rich liquors (Manso, Nunes, Raposo, & Lima-Costa, 2010). However, these liquors are also rich in sulfates and easy access to sulfur sources, and so they cannot be directly used for BDS.

In this context, Silva et al. (2013) described an optimized formulation of a culture medium, using carob pulp liquor as an alternative carbon source, for the production of *G. alkanivorans* strain 1B in a BDS process, aiming for a cost reduction associated to the process. In this work, they treated the carob pulp liquor with barium chloride ($BaCl_2$) for prior sulfate removal and demonstrated the potential of this agroindustrial byproduct as a renewable and cheaper alternative carbon source for BDS. Prior to the treatment with $BaCl_2$, no 2-HBP production by strain 1B was observed within 125 hours of the growth on carob pulp liquor, but in the optimized treated liquor strain 1B was able to produce 237 μM of 2-HBP, which is a similar result to what was obtained in the RPS dialyzed hydrolyzate (Alves, Marques, et al., 2008).

Sugar Beet Molasses

Sugar beet molasses (SBM) is a very abundant industrial byproduct very rich in sucrose (~50%, *w/v*). Because of its composition, SBM can be easily hydrolyzed to glucose and fructose by acidic or enzymatic hydrolysis, with the advantage of having a lower price of commercialization ($100–$150 per ton, which is 5 to 6 times less than sucrose).-

Similarly to carob pulp liquor, SBM contains a large concentration of readily available sulfur sources, which can completely inhibit the BDS process. So, before testing the SBM as a potential alternative carbon source it is necessary to perform a prior sulfate removal as described by Silva et al. (2013). Thus, to express the *dsz* operon of the strain 1B, Alves & Paixão (2014a) treated the sugar beet molasses (SBM_t) with 0.25% $BaCl_2$ (*w/v*) to reduce sulfates to residual levels prior to use it as a carbon source for DBT desulfurization. With the SBM_t, the strain 1B achieved a μ_{max} of 0.04 h^{-1} and maximum 2-HBP production rate of 2.56 μM h^{-1}, which resulted in a $q_{2\text{-HBP}}$ of 2.20 μmol g^{-1} h^{-1}. The desulfurization results were better than those obtained with commercial sucrose (1.91 μM (2-HBP) h^{-1}; $q_{2\text{-HBP}}$ = 0.718 μmol g^{-1} h^{-1} from Alves & Paixão, 2014b), probably due to other nutrients present in a complex feedstock such as SBM that can act as inducers.

Moreover, the DBT desulfurization process by *G. alkanivorans* strain 1B using SBM_t was further enhanced using a simultaneous saccharification and fermentation (SSF) approach with (1% *v/v*) *Z. bailii* strain Talf1 crude extract with invertase activity. The results obtained showed that this approach allowed a faster growth of strain 1B with a μ_{max} of 0.08 h^{-1}, 2-fold higher than in the absence of enzyme extract, and a 2-HBP production rate of 7.78 μM h^{-1} corresponding to a $q_{2\text{-HBP}}$ of 3.12 μmol g^{-1} h^{-1} (Alves & Paixão, 2014a). Comparing this $q_{2\text{-HBP}}$ value to others prior obtained with the same strain grown in pure commercial substrates, such as glucose (1.03 μmol g^{-1} h^{-1} from Alves et al. (2005)) and fructose (2.12 μmol g^{-1} h^{-1} from Alves & Paixão, 2014b), an enhancement of about 203 and 47% was achieved, respectively.

Arez, Alves, & Paixão (2014), also using the SSF approach with commercial sucrose and *Z. bailii* Talf1 enzymatic extract within a DBT desulfurization process by strain 1B, observed a μ_{max} of 0.07 h^{-1} and a 2-HBP productivity of 5.80 μM h^{-1}, corresponding to a $q_{2\text{-HBP}}$ of 2.60 μmol g^{-1} h^{-1}. These desulfurization values, in addition to those obtained by Alves & Paixão (2014a) in the SSF with SBM_t, translate the advantage of the SSF approach towards enhanced BDS when sucrose or sucrose-rich feedstocks are used as carbon source. The higher desulfurization obtained with the SBM_t highlights the potential of this renewable feedstock as a low-cost alternative carbon source to enhance BDS by *G. alkanivorans* strain 1B.

Jerusalem Artichoke

Jerusalem artichoke (JA) is a species of sunflower that has recently gained much attention as a renewable feedstock, especially towards biofuels (bioethanol, biodiesel), because its tubers present high levels of inulin reaching 50–80% of dry matter (Cheng et al., 2009; L. Guo, Zhang, Hu, Dy Ryu, & Bao, 2013; L. Li, Li, Wang, Du, & Qin, 2013; Liang et al., 2012). Inulin is a carbohydrate composed of linear chains of β-2,1-linked D-fructofuranose molecules terminated by a glucose residue through a sucrose-type linkage at the reducing end, which can be easily hydrolyzed into a fructose-rich hydrolyzate either using inulinases or acidic hydrolysis.

The potential of JA juice as an alternative carbon source, rich in fructose, towards BDS by the fructophilic bacterium *G. alkanivorans* strain 1B was exploited by two different approaches, since this bacterium is not able to metabolize pure inulin as carbon source. One approach used was the SSF with *Z. bailii* crude inulinases (Paixão, Teixeira, Silva, Teixeira, & Alves, 2013), in which a maximum 2-HBP production rate of 4.80 μM h^{-1} was attained from JA juice (~10 g l^{-1} of total reducing sugars). In this study, no sulfate removal procedure was used, which interfered with the final desulfurization productivity. In fact, JA, as a complex natural material, contains very high concentrations of sulfates (> 450 mg l^{-1}) that will inhibit the desulfurization and consequently decrease the 2-HBP production. So, the second approach was the application of acidic hydrolysis of the JA juice, combined with $BaCl_2$ precipitation for sulfate removal (Silva, Paixão, Roseiro, & Alves, 2014). In their study, Silva et al. (2014) first optimized the sulfate removal process for JA juice (0.5% (*w/v*) $BaCl_2$ at pH 8.73 and 30°C for 36 h with agitation) and then performed the DBT desulfurization process by *G. alkanivorans* strain 1B using the best treated JA juice (~25 g l^{-1} of total reducing sugars). In this process, strain 1B presented a μ_{max} of 0.06 h^{-1} with a total conversion of 400 μM DBT into 2-HBP in less than 90 h, attaining a 2-HBP maximum production rate of 28.2 μM h^{-1} and a $q_{2\text{-HBP}}$ of 5.06 μmol^{-1} g^{-1} h^{-1}, which are 3.6-fold and 1.6-fold higher, respectively, than that observed for SBM_t by Alves & Paixão (2014a). Moreover, this $q_{2\text{-HBP}}$ is similar to that reported by Mohebali et al. (2008) using optimized resting cells of another *G. alkanivorans*, the strain RIPI90A. In both studies the specific production rate was 2-fold higher than the $q_{2\text{-HBP}}$ of 2.54 μmol^{-1} g^{-1} h^{-1} recently predicted by Aggarwal et al. (2013) using *in silico* modelling for a *G. alkanivorans*, assuming an uptake of 20 mg per g (DCW) per h of glucose or fructose.

The results, using JA juice, highlight the efficiency of the treatment applied to JA juice in making this agromaterial a promising low-cost renewable feedstock for improved BDS by the fructophilic strain 1B. Once again, the results obtained using a cheaper alternative carbon source demonstrate the advantages of exploring renewable feedstocks as promoters of the metabolic activity towards a less-expensive biocatalyst production, which is a fundamental step for the application of BDS at the industrial level.

BDS OPERATIONAL CONDITIONS

After the selection of a microorganism with characteristics suitable for the desulfurization process and the determination of the optimal cultivation conditions, the operational settings are of critical importance for BDS scale-up. The industrial application and cost-effectiveness of the biodesulfurization process is highly dependent on optimization in four main areas: 1) biocatalyst application, 2) bioreactor design and parameters, 3) biomass recovery and water separation on the process downstream, and 4) integration on the refinery operation.

Biocatalyst Application

Process Setup

Biodesulfurization may use as biocatalyst whole cells, free cell extract or pure enzymes, each with its own advantages and disadvantages.

Some consider the use of pure enzymes as a good option since they require much less water than whole cells to play their catalytic role with only a small film being needed, which would be an interesting alternative for the scale-up process (Vazquez-Duhalt et al., 2002). However, enzymes purification is an expensive process. To surpass this problem, other researchers suggest the use of free cell extracts, which do not require the purification and maintain most enzymatic activity. But both to pure enzymes and cell free extracts, there is still a need of supplying NADH and $FMNH_2$ for the biochemical reactions making these alternatives more expensive and less practical. Moreover, the desulfurization activity has been reported to be lower when free cell extract is used (0.01 g (DBT removed) g^{-1} (protein) h^{-1}) than when whole cells are used (0.4 g (DBT removed) g^{-1} (dw biomass) h^{-1}) (Setti, Lanzarini, & Pifferi, 1997).

These setbacks make whole cells biodesulfurization the most consensual approach for the treatment of DBT (Soleimani, Bassi, & Margaritis, 2007). Biodesulfurization based on whole cells can take place either by growing or resting cells.

Growing cells are easily applied in a bioreactor, with microbial growth and desulfurization occurring simultaneously. However, these systems present a number of constraints. In general, the process is rather slow, achieving reduced desulfurization yields (Shennan, 1996) and requires the need for complex growth media, which can interfere with the product analysis due to the presence of a complex mixture of metabolites and medium components (J. Konishi, Ishii, Onaka, Okumura, & Suzuki, 1997). Moreover, the desulfurization activity is repressed in the presence of easy access sulfur sources or 2-HBP, the end-product of the 4S pathway that also inhibits cell growth (M. Z. Li et al., 1996; Nekodzuka, Nakajima-Kambe, Nomura, Lu, & Nakahara, 1997; Ohshiro et al., 1996).

Resting cells, non-growing cells that retain most of their enzymatic activity, have shown advantages over growing cells, such as greater resistance to 2-HBP inhibition. Kobayashi et al. (2001) observed that 80% of desulfurization activity of resting cells of *R. erythropolis* KA 2-5-1 remained even in the presence of 2-HBP at 10 mM in a model oil system, reinforcing the benefit of using these cells for a scale-up of the biodesulfurization process.

These cells can also be used in much higher concentration than growing cells (Borgne & Quintero, 2003). Many authors have focused on the production of high cell density cultures that can be used for biodesulfurization as resting cells resulting in a reduction of reaction volume (Honda, Sugiyama, Saito, & Kobayashi, 1998; M. Konishi et al., 2005; Wang & Krawiec, 1996). Honda et al. (1998) used a fed-batch reactor assembly to grow *R. erythropolis* IGTS8 in the presence of sulfate and achieved a cell concentration of 33 g l^{-1} in the best conditions, compared to only 1.1 g l^{-1} observed when DBT was used as a sulfur source, which could be attributed to the growth inhibition by 2-HBP. Chang et al. (2001), using a different biocatalyst, *Gordonia nitida* CYKS1, achieved a cell concentration of 92.6 g l^{-1}. In both cases desulfurization induction time was around 4 h.

However, high density resting cells production also has some disadvantages since this process may give rise to cross contaminations with C-C attacking microorganisms with higher growth rates. Moreover, economically relevant sized facilities are required to prepare the biocatalyst, with issues related to the maintenance of desulfurization activity and transport to desulfurization site needing to be addressed (Onodera-Yamada, Morimoto, & Tani, 2001).

This problem might be minimized using lyophilized cells. Patel, Killbane, & Webster (1997) tested biodesulfurization in an aqueous-hydrocarbon phase with freeze-dried cells of *R. erythropolis* IGTS8, revived in a minimal amount of water, and despite a loss of 20% of desulfurization activity in the freeze-drying process, there was only a 2.8% of activity reduction after 10 weeks at -80°C. This opens the possibility of producing the biocatalyst in a different location of the BDS refinery since a much smaller biomass volume allows for easier transportation. Furthermore, the desulfurization conditions of lyophilized cells are more flexible. Luo et al. (2003) reported that the desulfurization ability of lyophilized cells of *Pseudomonas delafieldii* R-8 was not affected by pH's ranging in 4.6-8.5, unlike growing cells in which the growth was inhibited at pH below 5.

Nano-Upgrading of Biocatalyst

Biocatalyst upgrading is another trend followed for increased efficiency of the process. Functionalization of the microorganism cell surface is a promising trend in bio-nanotechnology (Ansari, Grigoriev, Libor, Tothill, & Ramsden, 2009).

One of the limiting factors for biodesulfurization is the transfer of poly-aromatic sulfur heterocycles such as DBT, from oil to water and from water to cells (Guobin, Huaiying, Weiquan, Jianmin, & Huizhou, 2005). This constraint can be minimized by the use of nanosorbents such as γ-Al_2O_3 or Fe_3O_4 that selectively absorb those compounds from the organic phase at a higher rate than desulfurization, thus removing this limitation on the process. Guobin, Huaiying, Weiquan et al. (2005) reported that assembling γ-Al_2O_3 to microbial cells led to a two-fold increase in desulfurization rate of *P. delafieldii* R-8 compared to original cells, making DBT and 2-HBP transfer no longer a limiting factor.

The adsorption of Fe_3O_4 nanoparticles on the surface of *R. erythropolis* IGST8 led to a 56% increase in desulfurization, probably related to an increase in membrane permeability to DBT, and also allowed an easy separation of the microbial biomass given the super magnetic characteristics of these nanoparticles (Ansari et al., 2009).

Nano-shellization of desulfurizing bacteria with bio-hybrids based in amino acids was reported by Jiang, Ying, Liu, Shen, & Hu (2014). These self-encapsulated cells not only gained the ability to be reused, but these shells also served as a platform in which other inorganic compounds could be incorporated to upgrade desulfurization without affecting the cell metabolic activity. The authors reported a 62% increase of the desulfurization rate when cells were functionalized with TiO_2 nanoparticles, known for increased desulfurization efficiency under UV light. These authors further reported the post-functionalization of cells with magnetic nanoparticles, allowing for an easier mechanism of biomass recovery.

Biocatalyst Immobilization

Another constraint for a large scale BDS is related to the utilization of free cells, which end up being suspended on the oil/water mixture, requiring costly downstream processing for their removal and also a great control to avoid cell contaminations (X.-L. Guo, Deng, Xu, & Wang, 2006). Furthermore, free cells are prone to shear stress when applied in a bioreactor, due to the requirements of mixing the oil and water phases.

Given the high demand of water for biological processes, application of free cells to biodesulfurization scale-up would require massive amounts of water and reactors of an unimaginable scale, thus hindering the possibility to make the industrial application a reality. So, biocatalyst immobilization emerges as an interesting solution presenting marked advantages such as ease of recovery and re-use of cells, better stability to different reaction conditions and components, and lower biocatalyst contamination of final product (Zhang, Prabhu, & Lee, 2010).

Many immobilization methods have been successfully used for biocatalysis and biodegradation (Bardania et al., 2013; Dinamarca et al., 2014; Gunam, Sujaya, Aryanta, Tomita, & Asano, 2013; He et al., 2012; F. Li et al., 2005; Y.-G. Li, Gao, Li, Xing, & Liu, 2009; Wu, Lin, & Chan, 1994) and also for biodesulfurization (Nuhu, 2013). Immobilization studies are mainly focused on adsorption or entrapment/encapsulation. Adsorption uses inorganic supports, such as Al, Si, Ti, Celite or Sepiolite, or inert organic supports such as lignin, with specific area and porosity in which microbial biomass adheres (Chang, Chang, Ryu, & Chang, 2000; Dinamarca et al., 2010; Huaiying, Guobin, Huizhou, & Jianmin, 2007; Mukhopadhyaya, Chowdhury, & Bhattacharya, 2007). Biocatalytic immobilization by adsorption is advantageous in relation to entrapment when considering the better mass transfer and fewer steric problems with large sulfur-containing molecules (Guobin, Jianmin, Chen, Huizhou, & Jiayong, 2005). Despite these advantages, adsorption is not a strongly followed strategy as it is a simple physical process and thus cell attachment strength is not enough to avoid a quick detachment from the adsorbent (Dai, Shao, Qi, & Ding, 2014).

Although entrapment and encapsulation refer to different methodologies, with the first referring to the trapping within polymeric structures and the second to the formation of a continuous membrane around encapsulants that wholly coats it, entrapment may be used as a more general term (Willaert & Baron, 1996). Entrapment has obvious advantages for cell immobilization, in relation to adsorption, as a mean to prevent cell loss. Ideally, a carrier for entrapped cells should have a highly porous structure that allows for non-hindered diffusion of solutes and dissolved gas without the loss of cells. Several polymers have been tested in desulfurization, all with advantages and drawbacks. Agar and alginates have found deep application in industrial processes for their biocompatibility, simplicity and low cost (Blandino, Macías, & Cantero, 1999). However, their reduced mechanical strength leads to elevated cell leakage, thus compromising their durability. Naito et al. (2001) tested photo-crosslinked pre-polymers (ENT-4000 and ENTP-4000) and polyurethane (PU-3 and PU-6) in the entrapment of *R. erythropolis* KA 2-5-1 cells, as well as agar and alginates. Polyurethane showed high toxicity to the microorganism, but ENT-4000 showed promising results. Besides being able to retain the biomass, it increased the biocatalyst longevity to 900 h, in contrast with the 192 h of maximum longevity previously reported by Pacheco et al. (1999). Despite the lower desulfurization activity in relation to free cells (attributed to diffusion limitations) and the high cost of this pre-polymer, Naito et al.(2001) reported to have achieved these results with no water addition to the n-tetradecane model oil, which presents as a major breakthrough on the feasibility of biodesulfurization in an industrial scale greatly simplifying separation steps.

Polyvinyl alcohol (PVA) is a nontoxic, biologically compatible, low-cost polymer also being used for cell entrapment. It can be prepared with chemical reagents such as boric acid (Hashimoto & Furukawa, 1987) or with no chemical reagents by the freezing-thawing technique (Lozinsky & Plieva, 1998). Guobin, Jianmin, Chen et al. (2005) used this technique for the preparation of *P. delafieldii* R-8 magnetic beads. These authors report having been able to reuse PVA-R8 beads for 12 cycles, without a significant decrease in desulfurization activity and with much higher resistance to temperature.

The desulfurization activity of *Sphingomonas subarctica* T7b was studied either in free resting cells or immobilized in different PVA and sodium alginate mixtures by Gunam et al. (2013). They reported higher stability to pH and temperature variations in immobilized cells when compared to free cells. Furthermore, they also achieved slightly higher desulfurization rates than those obtained for free cells, surpassing one of the limitations of cell entrapment, and the biodesulfurization activity remained stable for 8 cycles with reactivation between cycles. These results indicate PVA as a good candidate for application in large scale biodesulfurization, given its low cost, stability and biocompatibility.

In overall, the integration of the immobilization techniques into BDS allows improved results and increase of process efficiency, besides enhancing separation and biocatalyst reutilization, with great impact in biocatalyst application costs.

Bioreactor System

Mass Transfer

The setup of a fossil fuels BDS process usually occurs in a byphasic system: an aqueous and an oil phase, which in fact can be considered a four phase system, if we consider a gas phase- the oxygen needed for the biocatalytic process; and a solid phase- the microbial biomass (Foght, 2004) One of the biggest challenges for the industrial application of BDS is related to the immiscibility of water (needed for microbial activity) with hydrocarbon feedstocks, which results in a low mass transfer between phases.

In this context, the desulfurizing microorganisms can be distinguished in relation to the character of their membranes, being either hydrophobic or hydrophilic. The hydrophobic membrane microorganisms, such as *Rhodococci* and *Gordoniae*, have better resistance when used in water-oil environments and easier access to sulfur-containing substrates dissolved in the organic phase, adhering to the oil-water interface (Abbad-Andaloussi et al., 2003). In contrast, the hydrophilic membrane microorganisms, such as *Pseudomonas*, need the substrates to be dissolved in the aqueous phase (Caro, Boltes, Letón, & García-Calvo, 2007), making the hydrophobic membrane microorganisms a better candidate for large scale biphasic systems (Monticello, 2000).

Oil/water (o/w) volume ratio is amongst the factors that most influence the cost and productivity of large scale BDS (Luo et al., 2003). In one hand, microorganisms require water for their biocatalytic activity and high oil ratios are reported to be toxic to the biocatalyst, with strong impact in BDS (Caro, Boltes, et al., 2007) also hindering oxygen diffusion when aqueous phase is below 70% (Guchhait, Biswas, Bhattacharya, & Chowdhury, 2005b). But on the other hand, very low ratios would imply huge costs on a scale-up process, both in reactor scale and separation costs. Furthermore, high water contents decrease the solubility and bioavailability of hydrophobic substrates (Vazquez-Duhalt et al., 2002) and hinder the dissolution of BDS products in the organic phase, thus limiting process productivity (Kobayashi et al., 2001).

Strategies such as encapsulation, as above stated, can drastically increase oil/water ratios, since the water needed for the catalytic activity is immobilized together with the biocatalyst, thus allowing to foresee a future BDS process without the need of biphasic systems.

Bioreactor Design

Knowing the importance of an efficient mass transfer, a good contact surface between phases is needed aiming at getting the BDS system scale-up and has been the focus of several bioreactor design and techniques, which have been optimized to overcome the mass transport limitations.

Most initial bioreactor studies were performed in stirred tank reactors (STR) (Pacheco et al., 1999; Chowdhury & Duttagupta, 2005), and its application has been reported in batch, fed-batch, and continuous approach, with the last showing promising results for BDS scale-up (Schilling, Alvarez, Wang, & Cooney, 2002). The optimization of mass transport in this type of reactor is highly dependent on impeller design and speed (Abín-Fuentes, 2013) since a high input of energy and impeller speed is needed in order to remove mass transfer constraints. Monticello (2000) introduced a conceptual design for the continuous desulfurization of fuels. This author emphasized the need for three reactors working in tandem in order to achieve very low sulfur concentration oil. Yang et al., (2007) presented an interesting design of a continuous stirred tank reactor where aqueous and organic phases were kept partitioned, allowing for an actual separation between these phases with the obvious downstream advantages.

The elevated operational costs in power input for mixing and aeration, the high shear stress and poor air-liquid contact of STR's make the utilization of internal loop airlift reactors an interesting alternative (Talvy, Cockx, & Line, 2005) and, despite few oil application studies, a simulation performed by Nandi (2010) reports the applicability towards ultra-low sulfur diesel. Despite the advantages of these reactors, achieving a fine emulsion still requires high power input to surpass mass transfer constraints, and so electrically driven phase contactors could be used to disperse fine droplets between immiscible phases (T. C. Scott & Wham, 1989). This would allow lower power consumption and a reduction of the water demand, as low as only 5% (Kaufman et al., 1997).

All these types of free cells reactors, however, lead to difficulties in separation of the biocatalyst. To avoid this problem many researchers have focused on immobilized cell reactors, which allow for a simpler downstream process and continuous operation. A fluidized bed reactor (FBR) configuration was reported by the Argonne National Laboratory (ANL), where biocatalyst was immobilized in beads and was fluidized by oxygenated gasoline, aerated in a separated unit (McFarland et al., 1998). An alternative reactor design with immobilized cells was also reported by Amin (2011). This author assembled a Vertical Rotating Immobilized Cell Reactor (VRICR) where cells were grown attached to Immobilized Biomass Units (IBUs) in a first stage and carried out desulfurization in a second one. This design achieved 100% sulfur removal of a model oil continuously for more than 120 h. A very different design was developed by Mukhopadhyaya, Chowdhury, & Bhattacharya (2006): a bio-trickling reactor. In this system, the bacterial cells were immobilized in spheres and then submitted to a stream of oil in a small reactor, with controlled pressure and aeration, successfully achieving a reduction of 99% sulfur content in the tested oil, without the need to separate neither water nor biocatalyst from the final product.

Emulsifiers

Another strategy to improve mass transfer between water and hydrocarbon phases of a BDS system is the addition of surfactants (W. Li & Jiang, 2013). Surfactants are surface-active amphipathic agents with both hydrophobic and hydrophilic components, which act in the interphase of fluids with different polarities such as water-hydrocarbon, reducing surface tension and aiding in the emulsion or solubilization between phases (Rouse, Sabatini, Suflita, & Harwell, 1994). Surfactants have been reported to increase BDS yield, increase DBT diffusion rate in the aqueous phase and avoid end-product accumulation that is related to BDS inhibition (Caro, Letón, García-Calvo, & Setti, 2007). Many of these compounds have been applied to desulfurization. Patel et al. (1997) applied several surfactants to *R. erythropolis* IGTS8 desulfurization of DBT, using hexadecane as hydrocarbon phase and obtained interesting results with oleic diethanolamine and Triton N101, achieving 2-fold higher 2-HBP production.

Improvements were also obtained with Tween-80 on DBT desulfurization in hexadecane by *Corynebacterium* sp. ZD-1, resulting in a 50% increase on 2-HBP production (M. Wang, Li, Shi, Wang, & Feng, 2006). The same agent improved sulfur removal of a previously hydrotreated diesel from 76.8% to 91.2% by the microorganism *R. erythropolis* 1awq (Feng et al., 2006).

Surfactants can also be used to decrease the viscosity of heavy oils that are scarcely used due to their high sulfur content, thus increasing BDS efficiency in this type of oils. Biodesulfurization of MFO380 heavy oil with a mixed culture had a sulfur removal increase from 2.88% to 51.7%, when using the surfactant Triton X-100 (W. Li & Jiang, 2013). Kaufman, Borole, Shong, Sides, & Juengst (1999) also reported that surfactants can minimize the stabilization of asphaltene groups in crude oils when mixed with buffer salts, needed for BDS, which would lead to increased viscosity. Synthetic surfactants, however, have some issues, namely their non-biodegradability and potential toxicity both to the environment and to the desulfurizing microorganism (Banat, Makkar, & Cameotra, 2000; Patel et al., 1997). Different strains of *Pseudomonas aeruginosa*, *Rhodococcus erythropolis*, *Arthrobacter* sp., and *Bacillus subtilis* have been described as natural producers of biological equivalents of chemical surfactants (Batista, Mounteer, Amorim, & Totola, 2006; Cameotra & Singh, 2009; Dehghan-Noudeh, Housaindokht, & Bazzaz, 2005).

These biosurfactants are characterized by better physicochemical characteristics and environmental compatibility (Desai & Banat, 1997).

Biodesulfurizing microorganisms have to survive in hydrophobic environments and many can naturally produce biosurfactants such as phospholipids, rhamnolipids, and glycolipids (Bandyopadhyay, Chowdhury, Bhattacharjee, & Pan, 2013). Furthermore, 2-HBP can also act as a biosurfactant and its production is related to increased emulsification on water-oil mixtures (Bandyopadhyay et al., 2013). Since biosurfactants are produced during microbial growth, they tend to take time to have a significant effect, and so an external addition of a biosurfactant could be provided to improve BDS efficiency on an industrial scale. Amin, Bazaid, & Abd El-Halim (2013) designed a two-stage bioreactor where BDS of a model oil by *R. erythropolis* ATCC 53968 was coupled with a second bioreactor in which *B. subtilis* strain BDCC-TUSA-3, an efficient producer of surfactin, was generating the biosurfactant that was fed to the BDS bioreactor. Surfactin improved *R. erythropolis* cell growth 33% and the BDS rate 66%, which was better than the results with synthetic surfactants, and the BDS of the model oil decreased sulfur content from 398 to <5 ppm.

Bioemulsifiers, such as exopolysaccharides, are extra-cellular products that are not surface-active as biosurfactants. These are associated with microorganisms that are able to degrade oil substrates but do not produce biosurfactants (Ta-Chen, Chang, & Young, 2008). These agents, when compared to chemical surfactants, present higher emulsification of crude oils (Martínez-Checa, Toledo, Vilchez, Quesada, & Calvo, 2002), making them an interesting alternative to solve emulsification problems in BDS.

Separation

The best conditions for good mass transfer in the upstream of BDS, lead to the biggest constraints for the downstream part of the process. The tight emulsion needed for oil-cell-water contact must be broken in order to recover the three components (Monticello, 2000). While desulfurized oil must be clean of microbial contaminants and water to proceed in the refinery, recycling of the microbial biomass is vital for the industrial process, and byproducts separation must also be quick and cost-effective.

Bioreactor design, biocatalyst choice and immobilization, and the increase of o/w ratios can greatly improve separation. The two-layer bioreactor design with minimal mixing between water and oil phases (Yang et al., 2007) can significantly reduce the mixture of phases. Designs, where biocatalysts are immobilized, are also promising for this step of the process, both due to the easier recovery of the biocatalyst and increase of o/w ratios (Amin, 2011). The choice of hydrophobic biocatalysts such as *Rhodococci* and the use of emulsifiers, despite the advantages in mass transfer, can create very fine and stable emulsions between the cells and the oil, with cells adherent to oil phases, adding the need of effective and cost-friendly methods for phase separation (Borgne & Quintero, 2003).

Nowadays in industrial processes, centrifugation is one of the most used recovery methods, but its application in three-phase emulsions achieves no relevant separation, which combined with the high energy input needed for this method, turns this into a non-viable separation method, at least as a single step (Choi, Cho, Ryu, & Chang, 2003). An alternative to centrifugation can be the use of hydrocyclones, conical tubes specially designed for phase separation. The fluid is fed to the widest part and the spinning of these hydrocyclones can separate denser (water) from lighter (oil) fractions, as the denser fraction is directed to the outside of the tube and the lighter to the middle. Microbial biomass is associated with the oil-water interface thus, in a water-in-oil emulsion, these cells are associated with the water droplets. This allows a separation of a clean oil fraction in the first phase of de-oiling. Inverting phase emulsion of the overflow, through the addition of fresh oil, can make cells adhere to the oil droplets and thus the running through a second hydrocyclone will lead to a dehydration of the mix with a remaining small fraction of concentrated cells mixed with oil and a clean water phase, which can be recycled back to the bioreactor (Folsom, Meyer, & Yu, 1998). This method appears like an inexpensive and quick alternative to high energy methods such as centrifugation, with great impact on capital and operational costs (Monticello, 2000).

Methods based in filtration can also be an interesting alternative for phase separation (Koltuniewicz, Field, & Arnot, 1995; K. Scott, Jachuck, & Hall, 2001). Chen at al. (1996) patented a method based on a dual filtration system. This system involved the use of wetting agents for the sequenced filters that were: 1) miscible with liquid fossil fuel but not an aqueous phase, and 2) miscible with aqueous phase but not with liquid fossil fuel. This filtration sequence allows obtaining a clean fossil fuel filtrate on the first step, followed by the filtration of the retentate to ensure a clean water filtrate, which could then be recycled back to the BDS tank. The final retentate would be mainly comprised of the biocatalyst that could be subjected to purification procedure and reactivation in order to be also recycled back to BDS tank (Chen and Monticello, 1996). A different apparatus proposed by M. Konishi, Kishimoto, Tamesui, et al. (2005), based on the use of a polytetrafluroethylene (PTFE) tubular membrane, was able to separate model oil from a mixture of water in oil (50%, *v/v*) containing the biocatalyst in a continuous operation. Filtration methods, although potentially useful, present some constraints for continuous industrial use, namely the need for backwashing in order to avoid flux rate decrease and the high dependence on emulsion viscosity.

Another technique used in industry for phase separations is settling, in which separation is based on the density difference between phases. Settling tanks are present in oil refineries, but emulsified mixtures are much harder to separate by settling. Laboratory scale settling of BDS effluent has been applied for a continuous desulfurizing process (Schilling et al., 2002), but large scale application of this technique is unlikely to be effective for emulsified oil-water-biocatalyst effluents.

Destabilization of oil-water emulsions is a crucial step in BDS effluent separation, and different techniques are widely applied, namely in bioremediation and oil spills (Nordvik, Simmons, & Bitting, 1996). Coagulant agents such as aluminum and ferric sulfates have been reported to destabilize this kind of emulsions, allowing for recovery by flocculation (Al-Shamrani, James, & Xiao, 2002). Applied to BDS, however, these pose an obstacle as their toxicity to the biocatalyst can disable its recycling. Demulsifiers such as ethanol, however, present an interesting alternative due to its biocompatibility with the biocatalyst. Despite having little effect in surface tension, its application can reduce the viscosity of the emulsion and thus improve phase separation methods, without affecting BDS ability of the biocatalyst when recycled (Choi et al., 2003). The biodemulsification, *i.e.* the use of microorganisms to brake emulsions, can present itself as a more eco-friendly alternative in the future (Singh, Singh, & Ward, 2012), but no application to BDS has been yet studied, hence the difficulty in evaluating its applicability and cost-effectiveness.

The need to recover the biocatalyst from the BDS effluent for reapplication limits the techniques and chemicals that can be used for phase separation. If the biocatalyst can be successfully removed in advance, oil-water emulsions can be subjected to more effective, harsher separation methods. Immobilization techniques, despite very effective for biocatalyst removal, are usually related with a decrease of desulfurization rates (Naito et al., 2001). The upgrade with magnetic nanoparticles (such as magnetite) of the biocatalyst could emerge as a cost-effective, simple and fast process for biomass recovery (Guobin, Jianmin, Huaiying, & Huizhou, 2005; Haukanes & Kvam, 1993), as it would only require the application of an external magnetic field for the recovery of the biocatalyst. Several authors have applied different techniques and nanoparticles to coat desulfurizing microbial cells with success (Ansari et al., 2009; Guobin, Jianmin, Huaiying, et al., 2005; Y.-G. Li et al., 2009), and in all studies the desulfurizing ability was either unaffected or even improved (Ansari et al., 2009).

The wide array of techniques developed for the BDS downstream or that can be applied to it opens good perspectives for this part of the process towards the industrial application of BDS. These techniques should not be seen as mutually exclusive, but the best setup of integrated techniques should be found as the most effective and better up-scalable, always having in account its cost-effectiveness.

BDS Added-Value

Industrial application of biodesulfurization, amongst other factors, is dependent on its economic viability. The production of added-value products in parallel with BDS can ease the process cost and give economic sense to the changes needed in the refinery process (Bandyopadhyay et al., 2013). One type of byproducts produced in BDS are the biosurfactants, needed for the microorganism to thrive in highly hydrophobic environments (Bandyopadhyay, Chowdhury, & Bhattacharjee, 2014). When compared to chemical surfactants, biosurfactants have important advantages such as biodegradability and effectiveness at extreme temperatures and pH, and moreover lower toxicity. Besides oil refining enhancement, their application ranges from bioremediation, to healthcare, food processing and cosmetics (Mulligan, 2005; Perfumo, Smyth, Marchant, & Banat, 2010; L. Rodrigues, Banat, Teixeira, & Oliveira, 2006; L. R. Rodrigues & Teixeira, 2010; Urum & Pekdemir, 2004). Biosurfactants cannot yet compete with chemical surfactants due to their high production cost (Cameotra & Makkar, 1998), and here may lie the interest in the BDS technology, as biosurfactants can be produced as a byproduct of fuels desulfurization.

The final product of the 4S pathway (2-HBP) is a hydrotrope that can act as a surfactant (Mohebali et al., 2007). Other hydroxyl and carboxyl compounds, phospholipids, rhamnolipids, and glycolipids, amongst others, also with surfactant and emulsifier characteristics, have been described as extracellular byproducts of BDS process and could be recovered from BDS effluents (Bandyopadhyay et al., 2014).

Energy BioSystems Corporation design for a BDS industrial application included the production of surfactants from low-value oil. This technology implied modification of the 4S pathway: the removal of *dszB* gene from the desulfurizing microorganism stopped the 4S pathway in the production of hydroxyphenyl benzene sulfonate (HPBS) byproducts, which could then be converted into derivative molecules with the application as surfactants (Monticello, 2000). In fact, at an industrial level, the production of high added-value biosurfactants coupled to BDS can greatly improve the overall process economics since their production costs will be included within the sulfur removal bioprocess, performed at mild operational conditions, from a low-value feedstock.

Another type of byproducts that can be exploited to turn BDS process more cost-effective are carotenoids (liposoluble pigments). Other than their colour and light harvesting properties, carotenoids have diverse bioactive and chemical properties. These compounds can present antioxidant activities, with specific roles or general health-promoting roles that reduce the risk or progression of diseases associated with oxidative stress. They can be precursors to vitamins, have antitumor and antimicrobial properties, and act as oxygen transporters, etc. (Berman et al., 2015; Kushwaha, Saini, Saraswat, Agarwal, & Saxena, 2014; Esteban, Moran, Becerril,

& García-Plazaola, 2015; Goswami, Chaudhuri, & Dutta, 2010; Ramachandran, Iqbal, & Amirul, 2014). Indeed, there is a vast market for carotenoids, generating interest in industries such as pharmaceutical, nutraceutical, food/feed additive, textile, cosmetics, and fine chemical sectors, which results in an estimated market value of about $2.0 billion in 2022 (McWilliams, 2018).

Carotenoids produced by microorganisms are nontoxic and the use of microorganisms (microalgae, yeasts, and bacteria) as pigment-producers have several advantages. Unlike plants, microorganisms have a short life cycle and are unaffected by the season/climate and geographical conditions, making it easier to produce carotenoids worldwide. Microorganisms can also produce colour shades that are not found in plants; have high productivity; and they are relatively easy of genetically manipulation towards improved fermentation and scale-up process. Moreover, microbial pigments are usually easy to extract (Calegari-Santos, Diogo, Fontana, & Bonfim, 2016; Kushwaha et al., 2014; Ramachandran et al., 2014; Venil, Aruldass, Dufossé, Zakaria, & Ahmad, 2014). Thus, microbial pigments are in increasing demand, since they are a promising natural and safe alternative source for various industrial applications.

Recently, Silva, Paixão, & Alves (2016) and Fernandes, Paixão, Silva, Roseiro, & Alves (2018) highlighted the great potential of a desulfurizing microorganism, G. alkanivorans strain 1B, as a new hyper-pigment producer to be exploited towards several applications.

Integration Into a Refinery Process

The incorporation of BDS into a refining process arises several issues, even if the constraints focused above are all dealt with. First and foremost, BDS has to operate at the same speed and reliability as other refining processes if it is to be integrated, which emerges as a challenge when using microorganisms in a large scale operation, not initially prepared for that purpose (Kilbane & Le Borgne, 2004).

The decision of where to insert and what type of oils BDS should treat in the refinery process is another issue to be dealt with. Ultra-low sulfur diesel (ULSD) production is one of the potential goals of BDS, knowing the very strict levels that international regulations have been establishing and the difficulty of HDS in removing the recalcitrant organosulfur compounds.

Some authors refer to the application of BDS before HDS as the best option for an integrated process into a refinery towards ultra-low sulfur fuels. This upstream approach permits to remove a great portion of the recalcitrant organosulfur compounds prior to HDS, making it more efficient and less energetically costly since will need lower hydrogen input (Stanislaus, Marafi, & Rana, 2010). In an upstream design, there are authors defending that the best application of BDS would be in crude oil since it has higher water content than diesel or gasoline (Zhou & Zhang, 2004). An integrated process enclosing BDS to upgrade crude oils might reduce their recalcitrance, before being sent to the refinery, improving the effectiveness of the current refinery processes and enabling the utilization of certain low-quality oils that cannot be treated with conventional technology (Kilbane, 2006).

In contrast, other authors defend that BDS has the potential of being developed as a viable technology downstream of classical HDS (Monticello, 2000; Gupta et al., 2005; Mukhopadhyaya et al., 2007; Yang, et al., 2007; Bandyopadhyay et al., 2013, 2014). In this approach, BDS will deal only with the recalcitrant molecules that the HDS is unable to remove.

There are very few reports on BDS process designs and cost analysis (Gupta et al., 2005). Recently, Alves, Paixão, Pacheco, Ferreira, & Silva (2015) reported a study where two BDS process designs were analyzed in terms of energy consumption, greenhouse gas emissions (GHG) and costs. This study pointed out for the application of the BDS downstream HDS as the best cost-effective conceptual design to apply to an oil refinery. Once it is able of desulfurizing HDS recalcitrant compounds selectively, BDS integration may be led to the accomplishment of the stringent European limit of <10 ppm for S-content on fuels, which otherwise may imply the necessity of more severe conditions within HDS units. Deep desulfurization is a very costly option and is not environmental friendly because implies higher GHG emissions and substantially increase the carbon footprint.

CONCLUSION

Although the new environmental regulations that limit the sulfur levels of diesel and other transportation fuels to very low levels are beneficial from an environmental point of view, meeting the required stringent specifications represent a major operational and economic challenge for the petroleum refining industry. The tightening of sulfur specifications of diesel/gasoline fuel to very low levels requires ultra-deep desulfurization of diesel/gasoline feed stream. The shift from normal to ultra-deep desulfurization is a very complicated technical problem thus demanding to find cost-effective ways for ultra-low sulfur fuel production.

BDS has drawn wide attention because of its green processing of fossil fuel. Bioprocesses can potentially provide a solution to the need for improved and expanded fuel upgrading worldwide because bioprocesses for fuel upgrading do not require hydrogen and produce far less greenhouse gas (GHG) emission than thermochemical processes. In the last decade, many reviews have been written on the topic of biotechnology for fossil fuels and biodesulfurization, but most have centered on the isolation of new microorganisms, the description of the metabolic pathways and the increase of BDS activity through genetic enhancing.

This work aimed to focus the current problems and efforts being made towards the BDS reduction of costs and increase of its efficiency, both required for an industrially feasible process. In this context, different formulations of culture media have been described and optimized for each biocatalyst studied, reducing the concentration of inhibitors, increasing desulfurization activity and biocatalyst longevity, and exploring cheaper alternatives for the current commercial reagents. Moreover, the utilization of desulfurizing microorganisms that can grow in low nutrient culture media without vitamins and other growth promoters, like yeast extract, peptone, triptone, etc, is an advantage for BDS upgrade since it may reduce the biocatalyst production costs significantly.

Process-wise, mass transfer, and consequently oil/water ratios, have been identified as two of the greatest problems of the biodesulfurization technology. To surpass them, many approaches have been explored, either by using different types of bioreactors, adding substances such as emulsifiers or biosurfactants to the reaction media, or by looking for better ways to apply the biocatalyst such as encapsulation.

In overall, for BDS to become a reality it is fundamental to achieve a good integration into the oil refinery process. In this field, the biggest setback commonly identified is the separation of the three components characteristic of this process, *i.e.* oil, water, and biocatalyst. Different mechanical solutions have been explored, such as filtration, centrifugation and the use of hydrocyclones. However, recently, other approaches have been more in focus including the modification of the biocatalyst through magnetic nanoparticles, or the use of immobilized microorganisms. These techniques may result in a cheaper process to separate and reutilize the biocatalyst.

The use of biotechnology to industrially desulfurize oil is still far from being economically appealing. Even after optimizing every drawback, probably, there will be a need to take advantage of the high added-value products resulting from the growth of the biocatalyst. In fact, many authors have speculated that the production of biosurfactants, characteristic of most desulfurizing microorganisms, could be the decisive factor to balance the economical scale-up of this process, and even make it more profitable than HDS.

Much work is still required to do. With the increasing pressure being made by the legislators towards the reductions of sulfur levels in fossil fuels, as well as the ever-increasing demand for energy, BDS should continue to be explored. It becomes fundamental for the industry to discover new and cheaper ways to process heavier and more contaminated oils. Biodesulfurization coupled with other biological treatments, such as biodenitrification, biological removal of metals and reduction of viscosity, will definitely play a crucial role in the future of sustainable energy from fossil fuels.

ACKNOWLEDGMENT

The present work was financed by FEDER funds through POFC-COMPETE and by national funds through FCT (Fundação para a Ciência e a Tecnologia) in the scope of project Carbon4Desulf - FCOMP-01-0124-FEDER-013932.

REFERENCES

Abbad-Andaloussi, S., Lagnel, C., Warzywoda, M., & Monot, F. (2003). Multi-criteria comparison of resting cell activities of bacterial strains selected for biodesulfurization of petroleum compounds. *Enzyme and Microbial Technology*, *32*(3), 446–454. doi:10.1016/S0141-0229(02)00320-4

Abín-Fuentes, A. (2013). *Mechanistic understanding of microbial desulfurization* (Doctoral dissertation). Massachusetts Institute of Technology.

Abo-state, M. A., El-Gendy, N. S., El-Temtamy, S. A., Mahdy, H. M., & Nassar, H. N. (2014). Modification of basal salts medium for enhancing dibenzothiophene biodesulfurization by *Brevibacillus invocatus* C19 and *Rhodococcus erythropolis* IGTS8. *World Applied Sciences Journal*, *30*(2), 133–140.

Aggarwal, S., Karimi, I., & Lee, D. Y. (2011). Reconstruction of a genome-scale metabolic network of *Rhodococcus erythropolis* for desulfurization studies. *Molecular BioSystems*, *7*(11), 3122–3131. doi:10.1039/c1mb05201b PMID:21912787

Aggarwal, S., Karimi, I. A., & Ivan, G. R. (2013). In silico modeling and evaluation of *Gordonia alkanivorans* for biodesulfurization. *Molecular BioSystems*, *9*(10), 2530–2540. doi:10.1039/c3mb70132h PMID:23921469

Aggarwal, S., Karimi, I. A., Kilbane, J. J. II, & Lee, D. Y. (2012). Roles of sulfite oxidoreductase and sulfite reductase in improving desulfurization by *Rhodococcus erythropolis*. *Molecular BioSystems*, *8*(10), 2724–2732. doi:10.1039/c2mb25127b PMID:22832889

Al-Shamrani, A. A., James, A., & Xiao, H. (2002). Destabilisation of oil–water emulsions and separation by dissolved air flotation. *Water Research*, *36*(6), 1503–1512. doi:10.1016/S0043-1354(01)00347-5 PMID:11996340

Alves, L., Marques, S., Matos, J., Tenreiro, R., & Gírio, F. M. (2008). Dibenzothiophene desulfurization by *Gordonia alkanivorans* strain 1B using recycled paper sludge hydrolyzate. *Chemosphere*, *70*(6), 967–973. doi:10.1016/j.chemosphere.2007.08.016 PMID:17897697

Alves, L., Matos, J., Tenreiro, R., & Gírio, F. M. (2008). Evidence for the role of zinc on the performance of dibenzothiophene desulfurization by *Gordonia alkanivorans* strain 1B. *Journal of Industrial Microbiology & Biotechnology, 35*(1), 69–73. doi:10.100710295-007-0278-5 PMID:18043954

Alves, L., & Paixão, S. M. (2014a). Enhancement of dibenzothiophene desulfurization by *Gordonia alkanivorans* strain 1B using sugar beet molasses as alternative carbon source. *Applied Biochemistry and Biotechnology, 172*(6), 3297–3305. doi:10.100712010-014-0763-z PMID:24519629

Alves, L., & Paixão, S. M. (2014b). Fructophilic behaviour of *Gordonia alkanivorans* strain 1B during dibenzothiophene desulfurization process. *New Biotechnology, 31*(1), 73–79. doi:10.1016/j.nbt.2013.08.007 PMID:24012483

Alves, L., Paixão, S. M., Pacheco, R., Ferreira, A. F., & Silva, C. M. (2015). Biodesulphurization of fossil fuels: Energy, emissions and cost analysis. *RSC Advances, 5*(43), 34047–34057. doi:10.1039/C4RA14216K

Alves, L., Salgueiro, R., Rodrigues, C., Mesquita, E., Matos, J., & Gírio, F. M. (2005). Desulfurization of dibenzothiophene, benzothiophene, and other thiophene analogs by a newly isolated bacterium, *Gordonia alkanivorans* strain 1B. *Applied Biochemistry and Biotechnology, 120*(3), 199–208. doi:10.1385/ABAB:120:3:199 PMID:15767694

Amin, G. A. (2011). Integrated two-stage process for biodesulfurization of model oil by vertical rotating immobilized cell reactor with the bacterium *Rhodococcus erythropolis. Journal of Petroleum & Environmental Biotechnology, 2*(1), 107.

Amin, G. A., Bazaid, S. A., & Abd El-Halim, M. (2013). A two-stage immobilized cell bioreactor with *Bacillus subtilis* and *Rhodococcus erythropolis* for the simultaneous production of biosurfactant and biodesulfurization of model oil. *Petroleum Science and Technology, 31*(21), 2250–2257. doi:10.1080/10916466.2011.565292

Ansari, F., Grigoriev, P., Libor, S., Tothill, I. E., & Ramsden, J. J. (2009). DBT degradation enhancement by decorating *Rhodococcus erythropolis* IGST8 with magnetic Fe_3O_4 nanoparticles. *Biotechnology and Bioengineering, 102*(5), 1505–1512. doi:10.1002/bit.22161 PMID:19012265

Arez, B. F., Alves, L., & Paixão, S. M. (2014). Production and characterization of a novel yeast extracellular invertase activity towards improved dibenzothiophene biodesulfurization. *Applied Biochemistry and Biotechnology, 174*(6), 2048–2057. doi:10.100712010-014-1182-x PMID:25163885

Babich, I. V., & Moulijn, J. A. (2003). Science and technology of novel processes for deep desulfurization of oil refinery streams: A review. *Fuel*, *82*(6), 607–631. doi:10.1016/S0016-2361(02)00324-1

Banat, I. M., Makkar, R. S., & Cameotra, S. S. (2000). Potential commercial applications of microbial surfactants. *Applied Microbiology and Biotechnology*, *53*(5), 495–508. doi:10.1007002530051648 PMID:10855707

Bandyopadhyay, S., Chowdhury, R., & Bhattacharjee, C. (2014). Production of biosurfactants through biodesulfurization of spent engine oil – an experimental study. *Green Chemistry Letters and Reviews*, *7*(3), 288–295. doi:10.1080/17518253.2014.946972

Bandyopadhyay, S., Chowdhury, R., Bhattacharjee, C., & Pan, S. (2013). Simultaneous production of biosurfactant and ULSD (ultra low sulfur diesel) using *Rhodococcus* sp. in a chemostat. *Fuel*, *113*, 107–112. doi:10.1016/j.fuel.2013.05.036

Bardania, H., Raheb, J., Mohammad-Beigi, H., Rasekh, B., & Arpanaei, A. (2013). Desulfurization activity and reusability of magnetite nanoparticle–coated *Rhodococcus erythropolis* FMF and *R. erythropolis* IGTS8 bacterial cells. *Biotechnology and Applied Biochemistry*, *60*(3), 323–329. doi:10.1002/bab.1090 PMID:23656694

Batista, S. B., Mounteer, A. H., Amorim, F. R., & Totola, M. R. (2006). Isolation and characterization of biosurfactant/bioemulsifier-producing bacteria from petroleum contaminated sites. *Bioresource Technology*, *97*(6), 868–875. doi:10.1016/j.biortech.2005.04.020 PMID:15951168

Bender, J., & Weigel, H.-J. (2011). Changes in atmospheric chemistry and crop health: A review. *Agronomy for Sustainable Development*, *31*(1), 81–89. doi:10.1051/agro/2010013

Berman, J., Zorrilla-López, U., Farré, G., Zhu, C., Sandmann, G., Twyman, R. M., ... Christou, P. (2015). Nutritionally important carotenoids as consumer products. *Phytochemistry Reviews*, *14*(5), 727–743.

Bhatia, S., & Sharma, D. K. (2012). Thermophilic desulfurization of dibenzothiophene and different petroleum oils by *Klebsiella* sp. 13T. *Environmental Science and Pollution Research International*, *19*(8), 3491–3497. doi:10.100711356-012-0884-2 PMID:22467237

Blandino, A., Macías, M., & Cantero, D. (1999). Formation of calcium alginate gel capsules: Influence of sodium alginate and $CaCl_2$ concentration on gelation kinetics. *Journal of Bioscience and Bioengineering*, *88*(6), 686–689. doi:10.1016/S1389-1723(00)87103-0 PMID:16232687

Boniek, D., Figueiredo, D., dos Santos, A. F. B., & Stoianoff, M. A. de R. (2014). Biodesulfurization: A mini review about the immediate search for the future technology. *Clean Technologies and Environmental Policy*.

Calegari-Santos, R., Diogo, R. A., Fontana, J. D., & Bonfim, T. M. B. (2016). Carotenoid production by halophilic Archaea under different culture conditions. *Current Microbiology*, *72*(5), 641–651.

Cameotra, S. S., & Makkar, R. S. (1998). Synthesis of biosurfactants in extreme conditions. *Applied Microbiology and Biotechnology*, *50*(5), 520–529. doi:10.1007002530051329 PMID:9866170

Cameotra, S. S., & Singh, P. (2009). Synthesis of rhamnolipid biosurfactant and mode of hexadecane uptake by *Pseudomonas* species. *Microbial Cell Factories*, *8*(1), 16. doi:10.1186/1475-2859-8-16 PMID:19284586

Caro, A., Boltes, K., Letón, P., & García-Calvo, E. (2007). Dibenzothiophene biodesulfurization in resting cell conditions by aerobic bacteria. *Biochemical Engineering Journal*, *35*(2), 191–197. doi:10.1016/j.bej.2007.01.013

Caro, A., Letón, P., García-Calvo, E., & Setti, L. (2007). Enhancement of dibenzothiophene biodesulfurization using β-cyclodextrins in oil-to-water media. *Fuel*, *86*(16), 2632–2636. doi:10.1016/j.fuel.2007.02.033

Carvalheiro, F., Moniz, P., Duarte, L. C., Esteves, M. P., & Gírio, F. M. (2011). Mannitol production by lactic acid bacteria grown in supplemented carob syrup. *Journal of Industrial Microbiology & Biotechnology*, *38*(1), 221–227. doi:10.100710295-010-0823-5 PMID:20820868

Chang, J. H., Chang, Y. K., Ryu, H. W., & Chang, H. N. (2000). Desulfurization of light gas oil in immobilized-cell systems of *Gordona* sp. CYKS1 and *Nocardia* sp. CYKS2. *FEMS Microbiology Letters*, *182*(2), 309–312. doi:10.1111/j.1574-6968.2000.tb08913.x PMID:10620684

Chang, J. H., Kim, Y. J., Lee, B. H., Cho, K., Ryu, H. W., Chang, Y. K., & Chang, H. N. (2001). Production of a desulfurization biocatalyst by two-stage fermentation and its application for the treatment of model and diesel oils. *Biotechnology Progress*, *17*(5), 876–880. doi:10.1021/bp0100676 PMID:11587578

Chen, J. C., & Monticello, D. J. (1996). *U.S. Patent No. 5,525,235*. Washington, DC: U.S. Patent and Trademark Office.

Cheng, Y., Zhou, W., Gao, C., Lan, K., Gao, Y., & Wu, Q. (2009). Biodiesel production from Jerusalem artichoke (*Helianthus tuberosus* L.) tuber by heterotrophic microalgae *Chlorella protothecoides*. *Journal of Chemical Technology and Biotechnology (Oxford, Oxfordshire)*, *84*(5), 777–781. doi:10.1002/jctb.2111

Choi, O. K., Cho, K. S., Ryu, H. W., & Chang, Y. K. (2003). Enhancement of phase separation by the addition of de-emulsifiers to three-phase (diesel oil / biocatalyst / aqueous phase) emulsion in diesel biodesulfurization. *Biotechnology Letters*, *25*(1), 73–77. doi:10.1023/A:1021750719337 PMID:12882310

Chowdhury, R., & Duttagupta, A. (2005). Mathematical Modeling of the transient behaviour of a chemostat undergoing bio-desulfurization of model organosulfur compounds and diesel using pure and isolated bacterial strains. *International Journal of Chemical Reactor Engineering*, *3*(1), 1–12.

Dai, Y., Shao, R., Qi, G., & Ding, B.-B. (2014). Enhanced dibenzothiophene biodesulfurization by immobilized cells of *Brevibacterium lutescens* in n-octane-water biphasic system. *Applied Biochemistry and Biotechnology*, *174*(6), 2236–2244. doi:10.100712010-014-1184-8 PMID:25173674

Dehghan-Noudeh, G., Housaindokht, M., & Bazzaz, B. S. F. (2005). Isolation, characterization, and investigation of surface and hemolytic activities of a lipopeptide biosurfactant produced by *Bacillus subtilis* ATCC 6633. *Journal Of Microbiology-Seoul*, *43*(3), 272. PMID:15995646

Del Olmo, C. H., Santos, V. E., Alcon, A., & Garcia-Ochoa, F. (2005). Production of a *Rhodococcus erythropolis* IGTS8 biocatalyst for DBT biodesulfurization: Influence of operational conditions. *Biochemical Engineering Journal*, *22*(3), 229–237. doi:10.1016/j.bej.2004.09.015

Denis, P. A. (2010). On the enthalpy of formation of thiophene. *Theoretical Chemistry Accounts*, *127*(5-6), 621–626. doi:10.100700214-010-0759-x

Derikvand, P., Etemadifar, Z., & Biria, D. (2014). Taguchi optimization of dibenzothiophene biodesulfurization by *Rhodococcus erythropolis* R1 immobilized cells in a biphasic system. *International Biodeterioration & Biodegradation*, *86*, 343–348. doi:10.1016/j.ibiod.2013.10.006

Desai, J. D., & Banat, I. M. (1997). Microbial production of surfactants and their commercial potential. *Microbiology and Molecular Biology Reviews*, *61*(1), 47–64. PMID:9106364

Dinamarca, M. A., Ibacache-Quiroga, C., Baeza, P., Galvez, S., Villarroel, M., Olivero, P., & Ojeda, J. (2010). Biodesulfurization of gas oil using inorganic supports biomodified with metabolically active cells immobilized by adsorption. *Bioresource Technology*, *101*(7), 2375–2378. doi:10.1016/j.biortech.2009.11.086 PMID:20034786

Dinamarca, M. A., Rojas, A., Baeza, P., Espinoza, G., Ibacache-Quiroga, C., & Ojeda, J. (2014). Optimizing the biodesulfurization of gas oil by adding surfactants to immobilized cell systems. *Fuel*, *116*, 237–241. doi:10.1016/j.fuel.2013.07.108

Esteban, R., Moran, J. F., Becerril, J. M., & García-Plazaola, J. I. (2015). Versatility of carotenoids: An integrated view on diversity, evolution, functional roles and environmental interactions. *Environmental and Experimental Botany*, *119*, 63–75.

Feng, J., Zeng, Y., Ma, C., Cai, X., Zhang, Q., Tong, M., ... Xu, P. (2006). The surfactant tween 80 enhances biodesulfurization. *Applied and Environmental Microbiology*, *72*(11), 7390–7393. doi:10.1128/AEM.01474-06 PMID:16980422

Fernandes, A. S., Paixão, S. M., Silva, T. P., Roseiro, J. C., & Alves, L. (2018). Influence of culture conditions towards optimal carotenoid production by Gordonia alkanivorans strain 1B. *Bioprocess and Biosystems Engineering*, *41*(2), 143–155.

Foght, J. M. (2004). Whole-cell bio-processing of aromatic compounds in crude oil and fuels. In R. Vazquez-Duhalt & R. Quintero-Ramirez (Eds.), Petroleum biotechnology: Developments and perspectives (pp. 145–176). Amsterdam: Elsevier. doi:10.1016/S0167-2991(04)80146-0

Folsom, B. R., Meyer, T. A., & Yu, L. Q. (1998). *U.S. Patent No. 5,772,901*. Washington, DC: U.S. Patent and Trademark Office.

Goswami, G., Chaudhuri, S., & Dutta, D. (2010). Effect of pH and temperature on pigment production from an isolated bacterium. *Chemical Engineering Transactions*, *20*, 127–132.

Gray, K., Mrachko, G., & Squires, C. (2003). Biodesulfurization of fossil fuels. *Current Opinion in Microbiology*, *6*(3), 229–235. doi:10.1016/S1369-5274(03)00065-1 PMID:12831898

Guchhait, S., Biswas, D., Bhattacharya, P., & Chowdhury, R. (2005a). Bio-desulfurization of model organo-sulfur compounds and hydrotreated diesel - Experiments and modelling. *Chemical Engineering Journal*, *112*(1-3), 145–151. doi:10.1016/j.cej.2005.05.006

Guchhait, S., Biswas, D., Bhattacharya, P., & Chowdhury, R. (2005b). Bi-phasic bio-conversion of sulfur present in model organo-sulfur compounds and hydro-treated diesel. *Catalysis Today*, *106*(1-4), 233–237. doi:10.1016/j.cattod.2005.07.137

Gunam, I. B. W., Sujaya, I. N., Aryanta, W. R., Tomita, F., & Asano, K. (2013). Biodesulfurization of dibenzothiophene and its derivatives using resting and immobilized cells of *Sphingomonas subarctica* T7b. *Journal of Microbiology and Biotechnology*, *23*(4), 473–482. doi:10.4014/jmb.1207.07070 PMID:23568201

Guo, L., Zhang, J., Hu, F., Dy Ryu, D., & Bao, J. (2013). Consolidated bioprocessing of highly concentrated Jerusalem artichoke tubers for simultaneous saccharification and ethanol fermentation. *Biotechnology and Bioengineering*, *110*(10), 2606–2615. doi:10.1002/bit.24929 PMID:23568827

Guo, X.-L., Deng, G., Xu, J., & Wang, M.-X. (2006). Immobilization of *Rhodococcus* sp. AJ270 in alginate capsules and its application in enantioselective biotransformation of trans-2-methyl-3-phenyl-oxiranecarbonitrile and amide. *Enzyme and Microbial Technology*, *39*(1), 1–5. doi:10.1016/j.enzmictec.2005.01.044

Guobin, S., Huaiying, Z., Jianmin, X., Guo, C., Wangliang, L., & Huizhou, L. (2006). Biodesulfurization of hydrodesulfurized diesel oil with *Pseudomonas delafieldii* R-8 from high density culture. *Biochemical Engineering Journal*, *27*(3), 305–309. doi:10.1016/j.bej.2005.07.003

Guobin, S., Huaiying, Z., Weiquan, C., Jianmin, X., & Huizhou, L. (2005). Improvement of biodesulfurization rate by assembling nanosorbents on the surfaces of microbial cells. *Biophysical Journal*, *89*(6), L58–L60. doi:10.1529/biophysj.105.073718 PMID:16258046

Guobin, S., Jianmin, X., Chen, G., Huizhou, L., & Jiayong, C. (2005). Biodesulfurization using *Pseudomonas delafieldii* in magnetic polyvinyl alcohol beads. *Letters in Applied Microbiology*, *40*(1), 30–36. doi:10.1111/j.1472-765X.2004.01617.x PMID:15612999

Guobin, S., Jianmin, X., Huaiying, Z., & Huizhou, L. (2005). Deep desulfurization of hydrodesulfurized diesel oil by *Pseudomonas delafieldii* R-8. *Journal of Chemical Technology & Biotechnology*, *80*, 420–424.

Gupta, N., Roychoudhury, P. K., & Deb, J. K. (2005). Biotechnology of desulfurization of diesel: Prospects and challenges. *Applied Microbiology and Biotechnology*, *66*(4), 356–366. doi:10.100700253-004-1755-7 PMID:15538557

Hashimoto, S., & Furukawa, K. (1987). Immobilization of activated sludge by PVA–boric acid method. *Biotechnology and Bioengineering*, *30*(1), 52–59. doi:10.1002/bit.260300108 PMID:18576582

Haukanes, B.-I., & Kvam, C. (1993). Application of magnetic beads in bioassays. *Nature Biotechnology*, *11*(1), 60–63. doi:10.1038/nbt0193-60 PMID:7763485

He, Y.-C., Zhou, Q., Ma, C.-L., Cai, Z.-Q., Wang, L.-Q., Zhao, X.-Y., ... Wang, X.-D. (2012). Biosynthesis of benzoylformic acid from benzoyl cyanide by a newly isolated *Rhodococcus* sp. CCZU10-1 in toluene–water biphasic system. *Bioresource Technology*, *115*, 88–95. doi:10.1016/j.biortech.2011.09.084 PMID:22033370

Hirasawa, K., Ishii, Y., Kobayashi, M., Koizumi, K., & Maruhashi, K. (2001). Improvement of desulfurization activity in *Rhodococcus erythropolis* KA2-5-1 by genetic engineering. *Bioscience, Biotechnology, and Biochemistry*, *65*(2), 239–246. doi:10.1271/bbb.65.239 PMID:11302154

Ho, K. F., & Lee, S. C. (2002). Identification of atmospheric volatile organic compounds (VOCs), polycyclic aromatic hydrocarbons (PAHs) and carbonyl compounds in Hong Kong. *The Science of the Total Environment*, *289*(1-3), 145–158. doi:10.1016/S0048-9697(01)01031-2 PMID:12049391

Honda, H., Sugiyama, H., Saito, I., & Kobayashi, T. (1998). High cell density culture of *Rhodococcus rhodochrous* by pH-stat feeding and dibenzothiophene degradation. *Journal of Fermentation and Bioengineering*, *85*(3), 334–338. doi:10.1016/S0922-338X(97)85685-1

Huaiying, Z., Guobin, S., & Huizhou, L., & JianMin, X. (2007). Surface modification of γ-Al_2O_3 nano-particles with gum arabic and its applications in adsorption and biodesulfurization. *Surface and Coatings Technology*, *201*(16), 6917–6921.

Jiang, N., Ying, G. L., Liu, S. Y., Shen, L., Hu, J., Dai, L. J., ... Su, B. L. (2014). Amino acid-based biohybrids for nano-shellization of individual desulfurizing bacteria. *Chemical Communications (Cambridge)*, *50*(97), 15407–15410. doi:10.1039/C4CC06323F PMID:25350878

Kaufman, E. N., Borole, A. P., Shong, R., Sides, J. L., & Juengst, C. (1999). Sulfur specificity in the bench-scale biological desulfurization of crude oil by *Rhodococcus*. *Journal of Chemical Technology and Biotechnology (Oxford, Oxfordshire)*, *74*(10), 1000–1004. doi:10.1002/(SICI)1097-4660(199910)74:10<1000::AID-JCTB134>3.0.CO;2-0

Kaufman, E. N., Harkins, J. B., & Borole, A. P. (1998). Comparison of Batch-Stirred and Electro-Spray Reactors for Biodesulfurization of dibenzothiophene in Crude Oil and Hydrocarbon Feedstocks. *Applied Biochemistry and Biotechnology*, *73*(2-3), 127–144. doi:10.1007/BF02785650

Kaufman, E. N., Harkins, J. B., Rodriguez, M. Jr, Tsouris, C., Selvaraj, P. T., & Murphy, S. E. (1997). Development of an electro-spray bioreactor for crude oil processing. *Fuel Processing Technology*, *52*(1), 127–144. doi:10.1016/S0378-3820(97)00022-2

Khedkar, S., & Shanker, R. (2014). Isolation and classification of a soil actinomycete capable of sulphur-specific biotransformation of dibenzothiophene, benzothiophene and thianthrene. *Journal of Applied Microbiology*. PMID:25319398

Kilbane, J. J. II. (2006). Microbial biocatalyst developments to upgrade fossil fuels. *Current Opinion in Biotechnology*, *17*(3), 305–314. doi:10.1016/j.copbio.2006.04.005 PMID:16678400

Kilbane, J. J. II, & Le Borgne, S. (2004). Petroleum biorefining: The selective removal of sulfur, nitrogen, and metals. *Studies in Surface Science and Catalysis*, *151*, 29–65. doi:10.1016/S0167-2991(04)80143-5

Kirimura, K., Harada, K., Iwasawa, H., Tanaka, T., Iwasaki, Y., Furuya, T., ... Kino, K. (2004). Identification and functional analysis of the genes encoding dibenzothiophene-desulfurizing enzymes from thermophilic bacteria. *Applied Microbiology and Biotechnology*, *65*(6), 703–713. doi:10.100700253-004-1652-0 PMID:15221222

Kobayashi, M., Horiuchi, K., Yoshikawa, O., Hirasawa, K., Ishii, Y., Fujino, K., ... Maruhashi, K. (2001). Kinetic analysis of microbial desulfurization of model and light gas oils containing multiple alkyl dibenzothiophenes. *Bioscience, Biotechnology, and Biochemistry*, *65*(2), 298–304. doi:10.1271/bbb.65.298 PMID:11302162

Kobayashi, M., Onaka, T., Ishii, Y., Konishi, J., Takaki, M., Okada, H., ... Suzuki, M. (2000). Desulfurization of alkylated forms of both dibenzothiophene and benzothiophene by a single bacterial strain. *FEMS Microbiology Letters*, *187*(2), 123–126. doi:10.1111/j.1574-6968.2000.tb09147.x PMID:10856644

Koltuniewicz, A. B., Field, R. W., & Arnot, T. C. (1995). Cross-flow and dead-end microfiltration of oily-water emulsion. Part I: Experimental study and analysis of flux decline. *Journal of Membrane Science*, *102*, 193–207. doi:10.1016/0376-7388(94)00320-X

Konishi, J., Ishii, Y., Onaka, T., Okumura, K., & Suzuki, M. (1997). Thermophilic carbon-sulfur-bond-targeted biodesulfurization. *Applied and Environmental Microbiology*, *63*(8), 3164–3169. PMID:16535672

Konishi, M., Kishimoto, M., Omasa, T., Katakura, Y., Shioya, S., & Ohtake, H. (2005). Effect of sulfur sources on specific desulfurization activity of *Rhodococcus erythropolis* KA2-5-1 in exponential fed-batch culture. *Journal of Bioscience and Bioengineering*, *99*(3), 259–263. doi:10.1263/jbb.99.259 PMID:16233786

Konishi, M., Kishimoto, M., Tamesui, N., Omasa, T., Shioya, S., & Ohtake, H. (2005). The separation of oil from an oil–water–bacteria mixture using a hydrophobic tubular membrane. *Biochemical Engineering Journal*, *24*(1), 49–54. doi:10.1016/j.bej.2004.11.012

Kushwaha, K., & Saini, A. Saraswa,t P., Agarwal, M. K., & Saxena, J. (2014). Colorful world of microbes: Carotenoids and their applications. Advances in Biology 2014, Article ID 837891.

Le Borgne, S., & Quintero, R. (2003). Biotechnological processes for the refining of petroleum. *Fuel Processing Technology*, *81*(2), 155–169. doi:10.1016/S0378-3820(03)00007-9

Lee, M. K., Senius, J. D., & Grossman, M. J. (1995). Sulfur-specific microbial desulfurization of sterically hindered analogs of dibenzothiophene. *Applied and Environmental Microbiology*, *61*(12), 4362–4366. PMID:16535189

Li, F., Xu, P., Feng, J., Meng, L., Zheng, Y., Luo, L., & Ma, C. (2005). Microbial desulfurization of gasoline in a *Mycobacterium goodii* X7B immobilized-cell system. *Applied and Environmental Microbiology*, *71*(1), 276–281. doi:10.1128/AEM.71.1.276-281.2005 PMID:15640198

Li, L., Li, L., Wang, Y., Du, Y., & Qin, S. (2013). Biorefinery products from the inulin-containing crop Jerusalem artichoke. *Biotechnology Letters*, *35*(4), 471–477. doi:10.100710529-012-1104-3 PMID:23242496

Li, M. Z., Squires, C. H., Monticello, D. J., & Childs, J. D. (1996). Genetic analysis of the dsz promoter and associated regulatory regions of *Rhodococcus erythropolis* IGTS8. *Journal ofBacteriology*, *178*(22), 6409–6418. doi:10.1128/jb.178.22.6409-6418.1996 PMID:8932295

Li, W., & Jiang, X. (2013). Enhancement of bunker oil biodesulfurization by adding surfactant. *World Journal of Microbiology & Biotechnology*, *29*(1), 103–108. doi:10.100711274-012-1162-7 PMID:22940903

Li, Y.-G., Gao, H.-S., Li, W.-L., Xing, J.-M., & Liu, H.-Z. (2009). *In situ* magnetic separation and immobilization of dibenzothiophene-desulfurizing bacteria. *Bioresource Technology, 100*(21), 5092–50966. doi:10.1016/j.biortech.2009.05.064 PMID:19541480

Liang, Z.-X., Li, L., Li, S., Cai, Y.-H., Yang, S.-T., & Wang, J.-F. (2012). Enhanced propionic acid production from Jerusalem artichoke hydrolysate by immobilized *Propionibacterium acidipropionici* in a fibrous-bed bioreactor. *Bioprocess and Biosystems Engineering, 35*(6), 915–921. doi:10.100700449-011-0676-y PMID:22228298

Linguist, L., & Pacheco, M. (1999). Enzyme-based diesel desulfurization process offers energy, CO_2 advantages. *Oil & Gas Journal, 97*(8), 45–50.

Lozinsky, V. I., & Plieva, F. M. (1998). Poly (vinyl alcohol) cryogels employed as matrices for cell immobilization. 3. Overview of recent research and developments. *Enzyme and Microbial Technology, 23*(3), 227–242. doi:10.1016/S0141-0229(98)00036-2

Luo, M. F., Xing, J. M., Gou, Z. X., Li, S., Liu, H. Z., & Chen, J. Y. (2003). Desulfurization of dibenzothiophene by lyophilized cells of *Pseudomonas delafieldii* R-8 in the presence of dodecane. *Biochemical Engineering Journal, 13*(1), 1–6. doi:10.1016/S1369-703X(02)00078-5

Ma, C., Feng, J., Zeng, Y., Cai, X., Sun, B., Zhang, Z., ... Xu, P. (2006). Methods for the preparation of a biodesulfurization biocatalyst using *Rhodococcus* sp. *Chemosphere, 65*(1), 165–169. doi:10.1016/j.chemosphere.2006.03.010 PMID:16624377

Manso, T., Nunes, C., Raposo, S., & Lima-Costa, M. E. (2010). Carob pulp as raw material for production of the biocontrol agent *P. agglomerans* PBC-1. *Journal of Industrial Microbiology & Biotechnology, 37*(11), 1145–1155. doi:10.100710295-010-0762-1 PMID:20593299

Maricq, M. M., Chase, R. E., Xu, N., & Laing, P. M. (2002). The effects of the catalytic converter and fuel sulfur level on motor vehicle particulate matter emissions: Light duty diesel vehicles. *Environmental Science & Technology, 36*(2), 283–289. doi:10.1021/es0109621 PMID:11827064

Martin, A. B., Alcon, A., Santos, V. E., & Garcia-ochoa, F. (2004). Production of a biocatalyst of *Pseudomonas putida* CECT5279 for dibenzothiophene (DBT) biodesulfurization for different media compositions. *Energy & Fuels, 18*(3), 851–857. doi:10.1021/ef030174c

Martinez, I., Santos, V. E., Alcon, A., & Garcia-Ochoa, F. (2015). Enhancement of the biodesulfurization capacity of *Pseudomonas putida* CECT5279 by co-substrate addition. *Process Biochemistry, 50*(1), 119–124. doi:10.1016/j.procbio.2014.11.001

Martínez-Checa, F., Toledo, F., Vilchez, R., Quesada, E., & Calvo, C. (2002). Yield production, chemical composition, and functional properties of emulsifier H28 synthesized by *Halomonas eurihalina* strain H-28 in media containing various hydrocarbons. *Applied Microbiology and Biotechnology, 58*(3), 358–363. doi:10.100700253-001-0903-6 PMID:11935188

McFarland, B. L. (1999). Biodesulfurization. *Current Opinion in Microbiology, 2*(3), 257–264. doi:10.1016/S1369-5274(99)80045-9 PMID:10383871

McFarland, B. L., Boron, D. J., Deever, W., Meyer, J. A., Johnson, A. R., & Atlas, R. M. (1998). Biocatalytic sulfur removal from fuels: Applicability for producing low sulfur gasoline. *Critical Reviews in Microbiology, 24*(2), 99–147. doi:10.1080/10408419891294208 PMID:9675512

McWilliams, A. (2015). The global market for carotenoids, Jun 2018, Report id: FOD025F, BCC research. https://www.bccresearch.com/market-research/food-and-beverage/the-global-market-for-carotenoids-fod025f.html. Accessed 1 March 2019.

Mendes, A., Guerra, P., Madeira, V., Ruano, F., da Silva, T. L., & Reis, A. (2007). Study of docosahexaenoic acid production by the heterotrophic microalga *Crypthecodinium cohnii* CCMP 316 using carob pulp as a promising carbon source. *World Journal of Microbiology & Biotechnology, 23*(9), 1209–1215. doi:10.100711274-007-9349-z

Mohebali, G., Ball, A., Kaytash, A., & Rasekh, B. (2007). Stabilization of water/gas oil emulsions by desulfurizing cells of *Gordonia alkanivorans* RIPI90A. *Microbiology, 153*(Pt 5), 1573–1581. doi:10.1099/mic.0.2006/002543-0 PMID:17464072

Mohebali, G., & Ball, A. S. (2008). Biocatalytic desulfurization (BDS) of petrodiesel fuels. *Microbiology, 154*(Pt 8), 2169–2183. doi:10.1099/mic.0.2008/017608-0 PMID:18667551

Mohebali, G., Ball, A. S., Kaytash, A., & Rasekh, B. (2008). Dimethyl sulfoxide (DMSO) as the sulfur source for the production of desulfurizing resting cells of *Gordonia alkanivorans* RIPI90A. *Microbiology, 154*(3), 878–885. doi:10.1099/mic.0.2007/013011-0 PMID:18310033

Monticello, D. J. (2000). Biodesulfurization and the upgrading of petroleum distillates. *Current Opinion in Biotechnology, 11*(6), 540–546. doi:10.1016/S0958-1669(00)00154-3 PMID:11102787

Monticello, D. J., & Finnerty, W. R. (1985). Microbial desulfurization of fossil fuels. *Annual Review of Microbiology*, *39*(1), 371–389. doi:10.1146/annurev.mi.39.100185.002103 PMID:3904605

Mukhopadhyaya, M., Chowdhury, R., & Bhattacharya, P. (2005). Two phase mathematical model for a bio-trickling reactor for the production of Ultra low sulfur diesel (ULSD) from deeply hydrodesulfurized diesel. *Catalysis Today*, *106*(1-4), 180–185. doi:10.1016/j.cattod.2005.07.128

Mukhopadhyaya, M., Chowdhury, R., & Bhattacharya, P. (2006). Biodesulfurization of hydrodesulfurized diesel in a trickle bed reactor - Experiments and modeling. *Journal of Scientific and Industrial Research*, *65*(5), 432–436.

Mukhopadhyaya, M., Chowdhury, R., & Bhattacharya, P. (2007). Trickle bed biodesulfurizer of diesel with backwash and recycle. *AIChE Journal. American Institute of Chemical Engineers*, *53*(8), 2188–2197. doi:10.1002/aic.11240

Mulligan, C. N. (2005). Environmental applications for biosurfactants. *Environmental Pollution*, *133*(2), 183–198. doi:10.1016/j.envpol.2004.06.009 PMID:15519450

Murphy, S. E., Amin, S., Coletta, K., & Hoffman, D. (1992). Rat liver metabolism of benzo[b]naphtho[2,1-d]thiophene. *Chemical Research in Toxicology*, *5*(4), 491–495. doi:10.1021/tx00028a006 PMID:1382648

Naito, M., Kawamoto, T., Fujino, K., Kobayashi, M., Maruhashi, K., & Tanaka, A. (2001). Long-term repeated biodesulfurization by immobilized *Rhodococcus erythropolis* KA 2-5-1 cells. *Applied Microbiology and Biotechnology*, *55*(3), 374–378. doi:10.1007002530000527 PMID:11341322

Nandi, S. (2010). Biodesulfurization of hydro-desulfurized diesel in airlift reactor. *Journal of Scientific and Industrial Research*, *69*(7), 543–547.

Nekodzuka, S., Nakajima-Kambe, T., Nomura, N., Lu, J., & Nakahara, T. (1997). Specific desulfurization of dibenzothiophene by *Mycobacterium sp*. strain G3. *Biocatalysis and Biotransformation*, *15*(1), 17–27. doi:10.3109/10242429709003607

Nordvik, A. B., Simmons, J. L., Bitting, K. R., Lewis, A., & Strøm-Kristiansen, T. (1996). Oil and water separation in marine oil spill clean-up operations. *Spill Science & Technology Bulletin*, *3*(3), 107–122. doi:10.1016/S1353-2561(96)00021-7

Nuhu, A. A. (2013). Bio-catalytic desulfurization of fossil fuels: A mini review. *Reviews in Environmental Science and Biotechnology*, *12*(1), 9–23. doi:10.100711157-012-9267-x

Ohshiro, T., & Izumi, Y. (1999). Microbial desulfurization of organic sulfur compounds in petroleum. *Bioscience, Biotechnology, and Biochemistry*, *63*(1), 1–9. doi:10.1271/bbb.63.1 PMID:10052116

Ohshiro, T., Suzuki, K., & Izumi, Y. (1996). Regulation of dibenzothiophene degrading enzyme activity of *Rhodococcus erythropolis* D-1. *Journal of Fermentation and Bioengineering*, *81*(2), 121–124. doi:10.1016/0922-338X(96)87588-X

Okada, H., Nomura, N., Nakahara, T., Saitoh, K., Uchiyama, H., & Maruhashi, K. (2003). Analyses of microbial desulfurization reaction of alkylated dibenzothiophenes dissolved in oil phase. *Biotechnology and Bioengineering*, *83*(4), 489–497. doi:10.1002/bit.10694 PMID:12800143

Oldfield, C., Pogrebinsky, O., Simmonds, J., Olson, E. S., & Kulpa, C. F. (1997). Elucidation of the metabolic pathway for dibenzothiophene desulphurization by *Rhodococcus* sp. strain IGTS8 (ATCC 53968). *Microbiology*, *143*(9), 2961–2973. doi:10.1099/00221287-143-9-2961 PMID:9308179

Onodera-Yamada, K., Morimoto, M., & Tani, Y. (2001). Degradation of dibenzothiophene by sulfate-reducing bacteria cultured in the presence of only nitrogen gas. *Journal of Bioscience and Bioengineering*, *91*(1), 91–93. doi:10.1016/S1389-1723(01)80119-5 PMID:16232954

Oral, J., Sikula, J., Puchyr, R., Hajny, Z., Stehlik, P., & Bebar, L. (2005). Processing of waste from pulp and paper plant. *Journal of Cleaner Production*, *13*(5), 509–515. doi:10.1016/j.jclepro.2003.09.005

Pacheco, M. A., Lange, E. A., Pienkos, P. T., Yu, L.-Q., Rouse, M. P., Lin, Q., & Linguist, L. K. (1999). *Recent advances in biodesulfurization of diesel fuel*. National Petrochemical And Refiners Association-Publications-All Series.

Paixão, S. M., Teixeira, P. D., Silva, T. P., Teixeira, A. V., & Alves, L. (2013). Screening of novel yeast inulinases and further application to bioprocesses. *New Biotechnology*, *30*(6), 598–606. doi:10.1016/j.nbt.2013.02.002 PMID:23419675

Patel, S. B., Killbane, J. J., & Webster, D. A. (1997). Biodesulphurisation of dibenzothiophene in hydrophobic media by *Rhodococcus* sp. strain IGTS8. *Journal of Chemical Technology and Biotechnology (Oxford, Oxfordshire)*, *69*(1), 100–106. doi:10.1002/(SICI)1097-4660(199705)69:1<100::AID-JCTB675>3.0.CO;2-9

Perfumo, A., Smyth, T. J. P., Marchant, R., & Banat, I. M. (2010). Production and roles of biosurfactants and bioemulsifiers in accessing hydrophobic substrates. In *Handbook of hydrocarbon and lipid microbiology* (pp. 1501–1512). Springer. doi:10.1007/978-3-540-77587-4_103

Pope, C. A. III, Burnett, R. T., Thun, M. J., Calle, E. E., Krewski, D., Ito, K., & Thurston, G. D. (2002). Lung cancer, cardiopulmonary mortality, and long-term exposure to fine particulate air pollution. *Journal of the American Medical Association, 287*(9), 1132–1141. doi:10.1001/jama.287.9.1132 PMID:11879110

Ramachandran, H., Iqbal, M. A., & Amirul, A. A. (2014). Identification and characterization of the yellow pigment synthesized by Cupriavidus sp. USMAHM13. *Applied Biochemistry and Biotechnology, 174*(2), 461–470.

Reichmuth, D. S., Hittle, J. L., Blanch, H. W., & Keasling, J. D. (2000). Biodesulfurization of dibenzothiophene in *Escherichia coli* is enhanced by expression of a *Vibrio harveyi* oxidoreductase gene. *Biotechnology and Bioengineering, 67*(1), 72–79. doi:10.1002/(SICI)1097-0290(20000105)67:1<72::AID-BIT8>3.0.CO;2-C PMID:10581437

Rivera, J. L., Navarro-Santos, P., Guerra-Gonzalez, R., & Lima, E. (2014). Interaction of refractory dibenzothiophenes and polymerizable structures. *International Journal of Polymer Science, 2014*, 1–11. doi:10.1155/2014/103945

Rodrigues, L., Banat, I. M., Teixeira, J., & Oliveira, R. (2006). Biosurfactants: Potential applications in medicine. *The Journal of Antimicrobial Chemotherapy, 57*(4), 609–618. doi:10.1093/jac/dkl024 PMID:16469849

Rodrigues, L. R., & Teixeira, J. A. (2010). *Biomedical and therapeutic applications of biosurfactants*. New York: Springer.

Rouse, J. D., Sabatini, D. A., Suflita, J. M., & Harwell, J. H. (1994). Influence of surfactants on microbial degradation of organic compounds. *Critical Reviews in Environmental Science and Technology, 24*(4), 325–370. doi:10.1080/10643389409388471

Sánchez-Segado, S., Lozano, L. J., de Los Ríos, A. P., Hernández-Fernández, F. J., Godínez, C., & Juan, D. (2012). Process design and economic analysis of a hypothetical bioethanol production plant using carob pod as feedstock. *Bioresource Technology, 104*, 324–328. doi:10.1016/j.biortech.2011.10.046 PMID:22100234

Schilling, B. M., Alvarez, L. M., Wang, D. I. C., & Cooney, C. L. (2002). Continuous desulfurization of dibenzothiophene with *Rhodococcus rhodochrous* IGTS8 (ATCC 53968). *Biotechnology Progress, 18*(6), 1207–1213. doi:10.1021/bp0200144 PMID:12467453

Scott, K., Jachuck, R. J., & Hall, D. (2001). Crossflow microfiltration of water-in-oil emulsions using corrugated membranes. *Separation and Purification Technology, 22*(1-2), 431–441. doi:10.1016/S1383-5866(00)00180-5

Scott, T. C., & Wham, R. M. (1989). An electrically driven multistage countercurrent solvent extraction device: The emulsion-phase contactor. *Industrial & Engineering Chemistry Research*, *28*(1), 94–97. doi:10.1021/ie00085a016

Setti, L., Lanzarini, G., & Pifferi, P. G. (1997). Whole cell biocatalysis for an oil desulfurization process. *Fuel Processing Technology*, *52*(1-3), 145–153. doi:10.1016/S0378-3820(97)00023-4

Shennan, J. L. (1996). Microbial attack on sulphur-containing hydrocarbons: Implications for the biodesulphurisation of oils and coals. *Journal of Chemical Technology and Biotechnology (Oxford, Oxfordshire)*, *67*(2), 109–123. doi:10.1002/(SICI)1097-4660(199610)67:2<109::AID-JCTB532>3.0.CO;2-Y

Silva, C. R., Silva, M. L. C., Kamida, H. M., Goes-Neto, A., & Koblitz, M. G. B. (2014). Lytic enzyme production optimization using low-cost substrates and its application in the clarification of xanthan gum culture broth. *Food Science & Nutrition*, *2*(4), 299–307. doi:10.1002/fsn3.87 PMID:25473487

Silva, T. P., Paixão, S. M., & Alves, L. (2016). Ability of Gordonia alkanivorans strain 1B for high added value carotenoids production. *RSC Advances*, *6*, 58055–58063.

Silva, T. P., Paixão, S. M., Roseiro, J. C., & Alves, L. (2015). Jerusalem artichoke as low-cost fructose-rich feedstock for fossil fuels desulfurization by a fructophilic bacterium. *Journal of Applied Microbiology*, *118*(3), 609–618. doi:10.1111/jam.12721 PMID:25494982

Silva, T. P., Paixão, S. M., Teixeira, A. V., Roseiro, J. C., & Alves, L. (2013). Optimization of low sulfur carob pulp liquor as carbon source for fossil fuels biodesulfurization. *Journal of Chemical Technology and Biotechnology (Oxford, Oxfordshire)*, *88*(5), 919–923. doi:10.1002/jctb.3921

Singh, A., Singh, B., & Ward, O. (2012). Potential applications of bioprocess technology in petroleum industry. *Biodegradation*, *23*(6), 865–880. doi:10.100710532-012-9577-2 PMID:22829348

Soleimani, M., Bassi, A., & Margaritis, A. (2007). Biodesulfurization of refractory organic sulfur compounds in fossil fuels. *Biotechnology Advances*, *25*(6), 570–596. doi:10.1016/j.biotechadv.2007.07.003 PMID:17716849

Srivastava, V. C. (2012). An evaluation of desulfurization technologies for sulfur removal from liquid fuels. *RSC Advances*, *2*(3), 759–783. doi:10.1039/C1RA00309G

Stanislaus, A., Marafi, A., & Rana, M. S. (2010). Recent advances in the science and technology of ultra low sulfur diesel (ULSD) production. *Catalysis Today*, *153*(1-2), 1–68. doi:10.1016/j.cattod.2010.05.011

Stoner, D. L., Wey, J. E., Barrett, K. B., Jolley, J. G., Wright, R. B., & Dugan, P. R. (1990). Modification of water-soluble coal-derived products by dibenzothiophene-degrading microorganisms. *Applied and Environmental Microbiology*, *56*(9), 2667–2676. PMID:16348275

Ta-Chen, L., Chang, J.-S., & Young, C.-C. (2008). Exopolysaccharides produced by *Gordonia alkanivorans* enhance bacterial degradation activity for diesel. *Biotechnology Letters*, *30*(7), 1201–1206. doi:10.100710529-008-9667-8 PMID:18286235

Takada, M., Nomura, N., Okada, H., Nakajima-Kambe, T., Nakahara, T., & Uchiyama, H. (2005). De-repression and comparison of oil-water separation activity of the dibenzothiophene desulfurizing bacterium, *Mycobacterium* sp. G3. *Biotechnology Letters*, *27*(12), 871–874. doi:10.100710529-005-6721-7 PMID:16086250

Talvy, S., Cockx, A., & Line, A. (2005). Global modelling of a gas–liquid–solid airlift reactor. *Chemical Engineering Science*, *60*(22), 5991–6003. doi:10.1016/j.ces.2005.04.067

Tanaka, Y., Yoshikawa, O., Maruhashi, K., & Kurane, R. (2002). The cbs mutant strain of *Rhodococcus erythropolis* KA2-5-1 expresses high levels of Dsz enzymes in the presence of sulfate. *Archives of Microbiology*, *178*(5), 351–357. doi:10.100700203-002-0466-7 PMID:12375103

Tang, K., & Hong, X. (2014). A characterization of bacterium for fcc diesel oil desulfurization. *Petroleum Science and Technology*, *32*(13), 1545–1552. doi:10.1080/10916466.2012.689795

Tuli, D., & Kumar, A. (2008). Non-conventional technologies for fuel desulphurization. In *19th World Petroleum Congress*. World Petroleum Congress.

Urum, K., & Pekdemir, T. (2004). Evaluation of biosurfactants for crude oil contaminated soil washing. *Chemosphere*, *57*(9), 1139–1150. doi:10.1016/j.chemosphere.2004.07.048 PMID:15504473

Vazquez-Duhalt, R., Torres, E., Valderrama, B., & Le Borgne, S. (2002). Will biochemical catalysis impact the petroleum refining industry? *Energy & Fuels*, *16*(5), 1239–1250. doi:10.1021/ef020038s

Venil, C. K., Aruldass, C. A., Dufossé, L., Zakaria, Z. A., & Ahmad, W. A. (2014). Current perspective on bacterial pigments: Emerging sustainable compounds with coloring and biological properties for the industry – an incisive evaluation. *RSC Advances*, *4*, 39523–39529.

Wang, M., Li, W., Shi, Y., Wang, D., & Feng, H. (2006). Effects of surfactant on biodesulfurization by *Corynebacterium* sp. ZD-1 in the presence of organic phase. *Journal of Zhejiang University. Science A*, *7*(S2), 371–375. doi:10.1631/jzus.2006.AS0371

Wang, P., & Krawiec, S. (1996). Kinetic analyses of desulfurization of dibenzothiophene by *Rhodococcus erythropolis* in batch and fed-batch cultures. *Applied and Environmental Microbiology*, *62*(5), 1670–1675. PMID:16535315

Wang, Z., Wang, D., Li, Q., Li, W., Tang, H., & Xing, J. (2011). Enhanced biodesulfurization by expression of dibenzothiophene uptake genes in *Rhodococcus erythropolis*. *World Journal of Microbiology & Biotechnology*, *27*(9), 1965–1970. doi:10.100711274-011-0656-z

Werther, J., Saenger, M., Hartge, E.-U., Ogada, T., & Siagi, Z. (2000). Combustion of agricultural residues. *Progress in Energy and Combustion Science*, *26*(1), 1–27. doi:10.1016/S0360-1285(99)00005-2

Willaert, R. G., & Baron, G. V. (1996). Gel entrapment and micro-encapsulation : Methods, applications and engineering principles. *Reviews in Chemical Engineering*, *12*(1), 1–205. doi:10.1515/REVCE.1996.12.1-2.1

Wu, S., Lin, J., & Chan, S. I. (1994). Oxidation of dibenzothiophene catalyzed by heme-containing enzymes encapsulated in sol-gel glass. *Applied Biochemistry and Biotechnology*, *47*(1), 11–20. doi:10.1007/BF02788671 PMID:8203869

Yang, J., Hu, Y., Zhao, D., Wang, S., Lau, P. C. K., & Marison, I. W. (2007). Two-layer continuous-process design for the biodesulfurization of diesel oils under bacterial growth conditions. *Biochemical Engineering Journal*, *37*(2), 212–218. doi:10.1016/j.bej.2007.04.012

Yoshikawa, O., Ishii, Y., Koizumi, K.-I., Ohshiro, T., Izumi, Y., & Maruhashi, K. (2002). Enhancement and stabilization of desulfurization activity of *Rhodococcus erythropolis* KA2-5-1 by feeding ethanol and sulfur components. *Journal of Bioscience and Bioengineering*, *94*(5), 447–452. doi:10.1016/S1389-1723(02)80223-7 PMID:16233332

Zhang, Y.-W., Prabhu, P., & Lee, J.-K. (2010). Alginate immobilization of recombinant *Escherichia coli* whole cells harboring L-arabinose isomerase for L-ribulose production. *Bioprocess and Biosystems Engineering*, *33*(6), 741–748. doi:10.100700449-009-0397-7 PMID:19946709

Zhou, Z., & Zhang, K. (2004). Development situation and prospect of oil fields in China. *Petroleum Exploration And Development*, *31*(1), 84–87.

Chapter 8
Environmental Concerns and the Importance of Desulfurization

Tawfik A. Saleh
King Fahd University of Petroleum and Minerals, Saudi Arabia

ABSTRACT

The increased utilization of fossil fuels and subsequent industrialization in most of the world has led to a remarkable increase in the atmospheric sulfur compounds concentrations. Pollution released by the use of petroleum-based fuels contributes immensely to the deterioration of air quality despite regulatory and technological advances in place. SOx, NOx, and particulate matter are constantly emitted to the environment which affects public health, ecosystem, and general wellbeing of the people living mostly in urban areas. Sulfur dioxide, which is the immediate sulfur compound found in the lower atmosphere after combustion of fuels, has a major role to play in the formation of acid rain, smog formation, and particulate aerosols. Each of these formations affects the healthy living of animals, plants, soils, water, and the general ecosystem. This chapter discusses the environmental issues of sulfur.

DOI: 10.4018/978-1-7998-2146-5.ch008

INTRODUCTION

Environmental Concerns

The increased utilization of fossil fuels and subsequent industrialization in most part of the world has led to a remarkable increase in the atmospheric sulfur compounds concentrations. Pollution released by the use of petroleum-based fuels contributes immensely to the deterioration of air quality despite regulatory and technological advances in place. SOx, NOx, and particulate matter are constantly emitted to the environment which affects public health, ecosystem and general wellbeing of the people living mostly in urban areas. Sulfur dioxide which is the immediate sulfur compound found in the lower atmosphere after combustion of fuels has a major role to play in the formation of acid rain, smog formation, and particulate aerosols. Each of these formations affects the healthy living of animals, plants, soils, water, and the general ecosystem.

The four-day event of "London smog" that took place in December 1952 claimed the life of over 4000 people is a noticeable example of the dangers of sulfur compounds in the atmosphere. A similar tragedy happened in London in 1957 with a recorded casualty of over 700 lives. The lower the sulfur content in fuel the better its quality and marketability (Song, 2003). The purpose of this chapter is to provide cheap and reliable nanomaterials that will help in reducing the amount of sulfur released from hydrocarbon-based fuels into the environment.

There are pieces of evidence of substantial the increase of sulfur compounds over the past few decades in our immediate environment (Georgii, 1970). These compounds are the major contaminants found in hydrocarbon fuels (Shahadat *et al*, 2013) and they are equally reactive. Sulfur compounds released mainly from the burning of fossil fuels results in breathing impairment, respiratory illnesses, alterations in lung defenses, and aggravation of the existing cardiovascular disease. These compounds equally have grievous consequences for asthmatic patients.

Many countries have laws for regulating the emission of sulfur from fuels e.g. In the US, USEPA enacted a law known as ultra-low sulfur diesel (ULSD) since 2006 for maintaining the sulfur content in diesel at 15ppm from the previous 500ppm law (low sulfur diesel) (USEPA, 2014). Diesel engines since then were produced with devices that minimize the emissions of sulfur to the environment. This had the immediate health benefit to the population and the environment. The expected results from the enactment of the law were that: 8300 premature deaths, 23100 cases of bronchitis, and 360,000 asthma attacks are expected to be prevented annually. In addition, 7100 hospital visits, 2400 emergency visits of asthmatic patients and 1.5 million days of productivity loss will be avoided (USEPA, 2006).

USEPA has a tier 3 motor vehicle and fuel standard approach to reduce the sulfur content in gasoline. It is an upgraded version of the tier 2 approach and it is one of the most effective air quality control measures in place (USEPA, 2014[a]). This rule is expected to start by 2017 and it will consider both the vehicle and the fuel as an integrated system; the expected sulfur level will be 10ppm on an annual average. Japan, Europe, Canada, and South Korea have already achieved the tier 3 limits of sulfur in gasoline (Dasgupta *et al*, 2013). Lower sulfur gasoline will aid in the development of better technologies that will further improve fuel economy, reduce greenhouse gas effect and improve public health (USEPA, 2014[b]). It is projected that by 2030, tier 3 will annually prevent: close to 2000 premature death, over 2000 hospital visits and asthma-related cases, and almost 1.5 million days of productivity loss (USEPA, 2014[a]). Therefore, it is required to reduce the sulfur content in the fuel itself during the refining process to protect the environment.

The removal of sulfur from fuels involves three different approaches: decomposing the sulfur compounds, removal of the compounds without decomposition and finally separating the compounds followed by decomposition (Babich *et al*, 2003). The conventional method used by refineries for the removal of sulfur from the fuel is hydrodesulphurization (HDS). It is efficient in the removal of most aliphatic sulfur compounds from fuels e.g. thiols. However, it is not efficient in the removal of aromatic and refractory sulfur compounds e.g. dibenzothiophene which poses a more serious danger to the environment. In addition, it requires high temperature, pressure and high dosage of the catalyst before achieving the desired objective (Song, 2003).

Other methods of desulfurization in use to curtail the problems of hydrodesulphurization include: oxidative desulfurization where all the refractory sulfur compounds are oxidized to less harmful polar derivatives that can be isolated easily at room temperature and pressure, ionic liquids desulfurization (Prashant *et al*, 2010), and bio-desulfurization (Soleimani *et al*, 2007) such methods are considered viable alternatives for the desulfurization. Catalyst is combined with most of the processes to speed up the rate of desulfurization. Adsorbent materials are currently used in desulfurization for the selective removal of sulfur from petroleum products to trace limits. It has advantages over other methods of desulfurization because it is cheap and does not have scaling abilities. It has ease of application, possesses the regenerative ability and has the potential to remove the sulfur compounds to ultra-deep limits (Kostas *et al*, 2014). The major drawback of this process is the absence of efficient adsorbent materials that will reduce the sulfur content of fuel to minimal concentrations and be easily recycled to minimize environmental concerns of its disposal.

Sulfur and Health Impacts

Generally, sulphur compounds may have adverse effects on human health i.e. carcinogenic or mutagenic, on exposure to a certain level (Braid et al., 2005). H_2S is malodorous having a strong irritating odor and on short time exposure causes headache but on long-duration exposure causes unconsciousness. Some thiols have also stinging repulsive odor and are strong irritants which can be very toxic at high concentrations. Thiols may cause nausea, vomiting, headache, coughing, lung irritation and inflammation of eyes, throat, and nose on exposure to high concentrations. Under severe conditions long term exposure to high concentrations of thiols may also leads to uncomfortable breathing, muscular spasms or unconsciousness, however, effects of irritation disappears soon as the affected individual is brought to breathe in the fresh air. Furthermore, as thiols are highly soluble in fats hence soon after contact they are readily absorbed in the skin. Like other sulfur compounds, lower sulfides are also malodorous. The dimethyl- and diethyl-disulphides are relatively less toxic and have less repulsive odor than thiols, but still, their smell can be observed at very low concentrations, up to a range of 0.0001 to 0.002 ppm. Under high concentration, sulfides can cause irritation of eyes, throat, nose and muscular paralysis.

Poisoning of Catalysts

Some of the disadvantages of the sulfur compounds in petroleum is that they deactivate the base metals and precious metal catalysts used in a number of refining processes for downstream processing of petroleum and other catalytic environmental operations (Ma et al., 1994). During the operation the sulphur species adsorbs selectively on the metals surfaces or active sites present on the catalysts, thereby causing to inactivate or reduce their activity. Sulfur compounds when deposit on the surface of Cu, Ni, Co, etc. forming metal sulfides, rendering them to lose activity. Sulphur impurity on ppm level in the feed is sufficient to reduce the activity of nickel catalyst many folds in just few days. Generally sulfur species are irreversible poisons for most systems catalysts e.g. heavy metals like Hg, Pb, As, hence the catalyst is difficult to regenerate by treatment with air, O_2 or steam (Froment et al., 1987), however in some cases like precious metals are poisoned reversibly, and can be regenerated by removing sulfur compounds from the gas stream (Kent, 2010).

Sulfur compounds, in combustion operations, produces SO_2 and SO_3, which absorbs on metal surfaces and form inactive metal sulfates (Kent, 2010). Pt supported on Al_2O_3 used for reduction of the combustion emissions is also deactivated in a similar manner; the SO_2 produced during combustion is oxidized to SO_3 by Pt which adsorbs on the carrier (Al_2O_3) forming sulfate. The sulfate slowly covers the surface of the carrier, cause to occlude Pt, due to which the catalysts lose its activity. In various processes e.g. catalytic hydrogenation, reforming, isomerization and alkylation, sulphur selectively adsorb on the catalyst surface as H_2S or organic sulfides, and blocks the adsorption of any other species, hence leads to partial or complete loss in the activity of the catalyst. The degree of deactivation of the catalyst depends on the extent of the surface coverage, which may be mechanistically categorized as geometric blocking or by long-range electronic effects. In case when the surface is partially covered then the deactivation is resulted by combined geometric blocking and long-range electronic effects, on the other hand under high surface coverage, loss in activity is resulted only by geometric blocking (Ahmad, 2013).

Generally, during catalytic hydrotreatment processes, the sulfur compounds present in the feed are changed into H_2S under a high partial pressure of H_2, which builds up to a high concentration in recycle gas stream, and leads to the deactivation of the catalyst unless it is removed. The H_2S is known to be a reversible poison for Pt and deactivates the catalyst used in recyclization and hydrogenation reactions. In the earlier designs of catalytic reforming reactors the removal of H_2S gas from the recycle stream was effectuated by absorption with steam and di-ethanolamine. In modern units, the sulfur from the feed is removed through catalytic HDS, and then feed is subjected to reforming, hence the problems of catalyst deactivation and corrosion due to H_2S are controlled. In most practices, sulfur in the feed is reduced to about 0.5 ppm or less, in case of using high sulfur feed problems of low catalysts activity and poor products yield are encountered. Similarly, the pretreatment of the feed is necessary to reduce its sulfur contents to few ppm before subjecting to catalytic reforming in case of using naphtha as feed and also for catalytic isomerization of C_4 to C_6 paraffins, in order to avoid deactivation of catalyst.

Pre-sulfidation of the catalyst may be used to overcome the deactivation problem up to some extent in certain processes. The catalyst is pre-sulfided with various sulfur precursors before the reaction, the so-called sulfided catalyst can efficiently catalyze hydrogenation reactions even if strong inhibitors like H_2S and NH_3 are present. In addition, the sulfided catalysts also resist early deactivation through coke deposition (Absi Halabi et al., 1996).

Additionally, in gasoline, sulfur compounds tend to resist the susceptibility to octane boosting by addition of the octane improvers like tetra-methyl-lead and tetra-ethyl-lead (Magyar et al., 2005). Similarly, in addition to high aromatics and olefins, the FCC gasoline also contain high sulfur content, and in order to eliminate sulfur through hydrodesulphurization, olefins and aromatic saturation occurs which leads to produce gasoline with low octane number (Babich et al., 2003), hence to meet the octane rating according to standard fuel specifications additional processing is required.

Impact of Sulphur on Environment

In petroleum, the sulfur compounds are objectionable because they are associated with a number of problems, and are therefore removed from petroleum products. Below is an overview of the major problems caused by sulfur compounds in petroleum.

The irritating pungent smell of sour petroleum crudes is due to the presence of sulfur compounds, particularly H_2S and mercaptans (McKetta, 1992.). These compounds are not only malodorous but also toxic, therefore handling and processing of the petroleum products with objectionable odor creates health hazards and a nuisance to the workers. Besides most often these compounds bear corrosive character, hence they also influence the fuel quality. The bad odor is removed from the crude petroleum through sweetening processes or by various refining processes in the refining industry (Hobson et al., 1984).

Sulfur and Atmospheric Pollution

Sulfur present in petroleum products is converted to SO_2 during combustion, which is liberated into the atmosphere where it causes serious environmental pollution like acid rain and contributes to photochemical smog. Also, SO_2 is directly harmful to animals and vegetation.

Direct Effects of SO2 in the Atmosphere

SO_2 is colorless gas having a piercing pungent odor and is very harmful to man, animals, and vegetations; its main source is the conventional fuels containing sulfur compounds, which are utilized for transportation and in industries. The effect of SO_2 on humans primarily appears on the respiratory tract causing its irritation and constriction, which leads to increased resistance in the air passage. These effects are more pronounced in persons suffering from respiratory weakness and sensitized asthma. The irritation effects are experienced when the individuals are exposed to SO_2 level of 1-2 ppm in air, SO_2 level of 5-10 ppm causes severe bronchial spasm, whereas the level of 500 ppm may lead to death (Manahan, 2010).

A combination of SO_2 and smoke in the air, which usually occurs when smog is formed by a temperature inversion layer, leads to a widespread disaster that affects not only humans but also animals. Several incidents happened across the world in the past when high SO_2 level was trapped by smog. In Meuse River Valley, Belgium such incident occurred on December 1930, when the SO_2 level in atmosphere raised to 38 ppm, which led to deaths of approximately 60 people and many cattle. Similarly in October 1948 in Donora, Pennsylvania, the SO_2 level reached to 2 ppm, and caused the deaths of about 20 people and the illness of more than 40% of the population. Heavy smog having 1.3 ppm SO_2 and 4 mg/m^3 of smoke, existed for five days, during which approximately 4000 deaths occurred. These deaths were mainly caused due to pneumonia, bronchitis and related respiratory problems. A similar accident occurred in London, but this time the number of the deaths was small i.e. about 700, which was due to the implementation of the Clean Air Act, in 1962 (Manahan, 2010).

High level of SO_2 is also injurious to plants, causing a condition called leaf necrosis in which the edge of the leaf and its petiole is damaged. In plants, the chronic effect of SO_2 appears as yellowing of the leaf which is called chlorosis. It has been found that low-level chronic effects of exposure to SO_2 are more damaging than exposure to high dose for short durations. Furthermore, the SO_2 also damages the plants and aquatic flora in rivers and lakes indirectly by causing acid rain.

Acid Rain

Gases such as SO_2 released during combustion of fossil fuels containing sulfur compounds is photo-catalytically oxidized to SO_3 in the atmosphere. In the atmosphere, under humid conditions, SO_3 reacts with water and form H_2SO_4.

$$2SO_2 + 2O_2 \rightarrow 2SO_3$$

$$SO_3 + H_2O \rightarrow H_2SO_4$$

In a similar way, the oxides of nitrogen forms HNO_3 in the atmosphere, and when these acids combine to cause acidic precipitation, this is termed as acid rain. Today the acid rain is a major environmental problem; due to high reactivity, it is not only hazardous to living beings but also damages the non-living objects. The acid rain adversely affects the biota and its surrounding. Acid rain primarily affects the soil by leaching action. Acid rain dissolves and washes away the nutrients materials from the soil which are needed for the plants' development and growth, thereby rendering soil less fertile. Furthermore, the acidic washing also dissolves the harmful substances present in certain soils i.e. Hg, Al, Pb, As, etc., thus drains out hazardous substances which further pollute the freshwater bodies, vegetation and finally reaches to the human food chain. Certain beneficial microbial cultures in soil that cannot survive under low pH conditions are also destroyed by acid rain (Harrison et al., 2007). The acid rain has damaging effects on forests as well (Manahan, 2010), due to decreasing the soil fertility by washing away the important nutrients, the growth and development of forest vegetation, particularly large trees are retarded. Furthermore, the acid rain damages the waxy coating on the leaves and needles of the large coniferous trees, causing dead brown spots on it, due to which they lose their photosynthetic activity. The infected trees weaken and become vulnerable to diseases, insect manifestations, cold weather, and draught. The trees at peaks or high altitude remain in direct contact with the acidic clouds, hence these are at more risk of adverse effects.

The acid rain runoff drains into freshwater bodies like lakes, streams, and marshes, where it disturbs the aquatic life. The freshwater is commonly neutral, with pH in the range of 6- 8, and is suitable for supporting the aquatic life. The aquatic flora and fauna are very sensitive to high or low pH, therefore the acid rain which lowers the pH of freshwater bodies greatly threaten their survival. Acidic water mainly affects the reproduction and growth of the organism living in the water (Singh et al., 2007). In addition, the damage of the structures made of metals or stones like statues, buildings, automobiles, etc. on exposure for long durations (Singh & Agrawal, 2007). The famous historic buildings and statues across the world like the Taj Mahal in Agra, India, the Parthenon in Athens and Greece, the stonework of Lincoln Castle in Lincolnshire, England, etc. are damaged by acid pollution. The human health is not harmed directly by acid rain but indirectly through the intoxication of the food chain. The acidic water leaches the toxic substances from the soil and adds them to the freshwater bodies, from where they find their way to the food chain by directly consuming the contaminated water or utilize the animals or plant grown therein (Singh & Agrawal, 2007). For example in Sweden, several lakes have been found contaminated with Hg, which is leached by acid rain from the surrounding soils, and the residents are abandoned to use fishes as food from these lakes.

Sulfur and Corrosion

Commonly, the corrosive nature of sour crude petroleum and light distillate fractions is due to some of the sulfur compounds like H_2S, elemental sulphur, lower aliphatic sulfides and mercaptans, which are highly reactive towards metals and thereby causes the corrosion in the distribution systems and equipment in the petroleum refineries (McKetta, 1992.; Wood et al., 1925). These compounds when occurring abundantly in crude oil or light distillate fractions causes corrosion by reacting directly on the metals surfaces. The corrosion damages the storage tanks, pipes, and equipment, which not only reduces their durability but also leads to high repair and maintenance costs (Ahmad, 2013). Such type of corrosion in petroleum products is assessed as a relative degree of corrosivity through the copper strip corrosion test (ASTM D 130).

In addition to the direct corrosion, sulfur may cause indirect corrosion by producing SO_x damaging the internal linings of pipes, boilers, and parts of the engine, boilers or piping (Ma et al., 2007). During combustion of fuel containing sulfur in the boilers or internal combustion engines, organic sulfur is oxidized and set free as SO_x, which leads to the formation of H_2SO_4 by combining with water vapors present in the flue gases. As the temperature decreases, the acid condenses on various metal surfaces like linings of tubes, cylinder, internal parts of the engine, etc. and causes the corrosion of these parts, thereby decreasing their durability (Ahmad, 2013). Various techniques are used to combat such corrosion; these include the application of metal plating as protective coatings, increasing the temperature of the metal to avoid the condensation of acid, using different additives in the combustion chamber (Hobson & Petroleum, 1984). Similarly, the addition of additives to the flue gas stream or increasing the temperature of incoming gases by auxiliary heating systems is also carried out to avoid the acid precipitation on the metal surfaces.

REFERENCES

Absi Halabi, M., Stanislaus, A., Qamra, A., & Chopra, S. (1996). Effect of presulfiding on the activity and deactivation of hydrotreating catalysts in processing Kuwait vacuum residue. In J. B. M. Absi-Halabi & A. Stanislaus (Eds.), Studies in Surface Science and Catalysis (Vol. 100, pp. 243-251). Elsevier. doi:10.1016/S0167-2991(96)80025-5

Ahmad, W. (2013). *Study on cost effective methodes for desulfurization of kerosene, diesel and furnace oil (PhD)*. Peshawar: University of Peshawar.

Babich, I. V., & Moulijin, J. A. (2003). Science and technology of novel processes for deep desulfurization of oil refinery streams: A review. *Fuel, 82*(6), 607–631. doi:10.1016/S0016-2361(02)00324-1

Babich, I. V., & Moulijn, J. A. (2003). Science and technology of novel processes for deep desulfurization of oil refinery streams: A review. *Fuel, 82*(6), 607–631. doi:10.1016/S0016-2361(02)00324-1

Braid, C., & Cann, M. (2005). Envoirnmental chemistry (3rd ed.). New York: W.H. Freeman and Company.

Dasgupta, S., Gupta, P., Nanoti, A., Goswami, A. N., Garg, M. O., Tangstad, E., & Stocker, M. (2013). Adsorptive desulfurization of diesel by regenerable nickel based adsorbents. *Fuel, 108*, 184–189. doi:10.1016/j.fuel.2012.12.060

Froment, G. F., & Delmon, B. (1987). *Catalyst Deactivation 1987*. Elsevier Science.

Georgii, H. W. (1970). Contribution to the atmospheric sulfur budget. *Journal of Geophysical Research, 75*(2), 2365–2371. doi:10.1029/JC075i012p02365

Harrison, R. M., & Chemistry, R. S. o. (2007). *Principles of Environmental Chemistry*. RSC Publishing. doi:10.1039/9781847557780

Hobson, G. D., & Petroleum, I. O. (1984). *Modern petroleum technology*. Wiley; doi:10.1201/9781482277197

Kent, J. A. (2010). *Kent and Riegel's Handbook of Industrial Chemistry and Biotechnology* (Vol. 1). Springer.

Kulkarni & Afonso. (2010). *Deep desulfurization of diesel fuel using ionic liquids: current status and future challenges*. Academic Press.

Ma, X., Sakanishi, K., & Mochida, I. (1994). Hydrodesulfurization reactivities of various sulfur compounds in diesel fuel. *Industrial & Engineering Chemistry Research*, *33*(2), 218–222. doi:10.1021/ie00026a007

Magyar, S., Hancsók, J., & Kalló, D. (2005). Hydrodesulfurization and hydroconversion of heavy FCC gasoline on PtPd/H-USY zeolite. *Fuel Processing Technology*, *86*(11), 1151–1164. doi:10.1016/j.fuproc.2004.11.012

Manahan, S. E. (2010). *Environmental Chemistry* (7th ed.). Taylor & Francis.

McKetta, J. J. (1992). *Petroleum processing hand book*. Taylor & Francis Inc.

Mehran Soleimani, M., Bassi, A., & Margaritis, A. (2007). Biodesulfurization of refractory organic sulfur compounds in fossil fuels. *Biotechnology Advances*, *25*(6), 570–596. doi:10.1016/j.biotechadv.2007.07.003 PMID:17716849

Shahadat, H. A. M. H., & Tatarchuk, B. J. (2013). Adsorptive desulfurization of jet and diesel fuels using Ag/TiOx–Al_2O_3 and Ag/TiOx–SiO_2 adsorbents. *Fuel*, *107*, 465–473. doi:10.1016/j.fuel.2012.11.030

Singh, A., & Agrawal, M. (2007). Acid rain and its ecological consequences. *Journal of Environmental Biology*, *29*(1), 15. PMID:18831326

Song, S. C. (2003). An Overview of new approaches to deep desulfurization for ultra-clean gasoline, diesel fuel and jet fuel. *Catalysis Today*, *86*(1-4), 211–263. doi:10.1016/S0920-5861(03)00412-7

Triantafyllidis, K. S., & Deliyanni, E. A. (2014). Desulfurization of diesel fuels: Adsorption of 4, 6-DMDBT on different origin and surface chemistry nanoporous activated carbons. *Chemical Engineering Journal*, *236*, 406–414. doi:10.1016/j.cej.2013.09.099

United States Environmental Protection Agency (USEPA). (2014b). *EPA sets tier 3 motor vehicle emission and fuel standards, strengthens clean cars program*. EPA-420-F-14-010. Retrieved from www.epa.gov/otaq/documents/tier3/420f14010.pdf

United States Environmental Protection Agency (USEPA). (2006). *Introduction of cleaner-burning diesel fuel enables advanced pollution control for cars, trucks and busses*. EPA420-F-06-064. Retrieved from www.epa.gov/otaq/highway-diesel/regs/420f06064.pdf

Unites States Environmental Protection Agency (USEPA). (2014a). *EPA sets tier 3 motor vehicle emission and fuel standards.* EPA-420-F-14-009. Retrieved from www.epa.gov/otaq/documents/tier3/420f14009.pdf

Wood, A. E., Sheely, C., & Trusty, A. W. (1925). Corrosion Effect of Naphtha Solutions of Sulfur and Sulfur Compounds. *Industrial & Engineering Chemistry, 17*(8), 798–802. doi:10.1021/ie50188a008

Compilation of References

Abbad-Andaloussi, S., Lagnel, C., Warzywoda, M., & Monot, F. (2003). Multi-criteria comparison of resting cell activities of bacterial strains selected for biodesulfurization of petroleum compounds. *Enzyme and Microbial Technology*, *32*(3), 446–454. doi:10.1016/S0141-0229(02)00320-4

Abdalla, Z. E. A., Li, B. S., & Tufail, A. (2009). Direct synthesis of mesoporous (C19H42N)(4) H-3(PW11O39)/SiO2 and its catalytic performance in oxidative desulfurization. *Colloids and Surfaces a-Physicochemical and Engineering Aspects, 341*(1-3), 86-92.

Abdalla, Z. E. A., & Li, B. S. (2012). Preparation of MCM-41 supported (Bu4N)(4)H-3(PW11O39) catalyst and its performance in oxidative desulfurization. *Chemical Engineering Journal, 200*, 113–121. doi:10.1016/j.cej.2012.06.004

Abdullah, W., Nazwanie, W., Bakar, W. A., Azelee, W., Rusmidah, A., & Zaidi, E. (2015). Oxidative desulfurization of commercial diesel catalyzed by *tert*-butyl hydroperoxide polymolybdate on alumina: Optimization by Box-Behnken design. *Clean Technologies and Environmental Policy, 17*(2), 433–441. doi:10.100710098-014-0797-5

Abín-Fuentes, A. (2013). *Mechanistic understanding of microbial desulfurization* (Doctoral dissertation). Massachusetts Institute of Technology.

Abo-state, M. A., El-Gendy, N. S., El-Temtamy, S. A., Mahdy, H. M., & Nassar, H. N. (2014). Modification of basal salts medium for enhancing dibenzothiophene biodesulfurization by *Brevibacillus invocatus* C19 and *Rhodococcus erythropolis* IGTS8. *World Applied Sciences Journal, 30*(2), 133–140.

Absi Halabi, M., Stanislaus, A., Qamra, A., & Chopra, S. (1996). Effect of presulfiding on the activity and deactivation of hydrotreating catalysts in processing Kuwait vacuum residue. In J. B. M. Absi-Halabi & A. Stanislaus (Eds.), Studies in Surface Science and Catalysis (Vol. 100, pp. 243-251). Elsevier. doi:10.1016/S0167-2991(96)80025-5

Agarwal, P., & Sharma, D. K. (2010). Comparative studies on the biodesulfurization of crude oil with other desulfurization techniques and deep desulfurization through integrated processes. *Energy & Fuels, 24*(1), 518–524. doi:10.1021/ef900876j

Aggarwal, S., Karimi, I. A., & Ivan, G. R. (2013). In silico modeling and evaluation of *Gordonia alkanivorans* for biodesulfurization. *Molecular BioSystems*, *9*(10), 2530–2540. doi:10.1039/c3mb70132h PMID:23921469

Aggarwal, S., Karimi, I. A., Kilbane, J. J. II, & Lee, D. Y. (2012). Roles of sulfite oxidoreductase and sulfite reductase in improving desulfurization by *Rhodococcus erythropolis*. *Molecular BioSystems*, *8*(10), 2724–2732. doi:10.1039/c2mb25127b PMID:22832889

Aggarwal, S., Karimi, I., & Lee, D. Y. (2011). Reconstruction of a genome-scale metabolic network of *Rhodococcus erythropolis* for desulfurization studies. *Molecular BioSystems*, *7*(11), 3122–3131. doi:10.1039/c1mb05201b PMID:21912787

Ahmad, W. (2013). *Study on cost effective methodes for desulfurization of kerosene, diesel and furnace oil (PhD)*. Peshawar: University of Peshawar.

Aida, T., & Yamamoto, D. (1994). Annual book of ASTM standards, 2005. Standard test method for determination of sulfur compounds in natural gas and gaseous fuels by gas chromatography and chemiluminescence. *Preprints of Papers- American Chemical Society. Division of Fuel Chemistry*, *39*(5), 623.

Aida, T., Yamamoto, D., Iwata, M., & Sakata, K. (2000). Development of oxidative desulfurization process for diesel. *Review on Heteroatom Chemistry*, *22*, 241–256.

Al Zubaidy, I. A. H., Tarsh, F. B., Darwish, N. N., Majeed, B. S. S. A., Al Sharafi, A., & Abu Chacra, L. (2013). *Adsorption Process of Sulfur Removal from Diesel Oil Using Sorbent Materials*. Academic Press.

Alfonso, D. R. (2008). First-principles studies of H_2S adsorption and dissociation on metal surfaces. *Surface Science*, *602*(16), 2758–2768. doi:10.1016/j.susc.2008.07.001

AL-Hammadi, S. A., Al-Amer, A. M., & Saleh, T. A. (2018). Alumina-carbon nanofiber composite as a support for MoCo catalysts in hydrodesulfurization reactions. *Chemical Engineering Journal*, *345*, 242–251. doi:10.1016/j.cej.2018.03.106

Allen, B. L., Kichambare, P. D., & Star, A. (2008). Synthesis, characterization, and manipulation of nitrogen-doped carbon nanotube cups. *ACS Nano*, *2*(9), 1914–1920. doi:10.1021/nn800355v PMID:19206432

Allen, B. L., Shade, C. M., Yingling, A. M., Petoud, S., & Star, A. (2009). Graphitic nanocapsules. *Advanced Materials*, *21*, 4692–4695.

Almarri, M., Ma, X., & Song, C. (2009). Role of surface oxygen-containing functional groups in liquid-phase adsorption of nitrogen compounds on carbon-based adsorbents. *Energy & Fuels*, *23*(8), 3940–3947. doi:10.1021/ef900051r

Al-Shamrani, A. A., James, A., & Xiao, H. (2002). Destabilisation of oil–water emulsions and separation by dissolved air flotation. *Water Research*, *36*(6), 1503–1512. doi:10.1016/S0043-1354(01)00347-5 PMID:11996340

Alves, L., Marques, S., Matos, J., Tenreiro, R., & Gírio, F. M. (2008). Dibenzothiophene desulfurization by *Gordonia alkanivorans* strain 1B using recycled paper sludge hydrolyzate. *Chemosphere*, *70*(6), 967–973. doi:10.1016/j.chemosphere.2007.08.016 PMID:17897697

Alves, L., Matos, J., Tenreiro, R., & Gírio, F. M. (2008). Evidence for the role of zinc on the performance of dibenzothiophene desulfurization by *Gordonia alkanivorans* strain 1B. *Journal of Industrial Microbiology & Biotechnology*, *35*(1), 69–73. doi:10.100710295-007-0278-5 PMID:18043954

Alves, L., & Paixão, S. M. (2014a). Enhancement of dibenzothiophene desulfurization by *Gordonia alkanivorans* strain 1B using sugar beet molasses as alternative carbon source. *Applied Biochemistry and Biotechnology*, *172*(6), 3297–3305. doi:10.100712010-014-0763-z PMID:24519629

Alves, L., & Paixão, S. M. (2014b). Fructophilic behaviour of *Gordonia alkanivorans* strain 1B during dibenzothiophene desulfurization process. *New Biotechnology*, *31*(1), 73–79. doi:10.1016/j.nbt.2013.08.007 PMID:24012483

Alves, L., Paixão, S. M., Pacheco, R., Ferreira, A. F., & Silva, C. M. (2015). Biodesulphurization of fossil fuels: Energy, emissions and cost analysis. *RSC Advances*, *5*(43), 34047–34057. doi:10.1039/C4RA14216K

Alves, L., Salgueiro, R., Rodrigues, C., Mesquita, E., Matos, J., & Gírio, F. M. (2005). Desulfurization of dibenzothiophene, benzothiophene, and other thiophene analogs by a newly isolated bacterium, *Gordonia alkanivorans* strain 1B. *Applied Biochemistry and Biotechnology*, *120*(3), 199–208. doi:10.1385/ABAB:120:3:199 PMID:15767694

Amin, G. A. (2011). Integrated two-stage process for biodesulfurization of model oil by vertical rotating immobilized cell reactor with the bacterium *Rhodococcus erythropolis*. *Journal of Petroleum & Environmental Biotechnology*, *2*(1), 107.

Amin, G. A., Bazaid, S. A., & Abd El-Halim, M. (2013). A two-stage immobilized cell bioreactor with *Bacillus subtilis* and *Rhodococcus erythropolis* for the simultaneous production of biosurfactant and biodesulfurization of model oil. *Petroleum Science and Technology*, *31*(21), 2250–2257. doi:10.1080/10916466.2011.565292

Amo-Ochoa, P., Hassanein, K., Gómez-García, C. J., Benmansour, S., Perles, J., Castillo, O., ... Zamora, F. (2015). Reversible stimulus-responsive Cu(I) iodide pyridine coordination polymer. *Chemical Communications (Cambridge)*, *51*(76), 14306–14309. doi:10.1039/C5CC04746C PMID:26264525

Andersson, J. T., Panda, S. K., & Schrader, W. (2007). Analysis of large sulfur aromatic compounds in fossil fuels by FT-ICR MS. In O. R. Koseoglu & S. Kasztelan (Eds.), *Hydrotreating/Hydrocracking Processing*. Washington, DC: American Chemical Society.

Andersson, J. T., & Schade, T. (2006). Speciation of alkylated dibenzothiophenes in a deeply desulfurized diesel fuel. *Energy & Fuels*, *20*(4), 1614–1620. doi:10.1021/ef0502507

Ania, C. O., & Bandosz, T. J. (2006). Metal-loaded polystyrene-based activated carbons as dibenzothiophene removal media via reactive adsorption. *Carbon*, *44*(12), 2404–2412. doi:10.1016/j.carbon.2006.05.016

Anisimov, A. V., Fedorova, E. V., Lesnugin, A. Z., Senyavin, V. M., Aslanov, L. A., Rybakov, V. B., & Tarakanova, A. V. (2003). Vanadium peroxocomplexes as oxidation catalysts of sulfur organic compounds by hydrogen peroxide in bi-phase systems. *Catalysis Today*, *78*(1-4), 319–325. doi:10.1016/S0920-5861(02)00311-5

An, K. H., Kim, W. S., Park, Y. S., Moon, J.-M., Bae, D. J., Lim, S. C., ... Lee, Y. H. (2001). Electrochemical Properties of High-Power Supercapacitors Using Single-Walled Carbon Nanotube Electrodes. Journal of Advanced Functional. *Materials (Basel)*, *11*, 387–392.

Anne Rodrigues, K. O., Ramos, J. E. T., Cavalcante, C. L., Rodriguez-Castellon, E., & Azevedo, D. C. S. (2014). Pd-loaded mesoporous silica as a robust adsorbent in adsorption/desorption desulfurization cycles. *Fuel*, *126*, 96–103. doi:10.1016/j.fuel.2014.02.019

Ansari, F., Grigoriev, P., Libor, S., Tothill, I. E., & Ramsden, J. J. (2009). DBT degradation enhancement by decorating *Rhodococcus erythropolis* IGST8 with magnetic Fe_3O_4 nanoparticles. *Biotechnology and Bioengineering*, *102*(5), 1505–1512. doi:10.1002/bit.22161 PMID:19012265

Anton, N. S., Mehdi, M. Y., Romaneh, J., Ouseph, P. J., Cohn, R. W., & Sumanasekera, G. U. (2007). Electrostatic deposition of graphene. *Nanotechnology*, *18*, 13530–13534. PMID:21730375

Arez, B. F., Alves, L., & Paixão, S. M. (2014). Production and characterization of a novel yeast extracellular invertase activity towards improved dibenzothiophene biodesulfurization. *Applied Biochemistry and Biotechnology*, *174*(6), 2048–2057. doi:10.100712010-014-1182-x PMID:25163885

Armatas, G. S. (2013). Heterogeneous Polyoxometalate-Containing Mesoporous Catalysts. In S. L. Suib (Ed.), *New and Future Developments in Catalysis* (pp. 311–342). Amsterdam: Elsevier. doi:10.1016/B978-0-444-53876-5.00014-3

Armatas, G. S., Bilis, G., & Louloudi, M. (2011). Highly ordered mesoporous zirconia-polyoxometalate nanocomposite materials for catalytic oxidation of alkenes. *Journal of Materials Chemistry*, *21*(9), 2997–3005. doi:10.1039/c0jm03395b

ATSDR. (1995). *Toxicological profile for fuel oils*. Atlanta, GA: U.S. Department of Health and Human Services, Public Health Service.

Avouris, P. (2002). Molecular electronics with carbon nanotubes. *Accounts of Chemical Research*, *35*(12), 1026–1034. doi:10.1021/ar010152e PMID:12484790

Babich, I. V., & Moulijin, J. A. (2003). Science and technology of novel processes for deep desulfurization of oil refinery streams: A review. *Fuel*, *82*(6), 607–631. doi:10.1016/S0016-2361(02)00324-1

Bacsa, J., Eve, D., & Dunbar, K. R. (2005). Catena-Poly[[diaquacobalt(II)]-μ-Oxalato]. *Acta Crystallographica. Section C, Crystal Structure Communications*, *61*(1), m58–m60. doi:10.1107/S0108270104030409 PMID:15640580

Baeza, P., Aguila, G., Gracia, F., & Araya, P. (2008). Desulfurization by adsorption with copper supported on zirconia. *Catalysis Communications*, *9*(9), 751–755. doi:10.1016/j.catcom.2007.08.020

Bai, S., & Shen, X. (2012). Graphene-inorganic nanocomposites. *Royal Society of Chemistry Advances*, *2*, 64–98.

Baker, R. T. K. (1995). A review of in-situ electron microscopy studies of metal/metal oxide graphite interactions. *Journal of Adhesion and Adhensives*, *52*(1-4), 13–40.

Baker, R. T. K., & Chludzinski, J. J. Jr. (1981). Catalytic gasification of graphite by chromium and copper in oxygen, steam and hydrogen. *Carbon*, *19*(2), 75–82. doi:10.1016/0008-6223(81)90110-X

Balula, S. S., Granadeiro, C. M., Barbosa, A. D. S., Santos, I. C. M. S., & Cunha-Silva, L. (2013). Multifunctional catalyst based on sandwich-type polyoxotungstate and MIL-101 for liquid phase oxidations. *Catalysis Today*, *210*, 142–148. doi:10.1016/j.cattod.2012.12.003

Banat, I. M., Makkar, R. S., & Cameotra, S. S. (2000). Potential commercial applications of microbial surfactants. *Applied Microbiology and Biotechnology*, *53*(5), 495–508. doi:10.1007002530051648 PMID:10855707

Bandyopadhyay, S., Chowdhury, R., & Bhattacharjee, C. (2014). Production of biosurfactants through biodesulfurization of spent engine oil – an experimental study. *Green Chemistry Letters and Reviews*, *7*(3), 288–295. doi:10.1080/17518253.2014.946972

Bandyopadhyay, S., Chowdhury, R., Bhattacharjee, C., & Pan, S. (2013). Simultaneous production of biosurfactant and ULSD (ultra low sulfur diesel) using *Rhodococcus* sp. in a chemostat. *Fuel*, *113*, 107–112. doi:10.1016/j.fuel.2013.05.036

Bansal, R. C., Donnet, J. B., & Stoeckli, F. (1988). *Active Carbon*. New York: Mercel Dekker.

Baran, E. J. (2014). Review: Natural oxalates and their analogous synthetic complexes. *Journal of Coordination Chemistry*, *67*(23-24), 3734–3768. doi:10.1080/00958972.2014.937340

Barbara, P., Rufino, M. N., Campos-Martin, J. M., & Fierro, J. L. G. (2011). Towards near zero-sulfur liquid fuels: A perspective review. *Catalysis Science & Technology*, *1*(1), 23–42. doi:10.1039/c0cy00049c

Bardania, H., Raheb, J., Mohammad-Beigi, H., Rasekh, B., & Arpanaei, A. (2013). Desulfurization activity and reusability of magnetite nanoparticle–coated *Rhodococcus erythropolis* FMF and *R. erythropolis* IGTS8 bacterial cells. *Biotechnology and Applied Biochemistry*, *60*(3), 323–329. doi:10.1002/bab.1090 PMID:23656694

Batista, S. B., Mounteer, A. H., Amorim, F. R., & Totola, M. R. (2006). Isolation and characterization of biosurfactant/bioemulsifier-producing bacteria from petroleum contaminated sites. *Bioresource Technology, 97*(6), 868–875. doi:10.1016/j.biortech.2005.04.020 PMID:15951168

Baughman, R. H., Zakhidov, A. A., & de Heer, W. A. (2002). Carbon nanotubes--the route towards applications. *Science, 297*(5582), 787–792. doi:10.1126cience.1060928 PMID:12161643

Baxendale, J. H., Evans, M. G., & Park, G. S. (1946). The mechanism and kinetics of the initiation of polymerisation by systems containing hydrogen peroxide. *Transactions of the Faraday Society, 42*, 155–169. doi:10.1039/tf9464200155

Bej. (2004). Revamping of diesel hydro-desulfurizers: Options available and future research needs. *Fuel Process. Technol, 85*(13), 1503–1517.

Ben Li, W. L., Wu, H., Yu, S., Cao, R., & Jiang, Z. (2012). Desulfurization of model gasoline by bioinspired oleophilic nanocomposite membranes. *Journal of Membrane Science, 415–416*, 278–287.

Bender, J., & Weigel, H.-J. (2011). Changes in atmospheric chemistry and crop health: A review. *Agronomy for Sustainable Development, 31*(1), 81–89. doi:10.1051/agro/2010013

Berman, J., Zorrilla-López, U., Farré, G., Zhu, C., Sandmann, G., Twyman, R. M., ... Christou, P. (2015). Nutritionally important carotenoids as consumer products. *Phytochemistry Reviews, 14*(5), 727–743.

Bernier, P., & Ajayan, P. M. (1994). Application of Carbon Nanotubes as Supports in Heterogeneous Catalysis. *Journal of the American Chemical Society, 116*(17), 7935–7936. doi:10.1021/ja00096a076

Bezeverkhyy, I., Ryzhikov, A., Gadacz, G., & Bellat, J. P. (2008). Kinetics of thiophene reactive adsorption on Ni/SiO_2 and Ni/ZnO. *Catalysis Today, 130*(1), 199–205. doi:10.1016/j.cattod.2007.06.038

Bezverkhyy, I., Safonova, O. V., Afanasiev, P., & Bellat, J. P. (2009). Reaction between thiophene and Ni nanoparticles supported on SiO_2 or ZnO: In situ synchrotron X-ray diffraction study. *The Journal of Physical Chemistry C, 113*(39), 17064–17069. doi:10.1021/jp903488p

Bhandari, V. M., Ko, C. H., Park, J. G., Han, S., Cho, S., & Kim, J. (2006). Desulfurization of diesel using ion-exchanged zeolites. *Chemical Engineering Science, 61*(8), 2599–2608. doi:10.1016/j.ces.2005.11.015

Bhatia, S., & Sharma, D. K. (2006). Emerging role of bio-refining of heavier crude oils and integration of bio-refining with petroleum refineries in the future. *Petroleum Science and Technology, 24*(10), 1125–1159. doi:10.1080/10916460600817447

Bhatia, S., & Sharma, D. K. (2012). Thermophilic desulfurization of dibenzothiophene and different petroleum oils by *Klebsiella* sp. 13T. *Environmental Science and Pollution Research International, 19*(8), 3491–3497. doi:10.100711356-012-0884-2 PMID:22467237

Bhutto, A. W., Abro, R., Gao, S., Abbas, T., Chen, X., & Yu, G. (2016). Oxidative desulfurization of fuels using ionic liquids: A review. *Journal of the Taiwan Institute of Chemical Engineers, 62*, 84–97. doi:10.1016/j.jtice.2016.01.014

Blanco-Brieva, G., Campos-Martin, J. M., Al-Zahrani, S. M., & Fierro, J. L. G. (2011). Effectiveness of metal–organic frameworks for removal of refractory organo-sulfur compound present in liquid fuels. *Fuel, 90*(1), 190–197. doi:10.1016/j.fuel.2010.08.008

Blanco-Brieva, G., Campos-Martin, J. M., Al-Zahrani, S. M., & Fierro, J. L. G. (2013). Thermal regeneration of the metal organic frameworks used in the adsorption of refractory organosulfur compounds from liquid fuels. *Fuel, 105*, 459–465. doi:10.1016/j.fuel.2012.08.003

Blanco-Lopez, M. C., Gutierrez-Fernandez, C., Lobo-Casta, Ç. M. J., Miranda-Ordieres, A. J., Tucon-Blanco, P. (2004). Electrochemical sensing with electrodes modified with molecularly imprinted polymer films. *Analytical and Bioanalytical Chemistry, 378*(8), 1922 –1928.

Blandino, A., Macías, M., & Cantero, D. (1999). Formation of calcium alginate gel capsules: Influence of sodium alginate and $CaCl_2$ concentration on gelation kinetics. *Journal of Bioscience and Bioengineering, 88*(6), 686–689. doi:10.1016/S1389-1723(00)87103-0 PMID:16232687

Boniek, D., Figueiredo, D., dos Santos, A. F. B., & Stoianoff, M. A. de R. (2014). Biodesulfurization: A mini review about the immediate search for the future technology. *Clean Technologies and Environmental Policy.*

Booth, B., Buglass, J. G., & Unsworth, J. F. (2001). Tailoring fuels for the new millennium. *Topics in Catalysis, 16-17*(1/4), 39–46. doi:10.1023/A:1016670528451

Braid, C., & Cann, M. (2005). Envoirnmental chemistry (3rd ed.). New York: W.H. Freeman and Company.

Brandon, N. P., Skinner, S., & Steele, B. C. (2003). Recent advances in materials for fuel cells. *Annual Review of Materials Research, 33*(1), 183–213. doi:10.1146/annurev.matsci.33.022802.094122

Brunauer, S., Emmett, P. H., & Teller, E. (1938). Adsorption of gasses in multimolecular layer. *Journal of the American Chemical Society, 60*(2), 309–319. doi:10.1021/ja01269a023

Buekens, A., & Zyaykina, N. N. (2009). *Adsorbents and adsorption processes for pollution control* (Vol. 2). Pollution Control Technologies.

Bu, J., Loh, G., Gwie, C. G., Dewiyanti, S., Tasrif, M., & Borgna, A. (2011). Desulfurization of diesel fuels by selective adsorption on activated carbons: Competitive adsorption of polycyclic aromatic sulfur heterocycles and polycyclic aromatic hydrocarbons. *Chemical Engineering Journal, 166*(1), 207–217. doi:10.1016/j.cej.2010.10.063

Calegari-Santos, R., Diogo, R. A., Fontana, J. D., & Bonfim, T. M. B. (2016). Carotenoid production by halophilic Archaea under different culture conditions. *Current Microbiology, 72*(5), 641–651.

Cameotra, S. S., & Makkar, R. S. (1998). Synthesis of biosurfactants in extreme conditions. *Applied Microbiology and Biotechnology, 50*(5), 520–529. doi:10.1007002530051329 PMID:9866170

Cameotra, S. S., & Singh, P. (2009). Synthesis of rhamnolipid biosurfactant and mode of hexadecane uptake by *Pseudomonas* species. *Microbial Cell Factories*, *8*(1), 16. doi:10.1186/1475-2859-8-16 PMID:19284586

Campestrini, S., Conte, V., Di Furia, F., Modena, G., & Bortolini, O. (1988). Metal catalysis in oxidation by peroxides. 30. Electrophilic oxygen transfer from anionic, coordinatively saturated molybdenum peroxo complexes. *The Journal of Organic Chemistry*, *53*(24), 5721–5724. doi:10.1021/jo00259a021

Campos-Martin, J. M., Capel-Sanchez, M. C., Perez-Presas, P., & Fierro, J. L. G. (2010). Oxidative processes of desulfurization of liquid fuels. *Journal of Chemical Technology and Biotechnology (Oxford, Oxfordshire)*, *85*(7), 879–890. doi:10.1002/jctb.2371

Cao, G. (2004). *Nanostructures and Nanomaterials: Synthesis, properties and applications*. Imperial College. doi:10.1142/p305

Caro, A., Boltes, K., Letón, P., & García-Calvo, E. (2007). Dibenzothiophene biodesulfurization in resting cell conditions by aerobic bacteria. *Biochemical Engineering Journal*, *35*(2), 191–197. doi:10.1016/j.bej.2007.01.013

Caro, A., Letón, P., García-Calvo, E., & Setti, L. (2007). Enhancement of dibenzothiophene biodesulfurization using β-cyclodextrins in oil-to-water media. *Fuel*, *86*(16), 2632–2636. doi:10.1016/j.fuel.2007.02.033

Carvalheiro, F., Moniz, P., Duarte, L. C., Esteves, M. P., & Gírio, F. M. (2011). Mannitol production by lactic acid bacteria grown in supplemented carob syrup. *Journal of Industrial Microbiology & Biotechnology*, *38*(1), 221–227. doi:10.100710295-010-0823-5 PMID:20820868

Castro, B., Whitcombe, M. J., Vulfson, E. N., Vazquez-Duhalt, R., & Barzana, E. (2001). Molecular imprinting for the selective adsorption of organosulfur compounds present in fuels. *Analytica Chimica Acta*, *435*(1), 83–90. doi:10.1016/S0003-2670(01)00799-1

Cerrato, G., Bordiga, S., Barbera, S., & Morterra, C. (1997). A surface study of monoclinic zirconia (m-ZrO_2). *Surface Science*, *377-379*(50-55).

Cerrato, G., Bordiga, S., Barbera, S., & Morterra, C. (1997). Surface Characterization of monoclinic ZrO_2 I morphology, FTIR spectral features and computer modelling. *Applied Surface Science*, *115*(1), 53–65. doi:10.1016/S0169-4332(96)00586-7

Chaichanawong, J., Yamamoto, T., Ohmori, T., & Endo, A. (2010). Adsorptive desulfurization of bioethanol using activated carbon loaded with zinc oxide. *Chemical Engineering Journal*, *165*(1), 218–224. doi:10.1016/j.cej.2010.09.020

Chamack, M., Mahjoub, A. R., & Aghayan, H. (2014). Cesium salts of tungsten-substituted molybdophosphoric acid immobilized onto platelet mesoporous silica: Efficient catalysts for oxidative desulfurization of dibenzothiophene. *Chemical Engineering Journal*, *255*, 686–694. doi:10.1016/j.cej.2014.06.054

Chang, G., Bao, Z., Ren, Q., Deng, S., Zhang, Z., Su, B., ... Yang, Y. (2014). Fabrication of cuprous nanoparticles in MIL-101: An efficient adsorbent for the separation of olefin–paraffin mixtures. *RSC Advances*, *4*(39), 20230–20233. doi:10.1039/C4RA02125H

Chang, J. H., Chang, Y. K., Cho, K. S., & Chang, H. N. (2000). Desulfurization of model and diesel oils by resting cells of Gordona sp. *Biotechnology Letters*, *22*(3), 193–196. doi:10.1023/A:1005637013879

Chang, J. H., Chang, Y. K., Ryu, H. W., & Chang, H. N. (2000). Desulfurization of light gas oil in immobilized-cell systems of *Gordona* sp. CYKS1 and *Nocardia* sp. CYKS2. *FEMS Microbiology Letters*, *182*(2), 309–312. doi:10.1111/j.1574-6968.2000.tb08913.x PMID:10620684

Chang, J. H., Kim, Y. J., Lee, B. H., Cho, K., Ryu, H. W., Chang, Y. K., & Chang, H. N. (2001). Production of a desulfurization biocatalyst by two-stage fermentation and its application for the treatment of model and diesel oils. *Biotechnology Progress*, *17*(5), 876–880. doi:10.1021/bp0100676 PMID:11587578

Chang, J. H., Rhee, S. K., Chang, Y. K., & Chang, H. N. (1998). Desulfurization of diesel oils by a newly isolated dibenzothiophene-degrading Norcadia sp. strain CYKS2. *Biotechnology Progress*, *14*(6), 851–855. doi:10.1021/bp9800788 PMID:9841646

Charlier, J. C. (2002). Defects in carbon nanotubes. *Accounts of Chemical Research*, *35*(12), 1063–1069. doi:10.1021/ar010166k PMID:12484794

Chen, J. C., & Monticello, D. J. (1996). *U.S. Patent No. 5,525,235*. Washington, DC: U.S. Patent and Trademark Office.

Cheng, G., & Guo, T. (2002). Surface segregation in Ni/Co bimetallic nanoparticles produced in single-walled carbon nanotube synthesis. *The Journal of Physical Chemistry B*, *106*(23), 5833–5839. doi:10.1021/jp025530r

Cheng, P., Zhang, S., Wang, P., Huang, S., & Tian, H. (2013). First-principles investigation of thiophene adsorption on Ni13 and Zn@Ni12 nanoclusters. *Computational & Theoretical Chemistry*, *1020*, 136–142. doi:10.1016/j.comptc.2013.07.044

Cheng, Y., Zhou, W., Gao, C., Lan, K., Gao, Y., & Wu, Q. (2009). Biodiesel production from Jerusalem artichoke (*Helianthus tuberosus* L.) tuber by heterotrophic microalgae *Chlorella protothecoides*. *Journal of Chemical Technology and Biotechnology (Oxford, Oxfordshire)*, *84*(5), 777–781. doi:10.1002/jctb.2111

Chen, H., & Hsieh, Y. L. (2004). Ultrafine hydrogel fibers with dual temperature- and pH-responsive swelling behaviors. *Journal of Polymer Science. Part A, Polymer Chemistry*, *42*(24), 6331–6339. doi:10.1002/pola.20461

Chen, Y., & Song, Y. F. (2014). Immobilization of LaW10 onto Ionic-Liquid-Modified Mesoporous Silica: Deep Desulfurization with Zero-Order Reaction Kinetics. *ChemPlusChem*, *79*(2), 304–309. doi:10.1002/cplu.201300323

Chen, Y., Yan, D., & Song, Y.-F. (2014). Tris(hydroxymethyl)aminomethane modified layered double hydroxides greatly facilitate polyoxometalate intercalation. *Dalton Transactions (Cambridge, England)*, *43*(39), 14570–14576. doi:10.1039/C4DT01827C PMID:25191827

Chen, Y., Zhao, S., & Song, Y. F. (2013). An efficient heterogeneous catalyst based on highly dispersed Na(7)H(2)LaW(10)O(36 center dot)32H(2)O nanoparticles on mesoporous silica for deep desulfurization. *Applied Catalysis A, General*, *466*, 307–314. doi:10.1016/j.apcata.2013.06.030

Chica, A., Corma, A., & Dómine, M. E. (2006). Catalytic oxidative desulfurization (ODS) of diesel fuel on a continuous fixed-bed reactor. *Journal of Catalysis*, *242*(2), 299–308. doi:10.1016/j.jcat.2006.06.013

Chica, A., Gatti, G., Moden, B., Marchese, L., & Iglesia, E. (2006). Selective catalytic oxidation of organosulfur compounds with *tert*-butylhydroperoxide. *Chemistry (Weinheim an der Bergstrasse, Germany)*, *12*(7), 1960–1967. doi:10.1002/chem.200500858 PMID:16385590

Choi, O. K., Cho, K. S., Ryu, H. W., & Chang, Y. K. (2003). Enhancement of phase separation by the addition of de-emulsifiers to three-phase (diesel oil / biocatalyst / aqueous phase) emulsion in diesel biodesulfurization. *Biotechnology Letters*, *25*(1), 73–77. doi:10.1023/A:1021750719337 PMID:12882310

Chong, E. J., Phan, T. T., Lim, I. J., Zhang, Y. Z., Bay, B. H., Ramakrishna, S., & Lim, C. T. (2007). Evaluation of electrospun PCL/gelatin nanofibrous scaffold for wound healing and layered dermal reconstitution. *Acta Biomaterialia*, *3*(3), 321–330. doi:10.1016/j.actbio.2007.01.002 PMID:17321811

Chowdhury, R., & Duttagupta, A. (2005). Mathematical Modeling of the transient behaviour of a chemostat undergoing bio-desulfurization of model organosulfur compounds and diesel using pure and isolated bacterial strains. *International Journal of Chemical Reactor Engineering*, *3*(1), 1–12.

Clean Fuel, I. I. (2019). Retrieved from http://www.sapia.org.za/key-issues/cleaner-fuels-ii

Cohen, S. M. (2012). Postsynthetic Methods for the Functionalization of Metal-Organic Frameworks. *Chemical Reviews*, *112*(2), 970–1000. doi:10.1021/cr200179u PMID:21916418

Collins, F. M., Lucy, A. R., & Sharp, C. (1997). Oxidative desulfurization of oils via hydrogen peroxide and heteropolyanion catalysis. *Journal of Molecular Catalysis A Chemical*, *117*(1-3), 397–403. doi:10.1016/S1381-1169(96)00251-8

Coronado, E., & Gómez-García, C. J. (1998). Polyoxometalate-Based Molecular Materials. *Chemical Reviews*, *98*(1), 273–296. doi:10.1021/cr970471c PMID:11851506

Crude oil reserves and spatial distribution. (n.d.). Retrieved September 13, 2013, from http://web2.mendelu.cz/af_291_projekty2/vseo/stranka.php?kod=312

Cui, R. L., Zhang, Y., Wang, J. Y., Zhou, W. W., & Li, Y. (2010). Comparison between copper and iron as catalyst for chemical vapor deposition of horizontally aligned ultralong single-walled carbon nanotubes on silicon substrates. *Journal of Physical Chemistry*, *114*(37), 15547–15552.

Cui, X., Yao, D., Li, H., Yang, J., & Hu, D. (2012). Nano-magnetic particles as multifunctional microreactor for deep desulfurization. *Journal of Hazardous Materials*, *205–206*(0), 17–23. doi:10.1016/j.jhazmat.2011.11.063 PMID:22244972

Cychosz, K. A., Wong-Foy, A. G., & Matzger, A. J. (2008). Liquid phase adsorption by microporous coordination polymers. *Journal of the American Chemical Society*, *130*(22), 6938–6943. doi:10.1021/ja802121u PMID:18465854

Czaja, A. U., Trkhan, N., & Muller, U. (2009). Industrial applications of metal organic frameworks. *Chemical Society Reviews*, *38*(5), 1284. doi:10.1039/b804680h PMID:19384438

Dai, H. (2002). Carbon nanotubes: Opportunities and challenges. *Surface Science*, *500*(1-3), 218–241. doi:10.1016/S0039-6028(01)01558-8

Daimler–Chrysler, A.G. (2000). *Methodfor desulfurization engine fuel on board a motor vehicle*. W.O 2000/20531.

Dai, Y., Shao, R., Qi, G., & Ding, B.-B. (2014). Enhanced dibenzothiophene biodesulfurization by immobilized cells of *Brevibacterium lutescens* in n-octane-water biphasic system. *Applied Biochemistry and Biotechnology*, *174*(6), 2236–2244. doi:10.100712010-014-1184-8 PMID:25173674

Darian, S. T., & Arabshshi, S. H. (1988). *Process for upgrading diesel oils*. US Patent 4746420.

Dasgupta, S., Gupta, P., Nanoti, A., Goswami, A. N., Garg, M. O., Tangstad, E., & Stocker, M. (2013). Adsorptive desulfurization of diesel by regenerable nickel based adsorbents. *Fuel*, *108*, 184–189. doi:10.1016/j.fuel.2012.12.060

Davis, F. J. (2004). Polymer chemistry-A practical approach. Oxford University Press.

Deck, C. P., & Vecchio, K. (2006). Prediction of carbon nanotube growth success by the analysis of carbon-catalyst binary phase diagrams. *Carbon*, *44*(2), 267–275. doi:10.1016/j.carbon.2005.07.023

Dehghan-Noudeh, G., Housaindokht, M., & Bazzaz, B. S. F. (2005). Isolation, characterization, and investigation of surface and hemolytic activities of a lipopeptide biosurfactant produced by *Bacillus subtilis* ATCC 6633. *Journal Of Microbiology-Seoul*, *43*(3), 272. PMID:15995646

Dekker, M., Yocum, R.H., & Nyquist, E.B. (1973). *Functional monomers; their preparation, polymerization, and application*. Academic Press.

Del Olmo, C. H., Santos, V. E., Alcon, A., & Garcia-Ochoa, F. (2005). Production of a *Rhodococcus erythropolis* IGTS8 biocatalyst for DBT biodesulfurization: Influence of operational conditions. *Biochemical Engineering Journal*, *22*(3), 229–237. doi:10.1016/j.bej.2004.09.015

Delgado, J. L., Herranz, M. A., & Martin, N. (2008). The nano-forms of carbon. *Journal of Materials Chemistry*, *18*(13), 1417–1426. doi:10.1039/b717218d

Demir, M. M., Yilgor, I., Yilgor, E., & Erman, B. (2002). Electrospinning of polyurethane fibers. *Polymer*, *43*(11), 3303–3309. doi:10.1016/S0032-3861(02)00136-2

Denis, P. A. (2010). On the enthalpy of formation of thiophene. *Theoretical Chemistry Accounts*, *127*(5-6), 621–626. doi:10.100700214-010-0759-x

Dercz, G., Prusik, K., & Pajak, L. (2008). X-ray and SEM studies on zirconia powders. *Journal of Mat & Manufacturing Eng*, *31*(2).

Deria, P., Mondloch, J. E., Karagiaridi, O., Bury, W., Hupp, J. T., & Farha, O. K. (2014). Beyond post-synthesis modification: Evolution of metal–organic frameworks *via* building block replacement. *Chemical Society Reviews*, *43*(16), 5896–5912. doi:10.1039/C4CS00067F PMID:24723093

Derikvand, P., Etemadifar, Z., & Biria, D. (2014). Taguchi optimization of dibenzothiophene biodesulfurization by *Rhodococcus erythropolis* R1 immobilized cells in a biphasic system. *International Biodeterioration & Biodegradation*, *86*, 343–348. doi:10.1016/j.ibiod.2013.10.006

Desai, J. D., & Banat, I. M. (1997). Microbial production of surfactants and their commercial potential. *Microbiology and Molecular Biology Reviews*, *61*(1), 47–64. PMID:9106364

De Volder, M. F. L., Tawfick, S. H., Baughman, R. H., & Hart, A. J. (2013). Carbon nanotubes: Present and future commercial applications. *Science*, *339*(6119), 535–539. doi:10.1126cience.1222453 PMID:23372006

Di Noto, V., Lavina, S., Wintersgill, M. C., & Fontanella, J. J. (2010). A formalism relating the conductivity of functionalized nanoparticles to constituent ligand molecules and application to water-containing silica. *Physical Chemistry Chemical Physics*, *12*(23), 5993–5997. doi:10.1039/b924811k PMID:20396820

Di-Giuseppe, A., Crucianelli, M., De Angelis, F., Crestini, C., & Saladino, R. (2009). Efficient oxidation of thiophene derivatives with homogeneous and heterogeneous MTO/H_2O_2 systems: A novel approach for, oxidative desulfurization (ODS) of diesel fuel. *Applied Catalysis B: Environmental*, *89*(1-2), 239–245. doi:10.1016/j.apcatb.2009.02.009

Dinamarca, M. A., Ibacache-Quiroga, C., Baeza, P., Galvez, S., Villarroel, M., Olivero, P., & Ojeda, J. (2010). Biodesulfurization of gas oil using inorganic supports biomodified with metabolically active cells immobilized by adsorption. *Bioresource Technology*, *101*(7), 2375–2378. doi:10.1016/j.biortech.2009.11.086 PMID:20034786

Dinamarca, M. A., Rojas, A., Baeza, P., Espinoza, G., Ibacache-Quiroga, C., & Ojeda, J. (2014). Optimizing the biodesulfurization of gas oil by adding surfactants to immobilized cell systems. *Fuel*, *116*, 237–241. doi:10.1016/j.fuel.2013.07.108

Ding, B., Kim, H. Y., Lee, S. C., Shao, C. L., Lee, D. R., Park, S. J., ... Choi, K. J. (2002). Preparation and characterization of a nanoscale poly(vinyl alcohol) fiber aggregate produced by an electrospinning method. *Journal of Polymer Science. Part B, Polymer Physics*, *40*(13), 1261–1268. doi:10.1002/polb.10191

Ding, B., Kimura, E., Sato, T., Fujita, S., & Shiratori, S. (2004). Fabrication of blend biodegradable nanofibrous nonwoven mats via multi-jet electrospinning. *Polymer*, *45*(6), 1895–1902. doi:10.1016/j.polymer.2004.01.026

Ding, F., Larsson, P., Larsson, J. A., Ahuja, R., Duan, H. M., Rosen, A., & Bolton, K. (2008). The importance of strong carbon-metal adhesion for catalytic nucleation of single-walled carbon nanotubes. *Nano Letters*, *8*(2), 63–68. doi:10.1021/nl072431m PMID:18162001

Ding, M., Tang, Y., & Star, A. (2012). Understanding Interfaces in Metal–Graphitic Hybrid Nanostructures. *The Journal of Physical Chemistry Letters*, *4*(1), 147–160. doi:10.1021/jz301711a PMID:26291227

Ding, Y., Zhu, W., Li, H., Jiang, W., Zhang, M., Duan, Y., & Chang, Y. (2011). Catalytic oxidative desulfurization with a hexatungstate/aqueous H2O2/ionic liquid emulsion system. *Green Chemistry*, *13*(5), 1210–1216. doi:10.1039/c0gc00787k

Djangkung, S., Murti, S., Yang, H., Choi, K., Kora, Y., & Mochida, I. (2003). Influences of nitrogen species on the hydrodesulfurization reactivity of a gas oil over sulfide catalysts of variable activity. *Applied Catalysis A, General*, *252*(2), 331–346. doi:10.1016/S0926-860X(03)00468-X

Doshi, J., & Reneker, D. H. (1995). Electrospinning process and applications of electrospun fibers. *Journal of Electrostatics*, *35*(2-3), 151–160. doi:10.1016/0304-3886(95)00041-8

Douglas, R. M. (1984). Principle of adsorption and adsorption processes. J. Wiley & Sons.

Dreizler, R. M., & Gross, E. K. U. (1990). *Density functional theory*. Berlin: Springer. doi:10.1007/978-3-642-86105-5

Dribinskii, A. V., Tarasevich, M. R., & Kazarinov, V. E. (1989). Electrocatalysis on carbon materials. *Materials Chemistry and Physics*, *22*(3-4), 377–400. doi:10.1016/0254-0584(89)90006-0

Duan, A., Peng, Z., Liang, Z., Ji, J., & Zhen, Z. (2014). *Computational simulation application in the research of desulfurization mechanism*. The 9th International conference on computer science & education (ICCSE 2014), Vancouver, Canada.

Duan, B., Dong, C. H., Yuan, X. Y., & Yao, K. D. (2004). Electrospinning of chitosan solutions in acetic acid with poly(ethylene oxide). *Journal of Biomaterials Science. Polymer Edition*, *15*(6), 797–811. doi:10.1163/156856204774196171 PMID:15255527

Du, D. Y., Qin, J. S., Li, S. L., Su, Z. M., & Lan, Y. Q. (2014). Recent advances in porous polyoxometalate-based metal-organic framework materials. *Chemical Society Reviews*, *43*(13), 4615–4632. doi:10.1039/C3CS60404G PMID:24676127

Dzenis, Y. (2004). Spinning continuous fibers for nanotechnology. *Science*, *304*, 1917–1919.

Ebbesen, T. W., & Ajayan, P. M. (1992). Large scale synthesis of carbon nanotubes. *Nature*, *358*(6383), 220–222. doi:10.1038/358220a0

Ebbesen, T. W., Hiura, H., Bisher, M. E., Treacy, M. M. J., Shreeve-Keyer, J. L., & Haushalter, R. C. (1996). Decoration of carbon nanotubes. *Advanced Materials*, *8*(2), 155–157. doi:10.1002/adma.19960080212

Eder, D. (2010). Carbon Nanotube–Inorganic Hybrids. *Chemical Reviews*, *110*(3), 1348–1385. doi:10.1021/cr800433k PMID:20108978

Edwards, R. S., & Coleman, K. S. (2013). Graphene film growth on polycrystalline metals. *Accounts of Chemical Research*, *46*(1), 23–30. doi:10.1021/ar3001266 PMID:22891883

El-Newehy, M.H., Al, D.S.S., Kenawy, E.R., & Abdel, M.A. (2011). Nanospider technology for the production of nylon 6-nanofibers for biomedical applications. *Journal of Nanomaterial*, 1-8.

Emtsev, K. V., Bostwick, A., Horn, K., Jobst, J., Kellogg, G. L., Ley, L., ... Seyller, T. (2009). Towards wafersize graphene layers by atmospheric pressure graphitization ofsilicon carbide. *Nature Materials*, *8*(3), 203–207. doi:10.1038/nmat2382 PMID:19202545

EPA sulfur distribution. (n.d.). Retrieved January 28, 2013, from http://www.eia.gov/todayinenergy/detail.cfm?id=7110

Esteban, R., Moran, J. F., Becerril, J. M., & García-Plazaola, J. I. (2015). Versatility of carotenoids: An integrated view on diversity, evolution, functional roles and environmental interactions. *Environmental and Experimental Botany*, *119*, 63–75.

Faghihian, H., & Naeemi, S. (2013). Application of a Novel Nanocomposite for Desulfurization of a Typical Organo Sulfur Compound. *Iran. J. Chem. Chem. Eng.*, *32*(3).

Falcaro, P., Ricco, R., Doherty, C. M., Liang, K., Hill, A. J., & Styles, M. J. (2014). MOF positioning technology and device fabrication. *Chemical Society Reviews*, *43*(16), 5513–5560. doi:10.1039/C4CS00089G PMID:24802634

Fallah, R. N., & Azizian, S. (2012). Removal of thiophenic compounds from liquid fuel by different modified activated carbon cloths. *Fuel Processing Technology*, *93*(1), 45–52. doi:10.1016/j.fuproc.2011.09.012

Fan, J., Wang, G., Sun, Y., Xu, C., Zhou, H., Zhou, G., & Gao, J. (2010). Research on reactive adsorption desulfurization over Ni/ZnO-SiO_2-Al_2O_3 adsorbent in a fixed-fluidized bed reactor. *Industrial & Engineering Chemistry Research*, *49*(18), 8450–8460. doi:10.1021/ie100923v

Fariba, G., Vasheghani-Farahani, S., & Vasheghani-Farahani, E. (2010). Theoretical description of hydrogel swelling: A review. *Iranian Polymer Journal*, *19*, 375–398.

Farsani, M. R., & Yadollahi, B. (2014). Synthesis, characterization and catalytic performance of a Fe polyoxometalate/silica composite in the oxidation of alcohols with hydrogen peroxide. *Journal of Molecular Catalysis A Chemical*, *392*, 8–15. doi:10.1016/j.molcata.2014.05.001

Feng, J., Zeng, Y., Ma, C., Cai, X., Zhang, Q., Tong, M., ... Xu, P. (2006). The surfactant tween 80 enhances biodesulfurization. *Applied and Environmental Microbiology*, *72*(11), 7390–7393. doi:10.1128/AEM.01474-06 PMID:16980422

Férey, G., Mellot-Draznieks, C., Serre, C., Millange, F., Dutour, J., Surblé, S., & Margiolaki, I. (2005). A Chromium Terephthalate-Based Solid with Unusually Large Pore Volumes and Surface Area. *Science*, *309*(5743), 2040–2042. doi:10.1126cience.1116275 PMID:16179475

Fernandes, A. S., Paixão, S. M., Silva, T. P., Roseiro, J. C., & Alves, L. (2018). Influence of culture conditions towards optimal carotenoid production by Gordonia alkanivorans strain 1B. *Bioprocess and Biosystems Engineering*, *41*(2), 143–155.

Ferrando-Soria, J., Ruiz-Garcia, R., Cano, J., Stiriba, S., Vallejo, J., Castro, I., ... Pardo, E. (2012). Reversible Solvatomagnetic Switching in a Spongelike Manganese(II)-Copper(II) 3D Open Framework with a Pillared Square/Octagonal Layer Architecture. *Chemistry (Weinheim an der Bergstrasse, Germany)*, *18*(6), 1608–1617. doi:10.1002/chem.201103308 PMID:22262643

Flavia, I. P. N., Lakshmi Prabha, V., & Aswathy Paul, D. (2014). Nirmal: Synthesis and dielectric studies of monoclinic nanosized zirconia. Advances in Condensed Matter Physics.

Foght, J. M. (2004). Whole-cell bio-processing of aromatic compounds in crude oil and fuels. In R. Vazquez-Duhalt & R. Quintero-Ramirez (Eds.), Petroleum biotechnology: Developments and perspectives (pp. 145–176). Amsterdam: Elsevier. doi:10.1016/S0167-2991(04)80146-0

Folsom, B. R., Meyer, T. A., & Yu, L. Q. (1998). *U.S. Patent No. 5,772,901*. Washington, DC: U.S. Patent and Trademark Office.

Fong, H., Chun, I., & Reneker, D. H. (1999). Beaded nanofibers formed during electrospinning. *Polymer*, *40*(16), 4585–4592. doi:10.1016/S0032-3861(99)00068-3

Forte, P. (1996). *Process for the Removal of Sulfur from Petroleum Fractions*. US Patent, 5582714.

Francisco, M., Arce, A., & Soto, A. (2010). Ionic liquids on desulfurization of fuel oils. *Fluid Phase Equilibria*, *294*(1-2), 39–48. doi:10.1016/j.fluid.2009.12.020

François-Xavier, C., & Fuchs, A. H. (2015). Computational characterization and prediction of metal–organic framework properties. *Coordination Chemistry Reviews*, *307*, 211–236.

Freitas, P. A. M., Iha, K., Felinto, M. C. F. C., & Suárez-Iha, M. E. V. (2008). Adsorption of di-2-pyridyl ketone salicyloylhydrazone on Amberlite XAD-2 and XAD-7 resins: Characteristics and isotherms. *J. Colloid Inter. Sci.*, *323*(1), 1–5. doi:10.1016/j.jcis.2008.04.001 PMID:18466914

Freundlich, H. M. F. (1906). Over the adsorption in solution. *Journal of Physical Chemistry*, *57*, 385–471.

Froment, G. F., & Delmon, B. (1987). *Catalyst Deactivation 1987*. Elsevier Science.

Funakoshi, I., & Aida, T. (1998). *Process for recovering organic sulfur compounds from fuel oil*. US Patent, 5753102.

Furimsky, E. (1996). Spent refinery catalysts: Environment, safety and utilization. *Catalysis Today*, *30*(4), 223–286. doi:10.1016/0920-5861(96)00094-6

García-Gutiérrez, J. L., Fuentes, G. A., Hernández-Terán, M. E., Murrieta, F., Navarrete, J., & Jiménez-Cruz, F. (2006). Ultra-deep oxidative desulfurization of diesel fuel with H2O2 catalyzed under mild conditions by polymolybdates supported on Al2O3. *Applied Catalysis A, General*, *305*(1), 15–20. doi:10.1016/j.apcata.2006.01.027

Ge, H., Tang, M., & Wen, X. (2016) Ni/ZnO Nano Sorbent for Reactive Adsorption Desulfurization of Refinery Oil Streams. In Applying Nanotechnology to the Desulfurization Process in Petroleum Engineering (pp. 216-239). IGI Global.

Geim, A. K. (2009). Graphene: Status and prospects. *Science*, *324*(5934), 1530–1534. doi:10.1126cience.1158877 PMID:19541989

Geim, A. K., & Novoselov, K. S. (2007). The rise of graphene. *Nature Materials*, *6*(3), 183–191. doi:10.1038/nmat1849 PMID:17330084

Geng, X. Y., Kwon, O. H., & Jang, J. H. (2005). Electrospinning of chitosan dissolved in concentrated acetic acid solution. *Biomaterials*, *26*(27), 5427–5432. doi:10.1016/j.biomaterials.2005.01.066 PMID:15860199

Georgii, H. W. (1970). Contribution to the atmospheric sulfur budget. *Journal of Geophysical Research*, *75*(2), 2365–2371. doi:10.1029/JC075i012p02365

Ghosh, P., Hickey, K. J., & Jaffe, S. B. (2006). Development of a Detailed Gasoline Composition-Based Octane Model. *Industrial & Engineering Chemistry Research*, *45*(1), 337–345. doi:10.1021/ie050811h

Goswami, G., Chaudhuri, S., & Dutta, D. (2010). Effect of pH and temperature on pigment production from an isolated bacterium. *Chemical Engineering Transactions*, *20*, 127–132.

Granadeiro, C. M., Barbosa, A. D. S., Ribeiro, S., Santos, I. C. M. S., de Castro, B., Cunha-Silva, L., & Balula, S. S. (2014). Oxidative catalytic versatility of a trivacant polyoxotungstate incorporated into MIL-101(Cr). *Catalysis Science & Technology*, *4*(5), 1416–1425. doi:10.1039/c3cy00853c

Granadeiro, C. M., Barbosa, A. D. S., Silva, P., Paz, F. A. A., Saini, V. K., Pires, J., ... Cunha-Silva, L. (2013). Monovacant polyoxometalates incorporated into MIL-101(Cr): Novel heterogeneous catalysts for liquid phase oxidation. *Applied Catalysis A, General*, *453*(0), 316–326. doi:10.1016/j.apcata.2012.12.039

Gray, K., Mrachko, G., & Squires, C. (2003). Biodesulfurization of fossil fuels. *Current Opinion in Microbiology*, *6*(3), 229–235. doi:10.1016/S1369-5274(03)00065-1 PMID:12831898

Guchhait, S., Biswas, D., Bhattacharya, P., & Chowdhury, R. (2005a). Bio-desulfurization of model organo-sulfur compounds and hydrotreated diesel - Experiments and modelling. *Chemical Engineering Journal*, *112*(1-3), 145–151. doi:10.1016/j.cej.2005.05.006

Guchhait, S., Biswas, D., Bhattacharya, P., & Chowdhury, R. (2005b). Bi-phasic bio-conversion of sulfur present in model organo-sulfur compounds and hydro-treated diesel. *Catalysis Today*, *106*(1-4), 233–237. doi:10.1016/j.cattod.2005.07.137

Gunam, I. B. W., Sujaya, I. N., Aryanta, W. R., Tomita, F., & Asano, K. (2013). Biodesulfurization of dibenzothiophene and its derivatives using resting and immobilized cells of *Sphingomonas subarctica* T7b. *Journal of Microbiology and Biotechnology*, *23*(4), 473–482. doi:10.4014/jmb.1207.07070 PMID:23568201

Guobin, S., Jianmin, X., Huaiying, Z., & Huizhou, L. (2005). Deep desulfurization of hydrodesulfurized diesel oil by *Pseudomonas delafieldii* R-8. *Journal of Chemical Technology & Biotechnology, 80*, 420–424.

Guobin, S., Huaiying, Z., Jianmin, X., Guo, C., Wangliang, L., & Huizhou, L. (2006). Biodesulfurization of hydrodesulfurized diesel oil with *Pseudomonas delafieldii* R-8 from high density culture. *Biochemical Engineering Journal, 27*(3), 305–309. doi:10.1016/j.bej.2005.07.003

Guobin, S., Huaiying, Z., Weiquan, C., Jianmin, X., & Huizhou, L. (2005). Improvement of biodesulfurization rate by assembling nanosorbents on the surfaces of microbial cells. *Biophysical Journal, 89*(6), L58–L60. doi:10.1529/biophysj.105.073718 PMID:16258046

Guobin, S., Jianmin, X., Chen, G., Huizhou, L., & Jiayong, C. (2005). Biodesulfurization using *Pseudomonas delafieldii* in magnetic polyvinyl alcohol beads. *Letters in Applied Microbiology, 40*(1), 30–36. doi:10.1111/j.1472-765X.2004.01617.x PMID:15612999

Guo, L., Zhang, J., Hu, F., Dy Ryu, D., & Bao, J. (2013). Consolidated bioprocessing of highly concentrated Jerusalem artichoke tubers for simultaneous saccharification and ethanol fermentation. *Biotechnology and Bioengineering, 110*(10), 2606–2615. doi:10.1002/bit.24929 PMID:23568827

Guo, T., Nikolaev, P., Thess, A., Colbert, D. T., & Smalley, R. E. (1995). Catalytic growth of single-walled nanotubes by laser vaporization. *Chemical Physics Letters, 243*(1-2), 49–54. doi:10.1016/0009-2614(95)00825-O

Guo, X.-L., Deng, G., Xu, J., & Wang, M.-X. (2006). Immobilization of *Rhodococcus* sp. AJ270 in alginate capsules and its application in enantioselective biotransformation of trans-2-methyl-3-phenyl-oxiranecarbonitrile and amide. *Enzyme and Microbial Technology, 39*(1), 1–5. doi:10.1016/j.enzmictec.2005.01.044

Gupta, N., Roychoudhury, P. K., & Deb, J. K. (2005). Biotechnology of desulfurization of diesel: Prospects and challenges. *Applied Microbiology and Biotechnology, 66*(4), 356–366. doi:10.100700253-004-1755-7 PMID:15538557

Gupta, P., Elkins, C., Long, T. E., & Wilkes, G. L. (2005). Electrospinning of linear homopolymers of poly(methyl methacrylate): Exploring relationships between fiber formation, viscosity, molecular weight and concentration in a good solvent. *Polymer, 46*(13), 4799–4810. doi:10.1016/j.polymer.2005.04.021

Gyanesh, K. (2001). Sorbent composition, process for producing and use in desulfurization. *PCT Int. Appl.* WO 0114051.

Hallensleben, M. L. (2000). *Polyvinyl compounds*, others. In *Ullmann's encyclopedia of industrial chemistry*. Weinheim: Wiley-VCH; doi:10.1002/14356007.a21_743

Hameed, B. H. (2009). Grass waste: A novel sorbent for the removal of basic dye from aqueous solution. *Journal of Hazardous Materials, 166*(1), 233–238. doi:10.1016/j.jhazmat.2008.11.019 PMID:19111987

Harrison, R. M., & Chemistry, R. S. o. (2007). *Principles of Environmental Chemistry*. RSC Publishing. doi:10.1039/9781847557780

Hart, A. J., & Slocum, A. H. (2006). Rapid growth and flow-mediated nucleation of millimeter-scale aligned carbon nanotube structures from a thin-film catalyst. *The Journal of Physical Chemistry B*, *110*(16), 8250–8257. doi:10.1021/jp055498b PMID:16623503

Hashimoto, S., & Furukawa, K. (1987). Immobilization of activated sludge by PVA–boric acid method. *Biotechnology and Bioengineering*, *30*(1), 52–59. doi:10.1002/bit.260300108 PMID:18576582

Haukanes, B.-I., & Kvam, C. (1993). Application of magnetic beads in bioassays. *Nature Biotechnology*, *11*(1), 60–63. doi:10.1038/nbt0193-60 PMID:7763485

Haupt, K. (2003). Molecularly imprinted polymers: The next generation. *Analytical Chemistry*, *75*(17), 376A–383A. doi:10.1021/ac031385h PMID:14632031

He, L., Li, H., Zhu, W., Guo, J., Jiang, X., Lu, J., & Yan, Y. (2008). Deep Oxidative Desulfurization of Fuels Using Peroxophosphomolybdate Catalysts in Ionic Liquids. *Industrial & Engineering Chemistry Research*, *47*(18), 6890–6895. doi:10.1021/ie800857a

Hernandez-Maldonado Arturo, J. (2005). Desulfurization of commercial fuels by p-complexation: Monolayer CuCl/g-Al_2O_3. *Applied Catalysis B: Environmental*, *61*, 212–218. doi:10.1016/j.apcatb.2005.05.003

Hernańdez-Maldonado, A. J., Stamatis, S. D., Yang, R. T., He, A. Z., & Cannella, W. (2004). New sorbents for desulfurization of diesel fuels via p-complexation: Layered beds and regeneration. *Industrial & Engineering Chemistry Research*, *43*(3), 769–776. doi:10.1021/ie034108+

Hernández-Maldonado, A. J., & Yang, R. T. (2004). Desulfurization of Diesel Fuels by Adsorption via p-Complexation with Vapor Phase Exchanged (VPIE) Cu (I)-Y Zeolites. *Journal of the American Chemical Society*, *126*(4), 992–993. doi:10.1021/ja039304m PMID:14746446

Hernandez-Maldonado, A. J., & Yang, R. T. (2004). Desulfurization of transportation fuels by adsorption. *Catalysis Reviews. Science and Engineering*, *46*(2), 111–150. doi:10.1081/CR-200032697

Hernandez-Maldonado, A. J., & Yang, R. T. (2004b). Desulfurization of diesel fuels via pi-complexation with nickel(JI)-exchanged X- and Y-zeolites. *Industrial & Engineering Chemistry Research*, *43*(4), 1081–1089. doi:10.1021/ie034206v

Hernandez-Maldonado, A. J., & Yang, R. T. (2004c). New sorbents for desulfurization of diesel fuels via pi-complexation. *The American Institute of Chemical Engineers Journal*, *50*(4), 791–801. doi:10.1002/aic.10074

Hernandez, S. P., Fino, D., & Russo, N. (2010). High performance sorbents for diesel oil desulfurization. *Chemical Engineering Science*, *65*(1), 603–609. doi:10.1016/j.ces.2009.06.050

Heunis, T. D. J., Botes, M., & Dicks, L. M. T. (2010). Encapsulation of *Lactobacillus plantarum* 423 and its bacteriocin in nanofibers. *Probiotics and Antimicrobial Proteins*, *2*(1), 46–51. doi:10.100712602-009-9024-9 PMID:26780900

Heunis, T. D. J., & Dicks, L. M. T. (2010). Nanofibers offer alternative ways to the treatment of skin infections. *Journal of Biomedicine & Biotechnology*, *2010*, 1–10. doi:10.1155/2010/510682 PMID:20798871

Heunis, T., Bshena, O., Klumperman, B., & Dicks, L. (2011). Release of bacteriocins from nanofibers prepared with combinations of poly(D,L-lactide) (PDLLA) and poly(ethylene oxide) (PEO). *International Journal of Molecular Sciences*, *12*(4), 2158–2173. doi:10.3390/ijms12042158 PMID:21731433

He, Y.-C., Zhou, Q., Ma, C.-L., Cai, Z.-Q., Wang, L.-Q., Zhao, X.-Y., ... Wang, X.-D. (2012). Biosynthesis of benzoylformic acid from benzoyl cyanide by a newly isolated *Rhodococcus* sp. CCZU10-1 in toluene–water biphasic system. *Bioresource Technology*, *115*, 88–95. doi:10.1016/j.biortech.2011.09.084 PMID:22033370

Hill, C. L., & Prosser-McCartha, C. M. (1995). Homogeneous catalysis by transition metal oxygen anion clusters. *Coordination Chemistry Reviews*, *143*(0), 407–455. doi:10.1016/0010-8545(95)01141-B

Hirasawa, K., Ishii, Y., Kobayashi, M., Koizumi, K., & Maruhashi, K. (2001). Improvement of desulfurization activity in *Rhodococcus erythropolis* KA2-5-1 by genetic engineering. *Bioscience, Biotechnology, and Biochemistry*, *65*(2), 239–246. doi:10.1271/bbb.65.239 PMID:11302154

Hizaddin, H. F., Hashim, M. A., & Anantharaj, R. (2013). Evaluation of molecular interaction in binary mixture of ionic liquids+ heterocyclic nitrogen compounds: Ab Initio method and COSMORS model. *Industrial & Engineering Chemistry Research*, *52*(50), 18043–18058. doi:10.1021/ie403032t

Hobson, G. D., & Petroleum, I. O. (1984). *Modern petroleum technology*. Wiley; doi:10.1201/9781482277197

Ho, K. F., & Lee, S. C. (2002). Identification of atmospheric volatile organic compounds (VOCs), polycyclic aromatic hydrocarbons (PAHs) and carbonyl compounds in Hong Kong. *The Science of the Total Environment*, *289*(1-3), 145–158. doi:10.1016/S0048-9697(01)01031-2 PMID:12049391

Holmes, A. B., & Stephenson, G. R. (1994). Molecular imprinting. *Chemistry & Industry*, *18*, 304.

Homs, N., Llorca, J., & de la Piscina, P. R. (2006). Low-temperature steam-reforming of ethanol over ZnO-supported Ni and Cu catalysts: The effect of nickel and copper addition to ZnO-supported cobalt-based catalysts. *Catalysis Today*, *116*(3), 361–366. doi:10.1016/j.cattod.2006.05.081

Honda, H., Sugiyama, H., Saito, I., & Kobayashi, T. (1998). High cell density culture of *Rhodococcus rhodochrous* by pH-stat feeding and dibenzothiophene degradation. *Journal of Fermentation and Bioengineering*, *85*(3), 334–338. doi:10.1016/S0922-338X(97)85685-1

Hou, X., & Zhuang, J. (Eds.). (2013). *Technology handbook of S-Zorb unit for desorption desulfurization of FCC gasoline*. Beijing, China: China Petrochemical Press.

Hsu, W. K., Terrones, M., Hare, J. P., Terrones, H., Kroto, H. W., & Walton, D. R. M. (1996). Electrochemical formation of carbon nanostructures. *Chemical Physics Letters*, *262*(1-2), 161–166. doi:10.1016/0009-2614(96)01041-X

Huaiying, Z., Guobin, S., & Huizhou, L., & JianMin, X. (2007). Surface modification of γ-Al_2O_3 nano-particles with gum arabic and its applications in adsorption and biodesulfurization. *Surface and Coatings Technology*, *201*(16), 6917–6921.

Huang, D., Wang, Y. J., Yang, L. M., & Luo, G. S. (2006). Chemical Oxidation of Dibenzothiophene with a Directly Combined Amphiphilic Catalyst for Deep Desulfurization. *Industrial & Engineering Chemistry Research*, *45*(6), 1880–1885. doi:10.1021/ie0513346

Huang, L., Ge, H., Yan, L., Wang, G., Qin, Z., & Wang, J. (2017). Competitive reactive adsorption desulfurization of dibenzothiophene and hydrogenation of naphthalene over Ni/ZnO. *Canadian Journal of Chemical Engineering*, *9999*, 1–8.

Huang, L., Nagapudi, K., Apkarian, R. P., & Chaikof, E. L. (2001). Engineered collagen-PEO nanofibers and fabrics. *Journal of Biomaterials Science. Polymer Edition*, *12*, 979–993. doi:10.1163/156856201753252516 PMID:11787524

Huang, L., Qin, Z., Wang, G., Du, M., Ge, H., Li, X., ... Wang, J. (2010b). A detailed study on the negative effect of residual sodium on the performance of Ni/ZnO adsorbent for diesel fuel desulfurization. *Industrial & Engineering Chemistry Research*, *49*(10), 4670–4675. doi:10.1021/ie100293h

Huang, L., Wang, G., Qin, Z., Dong, M., Du, M., Ge, H., ... Wang, J. (2011). In situ XAS study on the mechanism of reactive adsorption desulfurization of oil product over Ni/ZnO. *Applied Catalysis B: Environmental*, *106*(1-2), 26–38.

Huang, L., Wang, G., Qin, Z., Dong, M., Du, M., Ge, H., ... Wang, J. (2011). In situ XAS study on the mechanism of reactive adsorption desulfurization of oil product over Ni/ZnO. *Applied Catalysis B: Environmental*, *106*, 26–38.

Huang, L., Wang, G., Qin, Z., Du, M., Dong, M., Ge, H., ... Wang, J. (2010a). A sulfur K-edge XANES study on the transfer of sulfur species in the reactive adsorption desulfurization of diesel oil over Ni/ZnO. *Catalysis Communications*, *11*(7), 592–596. doi:10.1016/j.catcom.2010.01.001

Huang, M., Chang, G., Su, Ye., Xing, H., Su, B., Zhang, Z., ... Chen, B. (2015). A metal-organic framework with immobilized Ag(I) for highly efficient desulfurization of liquid fuels. *Chemical Communications*, *51*(61), 12205–12207. doi:10.1039/C5CC03648H PMID:26136210

Huang, W., Zhu, W., Li, H., Shi, H., Zhu, G., Liu, H., & Chen, G. (2010). Heteropolyanion-Based Ionic Liquid for Deep Desulfurization of Fuels in Ionic Liquids. *Industrial & Engineering Chemistry Research*, *49*(19), 8998–9003. doi:10.1021/ie100234d

Huang, X., Qi, X., Boey, F., & Zhang, H. (2012). Graphene-based composites. *Chemical Society Reviews*, *41*(2), 666–686. doi:10.1039/C1CS15078B PMID:21796314

Huang, Z. M., Zhang, Y. Z., Kotaki, M., & Ramakrishna, S. (2003). A review on polymer nanofibers by electrospinning and their applications in nanocomposites. *Composites Science and Technology*, *63*(15), 2223–2253. doi:10.1016/S0266-3538(03)00178-7

Hulea, V., Fajula, F., & Bousquet, J. (2001). Mild oxidation with H_2O_2 over Ti-containing molecular sieves-a very efficient method for removing aromatic sulfur compounds from fuels. *Journal of Catalysis*, *198*(2), 179–186. doi:10.1006/jcat.2000.3149

Hu, X., Lu, Y., Dai, F., Liu, C., & Liu, Y. (2013). Host–guest synthesis and encapsulation of phosphotungstic acid in MIL-101 via "bottle around ship": An effective catalyst for oxidative desulfurization. *Microporous and Mesoporous Materials*, *170*(0), 36–44. doi:10.1016/j.micromeso.2012.11.021

Ibrahima, M. H., Hayyan, M., Hashima, M. A., & Hayyan, A. (2017). Article. *Renewable & Sustainable Energy Reviews*, *76*, 1534–1549. doi:10.1016/j.rser.2016.11.194

Iijima, S. (1991). Helical microtubules of graphitic carbon. *Nature*, *354*(6348), 56–58. doi:10.1038/354056a0

Isarankura-Na-Ayudhya, C., Nantasenamat, C., Buraparuangsang, P., Piacham, T., Ye, L., Bülow, L., & Prachayasittikul, V. (2008). Computational insights on sulfonamide imprinted polymers. *Molecules (Basel, Switzerland)*, *13*(12), 3077–3091. doi:10.3390/molecules13123077 PMID:19078850

Ishihara, A., Wang, D. H., Dumeignil, F., Amano, H., Qian, E. W. H., & Kabe, T. (2005). Oxidative desulfurization and denitrogenation of a light gas oil using an oxidation/adsorption continuous flow process. *Applied Catalysis A, General*, *279*(1-2), 279–287. doi:10.1016/j.apcata.2004.10.037

Iwamoto, R., & Grimblot, J. (1997). Acidity and hydrogenation properties of Mo-P-alumina catalysts prepared by a sol-gel method. *Journal of Catalysis*, *172*(1), 252–255. doi:10.1006/jcat.1997.1855

Jacobs, C. B., Peairs, M. J., & Venton, B. J. (2010). Carbon nanotubes electrochemical sensors for biomolecules. *Analytica Chimica Acta*, *662*(2), 105–127. doi:10.1016/j.aca.2010.01.009 PMID:20171310

Jarusuwannapoom, T., Hongroijanawiwat, W., Jitjaicham, S., Wannatong, L., Nithitanakul, M., Pattamaprom, C., ... Supaphol, P. (2005). Effect of solvents on electro-spinnability of polystyrene solutions and morphological appearance of resulting electrospun polystyrene fibers. *European Polymer Journal*, *41*(3), 409–421. doi:10.1016/j.eurpolymj.2004.10.010

Javadli, R., & Klerk, A. (2012). Desulfurization of heavy oil. *Applied Petrochemical Research*, *1*(1-4), 3–19. doi:10.100713203-012-0006-6

Jeevanandam, P., Klabunde, K. J., & Tetzler, S. H. (2005). Adsorption of thiophenes out of hydrocarbons using metal impregnated nanocrystalline aluminum oxide. *Microporous and Mesoporous Materials*, *79*(1-3), 101–110. doi:10.1016/j.micromeso.2004.10.029

Jeon, H., Ko, C. H., Kim, S. H., & Kim, J. (2009). Removal of refractory sulphur compounds in diesel using activated carbon with controlled porosity. *Energy & Fuels*, *23*(5), 2537–2543. doi:10.1021/ef801050k

Jia, G. (2014). Retrieved from http://www.sinopecnews.com.cn/news/content/2014-02/17/content_1377530.shtml (in Chinese)

Jiang, C., Wang, J., Wang, S., Guan, H., Wang, X., & Huo, M. (2011). Oxidative desulfurization of dibenzothiophene with dioxygen and reverse micellar peroxotitanium under mild conditions. *Applied Catalysis B: Environmental*, *106*(3–4), 343–349. doi:10.1016/j.apcatb.2011.05.038

Jiang, H. L., Fang, D. F., Hsiao, B. S., Chu, B., & Chen, W. L. (2004). Optimization and characterization of dextran membranes prepared by electrospinning. *Biomacromolecules*, *5*(2), 326–333. doi:10.1021/bm034345w PMID:15002991

Jiang, N., Ying, G. L., Liu, S. Y., Shen, L., Hu, J., Dai, L. J., ... Su, B. L. (2014). Amino acid-based biohybrids for nano-shellization of individual desulfurizing bacteria. *Chemical Communications (Cambridge)*, *50*(97), 15407–15410. doi:10.1039/C4CC06323F PMID:25350878

Jiang, X., Li, H., Zhu, W., He, L., Shu, H., & Lu, J. (2009). Deep desulfurization of fuels catalyzed by surfactant-type decatungstates using H2O2 as oxidant. *Fuel*, *88*(3), 431–436. doi:10.1016/j.fuel.2008.11.010

Jiang, Z., Lu, H., Zhang, Y., & Li, C. (2011). Oxidative desulfurization of fuel oils. *Chinese Journal of Catalysis*, *32*(5), 707–715. doi:10.1016/S1872-2067(10)60246-X

Jia, S.-Y., Zhang, Y.-F., Liu, Y., Qin, F.-X., Ren, H.-T., & Wu, S.-H. (2013). Adsorptive removal of dibenzothiophene from model fuels over one-pot synthesized PTA@MIL-101(Cr) hybrid material. *Journal of Hazardous Materials*, *262*(0), 589–597. doi:10.1016/j.jhazmat.2013.08.056 PMID:24095999

Jiménez-Cruz, F., & Laredo, G. C. (2004). Molecular size evaluation of linear and branched paraffins from the gasoline pool by DFT quantum chemical calculations. *Fuel*, *83*(16), 2183–2188. doi:10.1016/j.fuel.2004.06.010

Jin, T., Yang, Q., Meng, C., Xu, J., Liu, H., Hu, J., & Ling, H. (2014). Promoting desulfurization capacity and separation efficiency simultaneously by the novel magnetic Fe_3O_4@PAA@MOF-199. *RSC Advances*, *4*(79), 41902–41909. doi:10.1039/C4RA06515H

Johnson, B. G., Slater, P., & Kidd, D. (2001). Application of Phillips's Zorb process to distillates-meeting the challenge. AM-01-14, New Orleans, LA.

Jorge, A., & Le Borgne, S. (2004). Selective adsorption of dibenzothiophene sulfone by an imprinted and stimuli-responsive chitosan hydrogel. *Macromolecules*, *37*(8), 2938–2943. doi:10.1021/ma049961e

Jourdain, V., & Bichara, C. (2013). Current understanding of the growth of carbon nanotubes in catalytic chemical vapour deposition. *Carbon, 58*, 2–39. doi:10.1016/j.carbon.2013.02.046

Journet, C., Maser, W. K., Bernier, P., Loiseau, A., de la Chapelle, M. L., Lefrant, S., ... Fischer, J. E. (1997). Large-scale production of single-walled carbon nanotubes by the electric-arc technique. *Nature, 388*(6644), 756–758. doi:10.1038/41972

Juan-Alcaniz, J., Gascon, J., & Kapteijn, F. (2012). Metal-organic frameworks as scaffolds for the encapsulation of active species: state of the art and future perspectives. *Journal of Materials Chemistry, 22*(20), 10102-10118.

Juan-Alcaniz, J., Ramos-Fernandez, E. V., Lafont, U., Gascon, J., & Kapteijn, F. (2010). Building MOF bottles around phosphotungstic acid ships: One-pot synthesis of bi-functional polyoxometalate-MIL-101 catalysts. *Journal of Catalysis, 269*(1), 229–241. doi:10.1016/j.jcat.2009.11.011

Julião, D., Gomes, A. C., Pillinger, M., Cunha-Silva, L., de Castro, B., Gonçalves, I. S., & Balula, S. S. (2015). Desulfurization of model diesel by extraction/oxidation using a zinc-substituted polyoxometalate as catalyst under homogeneous and heterogeneous (MIL-101(Cr) encapsulated) conditions. *Fuel Processing Technology, 131*(0), 78–86. doi:10.1016/j.fuproc.2014.10.030

Kappe, C. O. (2004). Controlled Microwave Heating in Modern Organic Synthesis. *Angewandte Chemie International Edition, 43*(46), 6250–6284. doi:10.1002/anie.200400655 PMID:15558676

Karas, L. J., Grey, R. A., & Lynch, M. W. (2008). *Desulfurization process*. US Patent 2008047875.

Karas, L. J., Han, Y. Z., & Leyshon, D. W. (2004). *Organosulfur oxidation process*. US Patent 2004178122.

Katsoulis, D. E. (1998). A survey of applications of polyoxometalates. *Chemical Reviews, 98*(1), 359–387. doi:10.1021/cr960398a PMID:11851510

Katz, H. E. (2004). Chemically Sensitive Field-Effect Transistors and Chemiresistors: New Materials and Device. Structures. *Journal of Electroanalytical Chemistry, 16*(22), 1837–1842. doi:10.1002/elan.200403071

Kauffman, D. R., Tang, Y., Kichambare, P. D., Jackovitz, J. F., & Star, A. (2010). Long-Term Performance of Pt-Decorated Carbon Nanotube Cathodes in Phosphoric Acid Fuel Cells. *Energy & Fuels, 24*(3), 1877–1881. doi:10.1021/ef100013v

Kaufman, E. N., Borole, A. P., Shong, R., Sides, J. L., & Juengst, C. (1999). Sulfur specificity in the bench-scale biological desulfurization of crude oil by *Rhodococcus*. *Journal of Chemical Technology and Biotechnology (Oxford, Oxfordshire), 74*(10), 1000–1004. doi:10.1002/(SICI)1097-4660(199910)74:10<1000::AID-JCTB134>3.0.CO;2-0

Kaufman, E. N., Harkins, J. B., & Borole, A. P. (1998). Comparison of Batch-Stirred and Electro-Spray Reactors for Biodesulfurization of dibenzothiophene in Crude Oil and Hydrocarbon Feedstocks. *Applied Biochemistry and Biotechnology, 73*(2-3), 127–144. doi:10.1007/BF02785650

Kaufman, E. N., Harkins, J. B., Rodriguez, M. Jr, Tsouris, C., Selvaraj, P. T., & Murphy, S. E. (1997). Development of an electro-spray bioreactor for crude oil processing. *Fuel Processing Technology*, *52*(1), 127–144. doi:10.1016/S0378-3820(97)00022-2

Kenta, K. (2017). Network polymers derived from the integration of flexible organic polymers and rigid metal–organic frameworks. *Polymer Journal*, *49*(4), 345–353. doi:10.1038/pj.2016.122

Kent, J. A. (2010). *Kent and Riegel's Handbook of Industrial Chemistry and Biotechnology* (Vol. 1). Springer.

Khan, A. (2012). The theory of adsorption equilibria analysis based on general equilibrium constant expression. *Turkish Journal of Chemistry*, *36*, 219–231.

Khan, N. A., Yoon, J. W., Chang, J., & Jhung, S. (2016). Enhanced adsorptive desulfurization with flexible metal–organic frameworks in the presence of diethyl ether and water. *Chemical Communications (Cambridge)*, *52*(56), 8667–8670. doi:10.1039/C6CC03976F PMID:27284597

Khare, G. P. (1999). Sorbent compositions. US Pat Appl. 4948830.

Khedkar, S., & Shanker, R. (2014). Isolation and classification of a soil actinomycete capable of sulphur-specific biotransformation of dibenzothiophene, benzothiophene and thianthrene. *Journal of Applied Microbiology*. PMID:25319398

Kholdeeva, O. A., Maksimchuk, N. V., & Maksimov, G. M. (2010). Polyoxometalate-based heterogeneous catalysts for liquid phase selective oxidations: Comparison of different strategies. *Catalysis Today*, *157*(1–4), 107–113. doi:10.1016/j.cattod.2009.12.016

Ki, C. S., Baek, D. H., Gang, K. D., Lee, K. H., Um, I. C., & Park, Y. H. (2005). Characterization of gelatin nanofiber prepared from gelatin-formic acid solution. *Polymer*, *46*(14), 5094–5102. doi:10.1016/j.polymer.2005.04.040

Kidoaki, S., Kwon, I. K., & Matsuda, T. (2005). Mesoscopic spatial designs of nano- and micron-fiber meshes for tissue-engineering matrix and scaffold based on newly devised multilayering and mixing electrospinning techniques. *Biomaterials*, *26*(1), 37–46. doi:10.1016/j.biomaterials.2004.01.063 PMID:15193879

Kilbane, J. J. II. (2006). Microbial biocatalyst developments to upgrade fossil fuels. *Current Opinion in Biotechnology*, *17*(3), 305–314. doi:10.1016/j.copbio.2006.04.005 PMID:16678400

Kilbane, J. J. II, & Le Borgne, S. (2004). Petroleum biorefining: The selective removal of sulfur, nitrogen, and metals. *Studies in Surface Science and Catalysis*, *151*, 29–65. doi:10.1016/S0167-2991(04)80143-5

Kim, B., Park, H., Lee, S. H., & Sigmund, W. M. (2005). Poly(acrylic acid) nanofibers by electrospinning. *Materials Letters*, *59*(7), 829–832. doi:10.1016/j.matlet.2004.11.032

Kim, H. S., Kim, K., Jin, H. J., & Chin, I. J. (2005). Morphological characterization of electrospun nano-fibrous membranes of biodegradable poly(L-lactide) and poly(lactideco-glycolide). *Macromolecular Symposia*, *224*(1), 145–154. doi:10.1002/masy.200550613

Kim, J. H., Ma, X., Zhou, A., & Song, C. (2006). Ultra-deep desulfurization and denitrogenation of diesel fuel by selective adsorption over three different adsorbents: A study on adsorptive selectivity and mechanism. *Catalysis Today*, *111*(1-2), 74–83. doi:10.1016/j.cattod.2005.10.017

King, D. L., & Li, L. (2006). Removal of sulfur components from low sulfur gasoline using copper exchanged zeolite Y at ambient temperature. *Catalysis Today*, *116*(4), 526–529. doi:10.1016/j.cattod.2006.06.026

Kinoshita, K., & Bett, J. A. S. (1973). Potentiodynamic analysis of surface oxides on carbon blacks. *Carbon*, *11*(4), 403–411. doi:10.1016/0008-6223(73)90080-8

Kirimura, K., Harada, K., Iwasawa, H., Tanaka, T., Iwasaki, Y., Furuya, T., ... Kino, K. (2004). Identification and functional analysis of the genes encoding dibenzothiophene-desulfurizing enzymes from thermophilic bacteria. *Applied Microbiology and Biotechnology*, *65*(6), 703–713. doi:10.100700253-004-1652-0 PMID:15221222

Kiselev, N. A., Sloan, J., Zakharov, D. N., Kukovitskii, E. F., Hutchison, J. L., Hammer, J., & Kotosonov, A. S. (1998). Carbon nanotubes from polyethylene precursors: Structure and structural changes as revealed by HREM. *Carbon*, *36*, 1149–1157. doi:10.1016/S0008-6223(98)00092-X

Klobes, P., Meyer, K., & Munro, R. G. (2006). Porosity and specific surface area measurements for solid materials. *National Institute of Standards and Technology Special Publication*, *89*, 960–917.

Knaebel Kent, S. (2004). *Adsorbent selection*. Dublin, OH: Adsorption Research, Inc.

Kobayashi, M., Horiuchi, K., Yoshikawa, O., Hirasawa, K., Ishii, Y., Fujino, K., ... Maruhashi, K. (2001). Kinetic analysis of microbial desulfurization of model and light gas oils containing multiple alkyl dibenzothiophenes. *Bioscience, Biotechnology, and Biochemistry*, *65*(2), 298–304. doi:10.1271/bbb.65.298 PMID:11302162

Kobayashi, M., Onaka, T., Ishii, Y., Konishi, J., Takaki, M., Okada, H., ... Suzuki, M. (2000). Desulfurization of alkylated forms of both dibenzothiophene and benzothiophene by a single bacterial strain. *FEMS Microbiology Letters*, *187*(2), 123–126. doi:10.1111/j.1574-6968.2000.tb09147.x PMID:10856644

Kocal, J. A., & Branvold, T. A. (2002). *Removal of sulfur-containing compounds from liquid hydrocarbon streams*. US Patent 6368495.

Kodama, K., Umehara, K., Shimizu, K., Nakatani, S., Minoda, Y., & Yamada, K. (1973). Identification of microbial products from dibenzothiophene and its proposed oxidation pathway. *Agricultural and Biological Chemistry*, *37*(1), 45–50. doi:10.1080/00021369.1973.10860640

Kohler, F., Roth, D., Kuhlmann, E., Wasserscheid, P., & Haumann, M. (2010). Continuous gas-phase desulfurisation using supported ionic liquid phase (SILP) materials. *Green Chemistry*, *12*(6), 979–984. doi:10.1039/c004883f

Koltuniewicz, A. B., Field, R. W., & Arnot, T. C. (1995). Cross-flow and dead-end microfiltration of oily-water emulsion. Part I: Experimental study and analysis of flux decline. *Journal of Membrane Science*, *102*, 193–207. doi:10.1016/0376-7388(94)00320-X

Komintarachat, C., & Trakarnpruk, W. (2006). Oxidative Desulfurization Using Polyoxometalates. *Industrial & Engineering Chemistry Research*, *45*(6), 1853–1856. doi:10.1021/ie051199x

Konishi, J., Ishii, Y., Onaka, T., Okumura, K., & Suzuki, M. (1997). Thermophilic carbon-sulfur-bond-targeted biodesulfurization. *Applied and Environmental Microbiology*, *63*(8), 3164–3169. PMID:16535672

Konishi, M., Kishimoto, M., Omasa, T., Katakura, Y., Shioya, S., & Ohtake, H. (2005). Effect of sulfur sources on specific desulfurization activity of *Rhodococcus erythropolis* KA2-5-1 in exponential fed-batch culture. *Journal of Bioscience and Bioengineering*, *99*(3), 259–263. doi:10.1263/jbb.99.259 PMID:16233786

Konishi, M., Kishimoto, M., Tamesui, N., Omasa, T., Shioya, S., & Ohtake, H. (2005). The separation of oil from an oil–water–bacteria mixture using a hydrophobic tubular membrane. *Biochemical Engineering Journal*, *24*(1), 49–54. doi:10.1016/j.bej.2004.11.012

Koriakin, A., Kim, Y.-H., & Lee, C.-H. (2012). Adsorptive Desulfurization of Natural Gas Using Lithium-Modified Mesoporous Silica. *Industrial & Engineering Chemistry Research*, *51*(44), 14489–14495. doi:10.1021/ie301066n

Koski, A., Yim, K., & Shivkumar, S. (2004). Effect of molecular weight on fibrous PVA produced by electrospinning. *Materials Letters*, *58*(3-4), 493–497. doi:10.1016/S0167-577X(03)00532-9

Kosynkin, D. V., Higginbotham, A. L., Sinitskii, A., Lomeda, J. R., Dimiev, A., Price, B. K., & Tour, J. M. (2009). Mechanically Assisted Exfoliation and Functionalization of Thermally Converted Graphene Sheets. *Nature*, *458*, 872–876. doi:10.1038/nature07872 PMID:19370030

Kruger, M., Buitelaar, M. R., Nussbaumer, T., Schonenberger, C., & Forro, L. (2001). Electrochemical carbon nanotube field-effect transistor. *Applied Physics Letters*, *78*(9), 1291–1293. doi:10.1063/1.1350427

Kruusma, J., Mould, N., Jurkschat, K., Crossley, A., & Banks, C. E. (2007). Single walled carbon nanotubes contain residual iron oxide impurities which can dominate their electrochemical activity. *Electrochemistry Communications*, *9*(9), 2330–2333. doi:10.1016/j.elecom.2007.06.024

Kuhlmann, E., Haumann, M., Jess, A., Seeberger, A., & Wasserscheid, P. (2009). Ionic Liquids in Refinery Desulfurization: Comparison between Biphasic and Supported Ionic Liquid Phase Suspension Processes. *ChemSusChem*, *2*(10), 969–977. doi:10.1002/cssc.200900142 PMID:19798713

Kulkarni & Afonso. (2010). *Deep desulfurization of diesel fuel using ionic liquids: current status and future challenges.* Academic Press.

Kumar, A. (2010). *Nanofibers*. Vukovar: Intech. doi:10.5772/45660

Kushwaha, K., & Saini, A. Saraswa,t P., Agarwal, M. K., & Saxena, J. (2014). Colorful world of microbes: Carotenoids and their applications. Advances in Biology 2014, Article ID 837891.

Kuznetsova, L., Detusheva, L., Kuznetsov, N., Duplyakin, V., & Likholobov, V. (2008). Liquid-phase oxidation of benzothiophene and dibenzothiophene by cumyl-hydroperoxide in the presence of catalysts based on supported metal oxides. *Kinetics and Catalysis*, *49*(5), 644–652. doi:10.1134/S002315840805008X

Kwon, J.-M., Moon, J.-H., Bae, Y.-S., Lee, D.-G., Sohn, H.-C., & Lee, C.-H. (2008). Adsorptive desulfurization and denitrogenation of refinery fuels using mesopourous silica adsorbents. *ChemSusChem*, *1*(4), 307–309. doi:10.1002/cssc.200700011 PMID:18605095

Kwon, T., Tsigdinos, G. A., & Pinnavaia, T. J. (1988). Pillaring of layered double hydroxides (LDH's) by polyoxometalate anions. *Journal of the American Chemical Society*, *110*(11), 3653–3654. doi:10.1021/ja00219a048

Lam, V., Li, G., Song, C., Chen, J., Fairbridge, C., Hui, R., & Zhang, J. (2012). A review of electrochemicaldesulfurization technologies for fossil fuels. *Fuel Processing Technology*, *98*, 30–38. doi:10.1016/j.fuproc.2012.01.022

Landau, M. V., Herskowitz, M., Agnihotri, R., & Kegerreis, J. E. (2009). Ultradeep hydrodesulfurization and adsorptive desulfurization of diesel fuel on metal-rich nickel phosphides. *Industrial & Engineering Chemistry Research*, *48*(11), 5239–5249. doi:10.1021/ie9000579

Langmuir, I. (1918). The adsorption of gases on plane surfaces of glass, mica and platinum. *Journal of the American Chemical Society*, *40*(9), 1361–1403. doi:10.1021/ja02242a004

Laplaze, D., Bernier, P., Maser, W. K., Flamant, G., Guillard, T., & Loiseau, A. (1998). carbon nanotubes: The solar approach. *Carbon*, *36*(5-6), 685–688. doi:10.1016/S0008-6223(98)00025-6

Le Borgne, S., & Quintero, R. (2003). Biotechnological processes for the refining of petroleum. *Fuel Processing Technology*, *81*(2), 155–169. doi:10.1016/S0378-3820(03)00007-9

Lee & Valla. (2019). Adsorptive desulfurization of liquid hydrocarbons using zeolite-based sorbents: a comprehensive review. *React. Chem. Eng*. DOI: doi:10.1039/c9re00036d

Lee, J. S., Choi, K. H., Do Ghim, H., Kim, S. S., Chun, D. H., Kim, H. Y., & Lyoo, W. S. (2004). Role of molecular weight of a tactic poly(vinyl alcohol) (PVA) in the structure and properties of PVA nanofabric prepared by electrospinning. *Journal of Applied Polymer Science*, *94*(4), 1638–1646. doi:10.1002/app.20602

Lee, M. K., Senius, J. D., & Grossman, M. J. (1995). Sulfur-specific microbial desulfurization of sterically hindered analogs of dibenzothiophene. *Applied and Environmental Microbiology*, *61*(12), 4362–4366. PMID:16535189

Lee, Y., Kim, J., & Ahn, W. (2013). Synthesis of metal-organic frameworks: A mini review. *Korean Journal of Chemical Engineering*, *30*(9), 1667–1680. doi:10.100711814-013-0140-6

Lefebvre, F. (2013). *Polyoxometalates Encapsulated in Inorganic Materials: Applications in Catalysis. In New and Future Developments in Catalysis: Hybrid Materials* (pp. 265–288). Composites, and Organocatalysts.

Leh & Chan. (1973). Sulfur compounds: Pollution, health effects and biological function. *J. Chem. Edu*, *50*(4), 246.

Li, Yue, Su, & Ga. (2011). Equilibrium and a two-stage batch adsorber design for reactive or disperse dye removal to minimize adsorbent amount. *Bioresource Technology*, *102*(9), 5290–5296.

Liang, Z.-X., Li, L., Li, S., Cai, Y.-H., Yang, S.-T., & Wang, J.-F. (2012). Enhanced propionic acid production from Jerusalem artichoke hydrolysate by immobilized *Propionibacterium acidipropionici* in a fibrous-bed bioreactor. *Bioprocess and Biosystems Engineering*, *35*(6), 915–921. doi:10.100700449-011-0676-y PMID:22228298

Li, C., Gao, J., Jiang, Z., Wang, S., Lu, H., Yang, Y., & Jing, F. (2005). Selective Oxidations on Recoverable Catalysts Assembled in Emulsions. *Topics in Catalysis*, *35*(1-2), 169–175. doi:10.100711244-005-3821-5

Li, C., Jiang, Z., Gao, J., Yang, Y., Wang, S., Tian, F., ... Han, C. (2004). Ultra-Deep Desulfurization of Diesel: Oxidation with a Recoverable Catalyst Assembled in Emulsion. *Chemistry (Weinheim an der Bergstrasse, Germany)*, *10*(9), 2277–2280. doi:10.1002/chem.200305679 PMID:15112217

Li, F., Xu, P., Feng, J., Meng, L., Zheng, Y., Luo, L., & Ma, C. (2005). Microbial desulfurization of gasoline in a *Mycobacterium goodii* X7B immobilized-cell system. *Applied and Environmental Microbiology*, *71*(1), 276–281. doi:10.1128/AEM.71.1.276-281.2005 PMID:15640198

Li, H., Zhang, P., Zhang, L., Zhou, T., & Hu, D. (2009). Composite microspheres with PAM microgel core and polymerisable surfactant/polyoxometalate complexes shell. *Journal of Materials Chemistry*, *19*(26), 4575–4586. doi:10.1039/b823075g

Li, K., Yang, X., Guo, Y., Ma, F., Li, H., Chen, L., & Guo Yihang, Y. (2010). Design of mesostructured H3PW12O40-titania materials with controllable structural orderings and pore geometries and their simulated sunlight photocatalytic activity towards diethyl phthalate degradation. *Applied Catalysis B: Environmental*, *99*(1-2), 364–375. doi:10.1016/j.apcatb.2010.07.012

Li, L., Li, L., Wang, Y., Du, Y., & Qin, S. (2013). Biorefinery products from the inulin-containing crop Jerusalem artichoke. *Biotechnology Letters*, *35*(4), 471–477. doi:10.100710529-012-1104-3 PMID:23242496

Li, M. Z., Squires, C. H., Monticello, D. J., & Childs, J. D. (1996). Genetic analysis of the dsz promoter and associated regulatory regions of *Rhodococcus erythropolis* IGTS8. *Journal of Bacteriology*, *178*(22), 6409–6418. doi:10.1128/jb.178.22.6409-6418.1996 PMID:8932295

Ling, L., Zhang, R., Han, P., & Wang, B. (2013). DFT study on the sulfurization mechanism during the desulfurization of H_2S on the ZnO desulfurizer. *Fuel Processing Technology*, *106*, 222–230. doi:10.1016/j.fuproc.2012.08.001

Linguist, L., & Pacheco, M. (1999). Enzyme-based diesel desulfurization process offers energy, CO_2 advantages. *Oil & Gas Journal*, *97*(8), 45–50.

Lin, T., Wang, H. X., Wang, H. M., & Wang, X. G. (2004). The charge effect of cationic surfactants on the elimination of fibre microspheres in the electrospinning of polystyrene. *Nanotechnology*, *15*(9), 1375–1381. doi:10.1088/0957-4484/15/9/044

Lissner, E., de Souza, W. F., Ferrera, B., & Dupont, J. (2009). Oxidative Desulfurization of Fuels with Task-Specific Ionic Liquids. *ChemSusChem*, *2*(10), 962–964. doi:10.1002/cssc.200900171 PMID:19795435

Liu, B. S., Xu, D. F., Chu, J. X., Liu, W., & Au, C. T. (2007). Deep desulfurization by the adsorption process of fluidized catalytic cracking (FCC) diesel over mesoporous Al-MCM- 41 materials. *Energy & Fuels*, *21*(1), 250–255. doi:10.1021/ef060249n

Liu, J., & Kumar, S. (2005). Microscopic polymer cups by electrospinning. *Polymer*, *46*(10), 3211–3214. doi:10.1016/j.polymer.2004.11.116

Liu, L., Zhang, Y., & Tan, W. (2014). Ultrasound-assisted oxidation of dibenzothiophene with phosphotungstic acid supported on activated carbon. *Ultrasonics Sonochemistry*, *21*(3), 970–974. doi:10.1016/j.ultsonch.2013.10.028 PMID:24326220

Liu, N., Luo, F., Wu, H., Liu, Y., Zhang, C., & Chen, J. (2008). One-step ionic-liquid-assisted electrochemical synthesis of ionic liquid functionalized graphene sheets directly from graphite. *Advanced Functional Materials*, *18*(10), 1518–1525. doi:10.1002/adfm.200700797

Liu, W., Wang, J., Li, Q., Sai, J., Zhang, T., & Ji, S. (2014). Synthesis of rare earth metal-organic frameworks (Ln-MOFs) and their properties of adsorption desulfurization. *Journal of Rare Earths*, *32*(2), 189–194. doi:10.1016/S1002-0721(14)60050-8

Liu, Y., Liu, S., Liu, S., Liang, D., Li, S., Tang, Q., ... Zheng, Z. (2013). Facile Synthesis of a Nanocrystalline Metal–Organic Framework Impregnated with a Phosphovanadomolybdate and Its Remarkable Catalytic Performance in Ultradeep Oxidative Desulfurization. *ChemCatChem*, *5*(10), 3086–3091. doi:10.1002/cctc.201300378

Li, W., & Jiang, X. (2013). Enhancement of bunker oil biodesulfurization by adding surfactant. *World Journal of Microbiology & Biotechnology*, *29*(1), 103–108. doi:10.100711274-012-1162-7 PMID:22940903

Li, W., Tang, H., Liu, Q., Xing, J., Li, Q., Wang, D., ... Liu, H. (2009). Deep desulfurization of diesel by integrating adsorption and microbial method. *Biochemical Engineering Journal*, *44*(2-3), 297–301. doi:10.1016/j.bej.2008.12.016

Li, X., Cai, W., An, J., Kim, S., Nah, J., Yang, D., ... Ruoff, R. S. (2009). Large-area synthesis of high-quality anduniform graphene films on copper foils. *Science*, *324*(5932), 1312–1314. doi:10.1126cience.1171245 PMID:19423775

Li, X., Nagaoka, K., Simon, L. J., Lercher, J. A., Wrabetz, S., Jentoft, F., ... Papp, H. (2005). Interaction between sulfated zirconia and alkanes: Prerequisite for active sites-formation and stability of reaction on intermediates. *Journal of Catalysis*, *230*(1), 214–225. doi:10.1016/j.jcat.2004.11.045

Li, X., Qin, Y., Picraux, S. T., & Guo, Z.-X. (2011). Noncovalent assembly of carbon nanotube-inorganic hybrids. *Journal of Materials Chemistry*, *21*(21), 7527–7547. doi:10.1039/c1jm10516g

Li, Y., Cui, R. L., Ding, L., Liu, Y., Zhou, W. W., Zhang, Y., ... Liu, J. (2010). How catalysts affect the growth of single-walled carbon nanotubes on substrates. *Advanced Materials*, *22*(13), 1508–1515. doi:10.1002/adma.200904366 PMID:20437500

Li, Y.-G., Gao, H.-S., Li, W.-L., Xing, J.-M., & Liu, H.-Z. (2009). *In situ* magnetic separation and immobilization of dibenzothiophene-desulfurizing bacteria. *Bioresource Technology*, *100*(21), 5092–50966. doi:10.1016/j.biortech.2009.05.064 PMID:19541480

Li, Y.-H., Wang, S., Cao, A., Zhao, D., Zhang, X., Xu, C., ... Wei, B. (2001). Adsorption of fluoride fromwater by amorphous alumina supported on carbon nanotubes. *Chemical Physics Letters*, *350*(5-6), 412–416. doi:10.1016/S0009-2614(01)01351-3

Lomeda, J. R., Doyle, C. D., Kosynkin, D. V., Hwang, W. F., & Tour, J. M. (2008). Diazonium functionalization of surfactant-wrapped chemically converted graphene sheets. *Journal of the American Chemical Society*, *130*(48), 16201–16206. doi:10.1021/ja806499w PMID:18998637

Loo, A. H., Bonanni, A., & Pumera, M. (2013). Biorecognition on graphene: Physical, covalent, and affinity immobilization methods exhibiting dramatic differences. *Asian Journal of Chemistry*, *8*(1), 198–203. doi:10.1002/asia.201200756 PMID:23090869

Lozinsky, V. I., & Plieva, F. M. (1998). Poly (vinyl alcohol) cryogels employed as matrices for cell immobilization. 3. Overview of recent research and developments. *Enzyme and Microbial Technology*, *23*(3), 227–242. doi:10.1016/S0141-0229(98)00036-2

Lu, L., Cheng, S.F., Gao, J.B., Gao, G.H., & He, M.Y. (2007). Title. *Energy and Fuels*, *21*, 383-384.

Lü, H., Gao, J., Jiang, Z., Jing, F., Yang, Y., Wang, G., & Li, C. (2006). Ultra-deep desulfurization of diesel by selective oxidation with [C18H37N(CH3)3]4[H2NaPW10O36] catalyst assembled in emulsion droplets. *Journal of Catalysis*, *239*(2), 369–375. doi:10.1016/j.jcat.2006.01.025

Lu, H., Gao, J., Jiang, Z., Yang, Y., Song, B., & Li, C. (2007). Oxidative desulfurization of dibenzothiophene with molecular oxygen using emulsion catalysis. *Chemical Communications*, (2): 150–152. doi:10.1039/B610504A PMID:17180229

Lü, H., Ren, W., Liao, W., Chen, W., Li, Y., & Suo, Z. (2013). Aerobic oxidative desulfurization of model diesel using a B-type Anderson catalyst [(C18H37)2N(CH3)2]3Co(OH)6Mo6O18·3H 2O. *Applied Catalysis B: Environmental*, *138–139*(0), 79–83. doi:10.1016/j.apcatb.2013.02.034

Lü, H., Ren, W., Wang, H., Wang, Y., Chen, W., & Suo, Z. (2013). Deep desulfurization of diesel by ionic liquid extraction coupled with catalytic oxidation using an Anderson-type catalyst [(C4H9)4N]4NiMo6O24H6. *Applied Catalysis A, General*, *453*(0), 376–382. doi:10.1016/j.apcata.2012.12.047

Lu, H., Zhang, Y., Jiang, Z., & Li, C. (2010). Aerobic oxidative desulfurization of benzothiophene, dibenzothiophene and 4,6-dimethyldibenzothiophene using an Anderson-type catalyst [(C18H37)2N(CH3)2]5 [IMo6O24]. *Green Chemistry*, *12*(11), 1954–1958. doi:10.1039/c0gc00271b

Luo, M. F., Xing, J. M., Gou, Z. X., Li, S., Liu, H. Z., & Chen, J. Y. (2003). Desulfurization of dibenzothiophene by lyophilized cells of *Pseudomonas delafieldii* R-8 in the presence of dodecane. *Biochemical Engineering Journal*, *13*(1), 1–6. doi:10.1016/S1369-703X(02)00078-5

Lu, W., Wei, Z., Gu, Z., Liu, T., Park, J., Park, J., ... Zhou, H. (2014). Tuning the structure and function of metal–organic frameworks *via* linker design. *Chemical Society Reviews*, *43*(16), 5561–5593. doi:10.1039/C4CS00003J PMID:24604071

Ma, C., Feng, J., Zeng, Y., Cai, X., Sun, B., Zhang, Z., ... Xu, P. (2006). Methods for the preparation of a biodesulfurization biocatalyst using *Rhodococcus* sp. *Chemosphere*, *65*(1), 165–169. doi:10.1016/j.chemosphere.2006.03.010 PMID:16624377

Maciuca, A.-L., Ciocan, C.-E., Dumitriu, E., Fajula, F., & Hulea, V. (2008). V-, Mo- and W-containing layered double hydroxides as effective catalysts for mild oxidation of thioethers and thiophenes with H2O2. *Catalysis Today*, *138*(1–2), 33–37. doi:10.1016/j.cattod.2008.04.031

Maes, M., Trekels, M., Boulhout, M., Schouteden, S., Vermoortele, F., Alaerts, L., ... De Vos, D. E. (2011). Selective Removal of N-Heterocyclic Aromatic Contaminants from Fuels by Lewis Acidic Metal-Organic Frameworks. *Angewandte Chemie International Edition*, *50*(18), 4210–4214. doi:10.1002/anie.201100050 PMID:21472941

Magyar, S., Hancsók, J., & Kalló, D. (2005). Hydrodesulfurization and hydroconversion of heavy FCC gasoline on PtPd/H-USY zeolite. *Fuel Processing Technology*, *86*(11), 1151–1164. doi:10.1016/j.fuproc.2004.11.012

Maleki, H., Duraes, L., & Portugal, A. (2014). Synthesis of lightweight polymers reinforced silica aerogels with improved mechanical and thermal insulation properties for space applications. *Microporous and Mesoporous Materials*, *197*, 116–129. doi:10.1016/j.micromeso.2014.06.003

Manahan, S. E. (2010). *Environmental Chemistry* (7th ed.). Taylor & Francis.

Mangnus, P. J., Poels, E. K., Van Langeveld, A. D., & Moulijn, J. A. (1995). Comparison of the sulfiding rate and mechanism of supported NiO and NiO particles. *Journal of Catalysis*, *137*(1), 92–101. doi:10.1016/0021-9517(92)90141-4

Manso, T., Nunes, C., Raposo, S., & Lima-Costa, M. E. (2010). Carob pulp as raw material for production of the biocontrol agent *P. agglomerans* PBC-1. *Journal of Industrial Microbiology & Biotechnology*, *37*(11), 1145–1155. doi:10.100710295-010-0762-1 PMID:20593299

Mansouri, A. (2014). Ultra-deep adsorptive desulfurization of a model diesel fuel on regenerable Ni–Cu/-Al2O3 at low temperatures in absence of hydrogen. *Journal of Hazardous Materials*, *271*, 120–130. doi:10.1016/j.jhazmat.2014.02.006 PMID:24632365

Marécot, P., Paraiso, E., Dumas, J. M., & Barbier, J. (1992). Deactivation of nickel catalysts by sulphur compounds: I. Benzene hydrogenation. *Applied Catalysis A*, *80*(1-2), 79–88. doi:10.1016/0926-860X(92)85109-O

Marı'n-Rosas, C., Ramı'rez-Verduzco, L. F., Murrieta-Guevara, F. R., Herna'ndez-Tapia, G., & Rodrı'guez-Otal, L. M. (2010). Desulfurization of Low Sulfur Diesel by Adsorption Using Activated Carbon: Adsorption Isotherms. *Industrial & Engineering Chemistry Research*, *49*(9), 4372–4376. doi:10.1021/ie901756b

Maria Angeles Larrubia, A. (2002). A FT-IR study of the adsorption of indole, carbazole, benzothiophene, dibenzothiophene and 4,6-dibenzothiophene over solid adsorbents and catalysts. *Applied Catalysis A, General*, *224*(1-2), 167–178. doi:10.1016/S0926-860X(01)00769-4

Maricq, M. M., Chase, R. E., Xu, N., & Laing, P. M. (2002). The effects of the catalytic converter and fuel sulfur level on motor vehicle particulate matter emissions: Light duty diesel vehicles. *Environmental Science & Technology*, *36*(2), 283–289. doi:10.1021/es010962l PMID:11827064

Martin, A. B., Alcon, A., Santos, V. E., & Garcia-ochoa, F. (2004). Production of a biocatalyst of *Pseudomonas putida* CECT5279 for dibenzothiophene (DBT) biodesulfurization for different media compositions. *Energy & Fuels*, *18*(3), 851–857. doi:10.1021/ef030174c

Martínez-Checa, F., Toledo, F., Vilchez, R., Quesada, E., & Calvo, C. (2002). Yield production, chemical composition, and functional properties of emulsifier H28 synthesized by *Halomonas eurihalina* strain H-28 in media containing various hydrocarbons. *Applied Microbiology and Biotechnology*, *58*(3), 358–363. doi:10.100700253-001-0903-6 PMID:11935188

Martinez, I., Santos, V. E., Alcon, A., & Garcia-Ochoa, F. (2015). Enhancement of the biodesulfurization capacity of *Pseudomonas putida* CECT5279 by co-substrate addition. *Process Biochemistry*, *50*(1), 119–124. doi:10.1016/j.procbio.2014.11.001

Mathieu, Y., Soulard, M., Patarin, J., & Moliere, M. (2012). Mesopourous materials for the removal of SO_2 from gas streams. *Fuel Processing Technology*, *99*, 35–42. doi:10.1016/j.fuproc.2012.02.005

Mattevi, C., Kim, H., & Chhowalla, M. (2011). A review of chemical vapour deposition of grapheme on copper. *Journal of Materials Chemistry*, *21*(10), 3324–3334. doi:10.1039/C0JM02126A

Ma, X. L., Sprague, M., Sun, L., & Song, C. S. (2002). Deep desulfurization of gasoline by SARS process using adsorbent for fuel cells. Prep. Am. Chem. Soc. DiV. *Fuel Chem.*, *47*, 452–453.

Ma, X., Lu, S., Yin, Z., & Song, C. (2001). New approaches to deep desulfurization of diesel fuel, jet fuel, and gasoline by adsorption for ultra-clean fuels and for fuel cell applications. *American Chemical Society. Division of Fuel Chemistry*, *46*(2), 648–649.

Ma, X., Sakanishi, K., Isoda, T., & Mochida, I. (1995). Quantum chemical calculation on the desulfurization reactivities of heterocyclic sulfur compounds. *Energy & Fuels*, *9*(1), 33–37. doi:10.1021/ef00049a005

Ma, X., Sakanishi, K., & Mochida, I. (1994). Hydrodesulfurization reactivities of various sulfur compounds in diesel fuel. *Industrial & Engineering Chemistry Research*, *33*(2), 218–222. doi:10.1021/ie00026a007

Ma, X., Sakanishi, K., & Mochida, I. (1996). Hydrodesulfurization reactivities of various sulfur compounds in vacuum gas oil. *Industrial & Engineering Chemistry Research*, *35*(8), 2487–2494. doi:10.1021/ie960137r

Ma, X., & Schobert, H. H. (2000). Molecular simulation on hydrodesulfurization of Thiophenic compounds over MoS_2 using ZINDO. *Journal of Molecular Catalysis A Chemical*, *160*(2), 409–427. doi:10.1016/S1381-1169(00)00278-8

Ma, X., Sun, L., & Song, C. (2002). A new approach to deep desulfurization of gasoline, diesel fuel and jet fuel by selective adsorption for ultra-clean fuels and for fuel cell applications. *Catalysis Today*, *77*(1-2), 107–116. doi:10.1016/S0920-5861(02)00237-7

McFarland, B. L. (1999). Biodesulfurization. *Current Opinion in Microbiology*, *2*(3), 257–264. doi:10.1016/S1369-5274(99)80045-9 PMID:10383871

McFarland, B. L., Boron, D. J., Deever, W., Meyer, J. A., Johnson, A. R., & Atlas, R. M. (1998). Biocatalytic sulfur removal from fuels: Applicability for producing low sulfur gasoline. *Critical Reviews in Microbiology*, *24*(2), 99–147. doi:10.1080/10408419891294208 PMID:9675512

McKetta, J. J. (1992). *Petroleum processing hand book*. Taylor & Francis Inc.

McWilliams, A. (2015). The global market for carotenoids, Jun 2018, Report id: FOD025F, BCC research. https://www.bccresearch.com/market-research/food-and-beverage/the-global-market-for-carotenoids-fod025f.html. Accessed 1 March 2019.

Meille, V., Shulz, E., Vrinat, M., & Lemaire, M. (1997). Hydrodesulfurization of alkyldibenzothiophenes over a NiMo/Al2O3 catalyst: Kinetics and mechanism. *Journal of Catalysis*, *170*(1), 29–36. doi:10.1006/jcat.1997.1732

Mendes, A., Guerra, P., Madeira, V., Ruano, F., da Silva, T. L., & Reis, A. (2007). Study of docosahexaenoic acid production by the heterotrophic microalga *Crypthecodinium cohnii* CCMP 316 using carob pulp as a promising carbon source. *World Journal of Microbiology & Biotechnology*, *23*(9), 1209–1215. doi:10.100711274-007-9349-z

Meng, X., Huang, H., & Shi, L. (2013). Reactive Mechanism and Regeneration Performance of NiZnO/Al_2O_3-Diatomite Adsorbent by Reactive Adsorption Desulfurization. *Industrial & Engineering Chemistry Research*, *52*(18), 6092–6100. doi:10.1021/ie303514y

Meyer, K., Klobes, P., & Röhl-Kuhn, B. (1997). Certification of reference material with special emphasis on porous solids. *Crystal Research and Technology*, *32*(1), 173–183. doi:10.1002/crat.2170320118

Milenkovic, A., Shulz, E., Meille, V., Lofferda, D., Forissier, M., Vrinat, M., ... Echegoyen, L. (2013). Preparation and characterization of soluble carbon nano-onions by covalent functionalization, employing a Na-K alloy. *Chemical Communications (Cambridge)*, *49*(24), 2406–2408. doi:10.1039/c3cc39077b PMID:23411670

Mit-Uppatham, C., Nithitanakul, M., & Supaphol, P. (2004). Ultrafine electrospun polyamide-6 fibers: Effect of solution conditions on morphology and average fiber diameter. *Macromolecular Chemistry and Physics*, *205*(17), 2327–2338. doi:10.1002/macp.200400225

Miyazaki, T., Takeda, Y., Akane, S., Itou, T., Hoshiko, A., & En, K. (2010). Role of boric acid for a poly (vinyl alcohol) film as a cross-linking agent: Melting behaviors of the films with boric acid. *Polymer*, *51*(23), 5539–5549. doi:10.1016/j.polymer.2010.09.048

Mizuno, N., & Misono, M. (1998). Heterogeneous Catalysis. *Chemical Reviews*, *98*(1), 199–218. doi:10.1021/cr960401q PMID:11851503

Mjalli, F. S., Ahmed, O. U., Al-Wahaibi, T., Al-Wahaibi, Y., & AlNashef, I. M. (2014). Deep oxidative desulfurization of liquid fuels. *Reviews in Chemical Engineering*, *30*(4), 337–378. doi:10.1515/revce-2014-0001

Mohebali, G., & Ball, A. S. (2008). Biocatalytic desulfurization (BDS) of petrodiesel fuels. *Microbiology*, *154*(Pt 8), 2169–2183. doi:10.1099/mic.0.2008/017608-0 PMID:18667551

Mohebali, G., Ball, A. S., Kaytash, A., & Rasekh, B. (2008). Dimethyl sulfoxide (DMSO) as the sulfur source for the production of desulfurizing resting cells of *Gordonia alkanivorans* RIPI90A. *Microbiology*, *154*(3), 878–885. doi:10.1099/mic.0.2007/013011-0 PMID:18310033

Mohebali, G., Ball, A., Kaytash, A., & Rasekh, B. (2007). Stabilization of water/gas oil emulsions by desulfurizing cells of *Gordonia alkanivorans* RIPI90A. *Microbiology*, *153*(Pt 5), 1573–1581. doi:10.1099/mic.0.2006/002543-0 PMID:17464072

Mokhtar, W. N. A. W., Bakar, W. A. W. A., Ali, R., & Abdul-Kadir, A. A. (2014). Deep desulfurization of model diesel by extraction with *N,N*-dimethylformamide: Optimization by Box–Behnken design. *Journal of the Taiwan Institute of Chemical Engineers*, *45*(4), 1542–1548. doi:10.1016/j.jtice.2014.03.017

Molecular sieves-Technical Information Bulletin: Sigma-Aldrich. (n.d.). Retrieved from www.sigmaaldrich.com/chemistry/chemical-synthesis/learning-center/technical-bulletins/al 1430/molecular-sieves.html

Monteiro, O. A. C. Jr, & Airoldi, C. (1999). Some studies of crosslinking chitosan–glutaraldehyde interaction in a homogeneous system. *International Journal of Biological Macromolecules*, *26*(2-3), 119–128. doi:10.1016/S0141-8130(99)00068-9 PMID:10517518

Monticello, D. J. (2000). Biodesulfurization and the upgrading of petroleum distillates. *Current Opinion in Biotechnology*, *11*(6), 540–546. doi:10.1016/S0958-1669(00)00154-3 PMID:11102787

Monticello, D. J., & Finnerty, W. R. (1985). Microbial desulfurization of fossil fuels. *Annual Review of Microbiology*, *39*(1), 371–389. doi:10.1146/annurev.mi.39.100185.002103 PMID:3904605

Mosbach, K. (1994). Molecular imprinting. *Trends in Biochemical Sciences*, *19*(1), 9–14. doi:10.1016/0968-0004(94)90166-X PMID:8140624

Mukhopadhyaya, M., Chowdhury, R., & Bhattacharya, P. (2005). Two phase mathematical model for a bio-trickling reactor for the production of Ultra low sulfur diesel (ULSD) from deeply hydrodesulfurized diesel. *Catalysis Today*, *106*(1-4), 180–185. doi:10.1016/j.cattod.2005.07.128

Mukhopadhyaya, M., Chowdhury, R., & Bhattacharya, P. (2006). Biodesulfurization of hydrodesulfurized diesel in a trickle bed reactor - Experiments and modeling. *Journal of Scientific and Industrial Research*, *65*(5), 432–436.

Mukhopadhyaya, M., Chowdhury, R., & Bhattacharya, P. (2007). Trickle bed biodesulfurizer of diesel with backwash and recycle. *AIChE Journal. American Institute of Chemical Engineers*, *53*(8), 2188–2197. doi:10.1002/aic.11240

Mulligan, C. N. (2005). Environmental applications for biosurfactants. *Environmental Pollution*, *133*(2), 183–198. doi:10.1016/j.envpol.2004.06.009 PMID:15519450

Murphy, S. E., Amin, S., Coletta, K., & Hoffman, D. (1992). Rat liver metabolism of benzo[b] naphtho[2,1-d]thiophene. *Chemical Research in Toxicology*, *5*(4), 491–495. doi:10.1021/tx00028a006 PMID:1382648

Muzic, M., Sertic-Bionda, K., Gomzi, Z., Podolski, S., & Telen, S. (2010). Study of diesel fuel desulfurization by adsorption. *Chemical Engineering Research & Design*, *8*(8), 487–495. doi:10.1016/j.cherd.2009.08.016

Mykola, S. (2009). Textural and chemical factors affecting adsorption capacity of activated carbon in highly efficient desulfurization of diesel fuel. *Carbon*, *47*(10), 2491–2500. doi:10.1016/j.carbon.2009.05.001

Nagai, M., Goto, Y., Ishii, H., & Omi, S. (2000). XPS and TPSR study of nitrided molybdena-alumina catalyst for the hydrodesulfurization of dibenzothiophene. *Applied Catalysis A, General*, *192*(2), 189–199. doi:10.1016/S0926-860X(99)00341-5

Nagai, Y., Shinjoh, H., & Yokota, K. (2002). Oxidation selectivity between n-hexane and sulfur dioxide in diesel simulated exhaust gas over platinum-supported zirconia catalyst. *Applied Catalysis B: Environmental*, *39*(2), 149–155. doi:10.1016/S0926-3373(02)00082-6

Nagarkar, S. S., Das, R., Poddar, P., & Ghosh, S. K. (2008). Bistable dynamic coordination polymer showing reversible structural and functional transformations. *Inorganic Chemistry*, *51*(15), 8317–8321. doi:10.1021/ic301481p PMID:22827720

Naidich, J. V. (1981). The wettability of solids by liquid metals. *Progress in Surface Membrane Science*, *14*, 353–484. doi:10.1016/B978-0-12-571814-1.50011-7

Naito, M., Kawamoto, T., Fujino, K., Kobayashi, M., Maruhashi, K., & Tanaka, A. (2001). Long-term repeated biodesulfurization by immobilized *Rhodococcus erythropolis* KA 2-5-1 cells. *Applied Microbiology and Biotechnology*, *55*(3), 374–378. doi:10.1007002530000527 PMID:11341322

Nandi, S. (2010). Biodesulfurization of hydro-desulfurized diesel in airlift reactor. *Journal of Scientific and Industrial Research*, *69*(7), 543–547.

Nantasenamat, C., Isarankura-Na-Ayudhya, C., Naenna, T., & Prachayasittikul, V. (2007). Quantitative structure-imprinting factor relationship of molecularly imprinted polymers. *Biosensors & Bioelectronics*, *22*(12), 3309–3317. doi:10.1016/j.bios.2007.01.017 PMID:17317143

Nantasenamat, C., Naenna, T., Isarankura Na Ayudhya, C., & Prachayasittikul, V. (2005). Quantitative prediction of imprinting factor of molecularly imprinted polymers by artificial neural network. *Journal of Computer-Aided Molecular Design*, *19*(7), 509–524. doi:10.100710822-005-9004-4 PMID:16244792

Nantasenamat, C., Tantimongcolwat, T., Naenna, T., Isarankura-Na-Ayudhya, C., & Prachayasittikul, V. (2006). Prediction of selectivity index of pentachlorophenol-imprinted polymers. *EXCLI Journal*, *5*, 150–163.

Narsi, R. B. (2004). Adsorption of Cr (VI) on activated rice husk carbon and activated alumina. *Bioresource Technology*, *91*(3), 305–307. doi:10.1016/S0960-8524(03)00204-9 PMID:14607491

Nekodzuka, S., Nakajima-Kambe, T., Nomura, N., Lu, J., & Nakahara, T. (1997). Specific desulfurization of dibenzothiophene by *Mycobacterium sp*. strain G3. *Biocatalysis and Biotransformation*, *15*(1), 17–27. doi:10.3109/10242429709003607

Ngah, W. S. W., & Liang, K. H. (1999). Adsorption of gold(III) ions onto chitosan and *N*-carboxymethyl chitosan: Equilibrium studies. *Industrial & Engineering Chemistry Research*, *38*(4), 1411–1414. doi:10.1021/ie9803164

Nicholls, I. A., Adbo, K., Andersson, H. S., Andersson, P. O., Ankarloo, J., Hedin-Dahlström, J., ... Wikman, S. (2001). Can we rationally design molecularly imprinted polymers? *Analytica Chimica Acta*, *435*(1), 9–18. doi:10.1016/S0003-2670(01)00932-1

Nie, Y., Dong, Y., Bai, L., Dong, H., & Zhang, X. (2013). Fast oxidative desulfurization of fuel oil using dialkylpyridinium tetrachloroferrates ionic liquids. *Fuel*, *103*(0), 997–1002. doi:10.1016/j.fuel.2012.07.071

Nisar, A., Lu, Y., Zhuang, J., & Wang, X. (2011). Polyoxometalate Nanocone Nanoreactors: Magnetic Manipulation and Enhanced Catalytic Performance. *Angewandte Chemie International Edition*, *50*(14), 3187–3192. doi:10.1002/anie.201006155 PMID:21370361

Nisar, A., Zhuang, J., & Wang, X. (2011). Construction of Amphiphilic Polyoxometalate Mesostructures as a Highly Efficient Desulfurization Catalyst. *Advanced Materials*, *23*(9), 1130–1135. doi:10.1002/adma.201003520 PMID:21360767

Nogueira, L. S., Ribeiro, S., Granadeiro, C. M., Pereira, E., Feio, G., Cunha-Silva, L., & Balula, S. S. (2014). Novel polyoxometalate silica nano-sized spheres: Efficient catalysts for olefin oxidation and the deep desulfurization process. *Dalton Transactions (Cambridge, England)*, *43*(25), 9518–9528. doi:10.1039/C3DT53444H PMID:24829931

Nolte, T., Posch, T. N., Huhn, C., & Andersson, J. T. (2013). Desulfurized fuels from Athabasca bitumen and their polycyclic aromatic sulfur heterocycles. Analysis based on capillary electrophoresis coupled with TOF MS. *Energy & Fuels*, *27*(1), 97–107. doi:10.1021/ef301424d

Nordvik, A. B., Simmons, J. L., Bitting, K. R., Lewis, A., & Strøm-Kristiansen, T. (1996). Oil and water separation in marine oil spill clean-up operations. *Spill Science & Technology Bulletin*, *3*(3), 107–122. doi:10.1016/S1353-2561(96)00021-7

Nuhu, A. A. (2013). Bio-catalytic desulfurization of fossil fuels: A mini review. *Reviews in Environmental Science and Biotechnology*, *12*(1), 9–23. doi:10.100711157-012-9267-x

Ogunlaja, A. S., Coombes, M. J., Torto, N., & Tshentu, Z. R. (2014). The adsorptive extraction of oxidised sulfur-containing compounds from fuels by using molecularly imprinted chitosan materials. *Reactive & Functional Polymers*, *81*, 61–76. doi:10.1016/j.reactfunctpolym.2014.04.006

Ogunlaja, A. S., du Sautoy, C. W., Torto, N., & Tshentu, Z. R. (2014). Design, fabrication and evaluation of intelligent sulfone-selective polybenzimidazole nanofibers. *Talanta*, *126*, 61–72. doi:10.1016/j.talanta.2014.03.035 PMID:24881535

Ogunlaja, A. S., Khene, S., Antunes, E., Nyokong, T., Torto, N., & Tshentu, Z. R. (2013). The development of catalytic oxovanadium(IV)-containing microspheres for the oxidation of various organosulfur compounds. *Applied Catalysis A, General*, *462-463*, 157–167. doi:10.1016/j.apcata.2013.05.004

Ogunlaja, A. S., Walmsley, R. S., du Sautoy, C. W., Torto, N., & Tshentu, Z. R. (2013). Development of a continuous flow system for the oxidative desulfurization of refractory organosulfur compounds in hydrotreated diesel. *Energy & Fuels*, *27*(12), 7714–7723. doi:10.1021/ef401708f

Ohshiro, T., & Izumi, Y. (1999). Microbial desulfurization of organic sulfur compounds in petroleum. *Bioscience, Biotechnology, and Biochemistry*, *63*(1), 1–9. doi:10.1271/bbb.63.1 PMID:10052116

Ohshiro, T., Suzuki, K., & Izumi, Y. (1996). Regulation of dibenzothiophene degrading enzyme activity of *Rhodococcus erythropolis* D-1. *Journal of Fermentation and Bioengineering*, *81*(2), 121–124. doi:10.1016/0922-338X(96)87588-X

Okada, H., Nomura, N., Nakahara, T., Saitoh, K., Uchiyama, H., & Maruhashi, K. (2003). Analyses of microbial desulfurization reaction of alkylated dibenzothiophenes dissolved in oil phase. *Biotechnology and Bioengineering*, *83*(4), 489–497. doi:10.1002/bit.10694 PMID:12800143

Okada, T., Miyamoto, K., Sakai, T., & Mishima, S. (2014). Encapsulation of a polyoxometalate into an organosilica microcapsule for highly active solid acid catalysis. *ACS Catalysis*, *4*(1), 73–78. doi:10.1021/cs400923q

Oldfield, C., Pogrebinsky, O., Simmonds, J., Olson, E. S., & Kulpa, C. F. (1997). Elucidation of the metabolic pathway for dibenzothiophene desulphurization by *Rhodococcus* sp. strain IGTS8 (ATCC 53968). *Microbiology*, *143*(9), 2961–2973. doi:10.1099/00221287-143-9-2961 PMID:9308179

Omwoma, S., Chen, W., Tsunashima, R., & Song, Y.-F. (2014). Recent advances on polyoxometalates intercalated layered double hydroxides: From synthetic approaches to functional material applications. *Coordination Chemistry Reviews*, *258–259*(0), 58–71. doi:10.1016/j.ccr.2013.08.039

Onodera-Yamada, K., Morimoto, M., & Tani, Y. (2001). Degradation of dibenzothiophene by sulfate-reducing bacteria cultured in the presence of only nitrogen gas. *Journal of Bioscience and Bioengineering*, *91*(1), 91–93. doi:10.1016/S1389-1723(01)80119-5 PMID:16232954

Oral, J., Sikula, J., Puchyr, R., Hajny, Z., Stehlik, P., & Bebar, L. (2005). Processing of waste from pulp and paper plant. *Journal of Cleaner Production*, *13*(5), 509–515. doi:10.1016/j.jclepro.2003.09.005

Orita, H., Uchida, K., & Itoh, N. (2004). A volcano-type relationship between the adsorption energy of thiophene on promoted MoS2 cluster-model catalysts and the experimental HDS activity: Ab initio density functional study. *Applied Catalysis A, General*, *258*(1), 115–120. doi:10.1016/j.apcata.2003.08.016

Otsuki, S., Nonaka, T., Qian, W., Ishihara, A., & Kabe, T. (1999). Oxidative desulfurization of middle distillate using ozone. *Journal of the Japan Petroleum Institute*, *42*(5), 315–320. doi:10.1627/jpi1958.42.315

Otsuki, S., Nonaka, T., Takashima, N., Qian, W., Ishihara, A., Imai, T., & Kabe, T. (2000). Oxidative desulfurization of light gas oil and vacuum gas oil by oxidation and solvent extraction. *Energy & Fuels*, *14*(6), 1232–1239. doi:10.1021/ef000096i

Pacheco, M. A., Lange, E. A., Pienkos, P. T., Yu, L.-Q., Rouse, M. P., Lin, Q., & Linguist, L. K. (1999). *Recent advances in biodesulfurization of diesel fuel*. National Petrochemical And Refiners Association-Publications-All Series.

Paixão, S. M., Teixeira, P. D., Silva, T. P., Teixeira, A. V., & Alves, L. (2013). Screening of novel yeast inulinases and further application to bioprocesses. *New Biotechnology*, *30*(6), 598–606. doi:10.1016/j.nbt.2013.02.002 PMID:23419675

Palomeque, J., Clacens, J. M., & Figueras, F. (2002). Oxidation of dibenzothiophene by hydrogen peroxide catalysed by solid bases. *Journal of Catalysis*, *211*(1), 103–108. doi:10.1016/S0021-9517(02)93721-8

Palomino, J. M., Tran, D. T., Hausser, J. L., Dong, H., & Oliver, S. R. J. (2014). Mesopourous Silica nanoparticles for high Capacity adsorptive desulfurization. *J. Matter. Chem A*, *2*(36), 14890–14895. doi:10.1039/C4TA02570A

Pandiyarajan, C., Pandikumar, A., & Ramaraj, R. (2013). Photoelectrocatalytic performance of a titania-keggin type polyoxometalate-gold nanocomposite modified electrode in methanol oxidation. *Nanotechnology*, *24*(43), 435401. doi:10.1088/0957-4484/24/43/435401 PMID:24077520

Paolo, D. F., & Scarsella, M. (2008). Functionalized hexagonal mesoporous silica as an oxidizing agent for the oxidative desulfurization of organosulfur compounds. *Industrial & Engineering Chemistry Research*, *47*(3), 973–975. doi:10.1021/ie071057y

Papirer, E. (2000). *Adsorption on silica surfaces*. CRC. doi:10.1201/9781482269703

Park, J. G., Ko, C. H., Yi, K. B., Park, J. H., Han, S. S., Cho, S. H., & Kim, J. N. (2008). Reactive adsorption of sulfur compounds in diesel on nickel supported on mesoporous silica. *Applied Catalysis B: Environmental*, *81*(3-4), 244–250. doi:10.1016/j.apcatb.2007.12.014

Park, S., & Ruoff, R. S. (2009). Chemical methods for the production of graphenes. *Nature Nanotechnology*, *4*(4), 217–224. doi:10.1038/nnano.2009.58 PMID:19350030

Parr, R. G., & Yang, W. (1989). *Density functional theory of atoms and molecules*. Oxford, UK: Oxford University Press.

Patel, S. B., Killbane, J. J., & Webster, D. A. (1997). Biodesulphurisation of dibenzothiophene in hydrophobic media by *Rhodococcus* sp. strain IGTS8. *Journal of Chemical Technology and Biotechnology (Oxford, Oxfordshire)*, *69*(1), 100–106. doi:10.1002/(SICI)1097-4660(199705)69:1<100::AID-JCTB675>3.0.CO;2-9

Patil, S. V., Sorokhaibam, L. G., Bhandari, V. M., Killedar, D. J., & Ranade, V. V. (2014). Investigating role of sulphur specific carbon adsorbents in deep desulphurization. *Journal of Environmental Chemical Engineering*, *2*(3), 1495–1505. doi:10.1016/j.jece.2014.07.009

Pawelec, B., Navarro, R.M., Campos-Martin, J.M., & Fierro, J.L.G. (2011). *Towards near zero-sulfur liquid fuels: a perspective review*. Academic Press.

Paybarah, A., Bone, R. L., & Corcoran, W. H. (1982). Selective oxidation of dibenzothiophene by peroxybenzoic acid formed in situ. *Industrial & Engineering Chemistry Product Research and Development*, *21*, 426–431.

Pearson, R. G. (1963). Hard and Soft Acids and Bases. *Journal of the American Chemical Society*, *85*(22), 3533–3539. doi:10.1021/ja00905a001

Pearson, R. G. (1986). Absolute electronegativity and hardness correlated with molecular orbital theory. *Proceedings of the National Academy of Sciences of the United States of America*, *83*(22), 8440–8441. doi:10.1073/pnas.83.22.8440 PMID:16578791

Peng, Y., Liu, J., Zhang, H., Luo, D., & Li, D. (2018). A size-matched POM@MOF composite catalyst for highly efficient and recyclable ultra-deep oxidative fuel desulfurization. *Inorganic Chemistry Frontiers*, *5*(7), 1563–1569. doi:10.1039/C8QI00295A

Peralta, D., Chaplais, G., Simon-Masseron, A., Barthelet, K., & Pirngruber, G. D. (2012). Separation of C6 Paraffins Using Zeolitic Imidazolate Frameworks: Comparison with Zeolite 5A. *Energy & Fuels*, *26*, 4953–4960. doi:10.1021/ef300762z

Perathoner, S., Gangeri, M., Lanzafame, P., & Centi, G. (2007). Nanostructured electrocatalytic Pt-carbon materials for fuel cells and CO2conversion. *Kinetics and Catalysis*, *48*(6), 877–883. doi:10.1134/S0023158407060171

Perfumo, A., Smyth, T. J. P., Marchant, R., & Banat, I. M. (2010). Production and roles of biosurfactants and bioemulsifiers in accessing hydrophobic substrates. In *Handbook of hydrocarbon and lipid microbiology* (pp. 1501–1512). Springer. doi:10.1007/978-3-540-77587-4_103

Persano, L., Camposeo, A., Tekmen, C., & Pisignano, D. (2013). Industrial upscaling of electrospinning and applications of polymer nanofibers: A review. *Macromolecular Materials and Engineering*, *298*(5), 504–520. doi:10.1002/mame.201200290

Pham, Q. P., Sharma, U., & Mikos, A. G. (2006). Electrospinning of polymeric nanofibers for tissue engineering applications: A review. *Tissue Engineering*, *12*(5), 1197–1211. doi:10.1089/ten.2006.12.1197 PMID:16771634

Picconi, C. (2014). *Saverio Giovanni Condo, Tomaz Kosmae: Alumina and Zirconia based ceramics for load bearing applications*. Butterworth-Heinemann.

Pilloni, M., Padella, F., Ennas, G., Lai, S., Bellusci, M., Rombi, E., ... Ferino, I. (2015). Liquid-assisted mechanochemical synthesis of an iron carboxylate metal organic framework and its evaluation in diesel fuel desulfurization. *Microporous and Mesoporous Materials*, *213*, 14–21. doi:10.1016/j.micromeso.2015.04.005

Pires, J., Pinto, M. L., Granadeiro, C. M., Barbosa, A. D. S., Cunha-Silva, L., Balula, S. S., & Saini, V. K. (2014). Effect on selective adsorption of ethane and ethylene of the polyoxometalates impregnation in the metal-organic framework MIL-101. *Adsorption*, *20*(4), 533–543. doi:10.100710450-013-9592-6

Pope, C. A. III, Burnett, R. T., Thun, M. J., Calle, E. E., Krewski, D., Ito, K., & Thurston, G. D. (2002). Lung cancer, cardiopulmonary mortality, and long-term exposure to fine particulate air pollution. *Journal of the American Medical Association*, *287*(9), 1132–1141. doi:10.1001/jama.287.9.1132 PMID:11879110

Popov, V. N. (2004). Carbon nanotubes: Properties and applications. *Materials Science and Engineering R Reports*, *43*(3), 61–102. doi:10.1016/j.mser.2003.10.001

Prasad, V. V. D. N., Jeong, K. E., Chae, H. J., Kim, C. U., & Jeong, S. Y. (2008). Oxidative desulfurization of 4,6-dimethyl dibenzothiophene and light cycle oil over supported molybdenum oxide catalysts. *Catalysis Communications*, *9*(10), 1966–1969. doi:10.1016/j.catcom.2008.03.021

Proust, A., Thouvenot, R., & Gouzerh, P. (2008). Functionalization of polyoxometalates: Towards advanced applications in catalysis and materials science. *Chemical Communications*, (16): 1837–1852. doi:10.1039/b715502f PMID:18401495

Pumera, M. (2009). The electrochemistry of carbon nanotubes: Fundamentals and applications. *Chemistry (Weinheim an der Bergstrasse, Germany)*, *15*(20), 4970–4978. doi:10.1002/chem.200900421 PMID:19360829

Pumera, M. (2010). Graphene-based nanomaterials and their electrochemistry. *Chemical Society Reviews*, *39*(11), 4146–4157. doi:10.1039/c002690p PMID:20623061

Qiu, J., Wang, G., Zeng, D., Tang, Y., Wang, M., & Li, Y. (2009). Oxidative desulfurization of diesel fuel using amphiphilic quaternary ammonium phosphomolybdate catalysts. *Fuel Processing Technology*, *90*(12), 1538–1542. doi:10.1016/j.fuproc.2009.08.001

Qiu, L., Zou, K., & Xu, G. (2013). Investigation on the sulfur state and phase transformation of spent and regenerated S zorb sorbents using XPS and XRD. *Applied Surface Science*, *266*, 230–234. doi:10.1016/j.apsusc.2012.11.156

Rafiee, E., & Eavani, S. (2013). Organic–inorganic polyoxometalate based salts as thermoregulated phase-separable catalysts for selective oxidation of thioethers and thiophenes and deep desulfurization of model fuels. *Journal of Molecular Catalysis A Chemical*, *380*(0), 18–27. doi:10.1016/j.molcata.2013.09.009

Ramachandran, H., Iqbal, M. A., & Amirul, A. A. (2014). Identification and characterization of the yellow pigment synthesized by Cupriavidus sp. USMAHM13. *Applied Biochemistry and Biotechnology*, *174*(2), 461–470.

Rana, M. S., Sa'mano, V., Ancheyta, J., & Diaz, J. A. I. (2007). A review of recent advances on processing technologies for upgrading of heavy oils and residual. *Fuel*, *86*(9), 1216–1231. doi:10.1016/j.fuel.2006.08.004

Rao, C. N. R., Sood, A. K., Subrahmanyam, K. S., & Govindaraj, A. (2009). Graphene: The new two-dimensional nanomaterial. *Angewandte Chemie International Edition*, *48*(42), 7752–7777. doi:10.1002/anie.200901678 PMID:19784976

Reddy, B. M., Sreekanth, P. M., & Lakshmanan, P. (2005). Sulfated zirconia as an efficient catalyst for organic synthesis and transformation reactions. *Journal of Molecular Catalysis A Chemical*, *237*(1-2), 93–100. doi:10.1016/j.molcata.2005.04.039

Reichmuth, D. S., Hittle, J. L., Blanch, H. W., & Keasling, J. D. (2000). Biodesulfurization of dibenzothiophene in *Escherichia coli* is enhanced by expression of a *Vibrio harveyi* oxidoreductase gene. *Biotechnology and Bioengineering*, *67*(1), 72–79. doi:10.1002/(SICI)1097-0290(20000105)67:1<72::AID-BIT8>3.0.CO;2-C PMID:10581437

Reina, A., Jia, X., Ho, J., Nezich, D., Son, H., Bulovic, V., ... Kong, J. (2009). Large area, few-layer graphene films on arbitrary substrates by chemical vapor deposition. *Nano Letters*, *9*(1), 30–35. doi:10.1021/nl801827v PMID:19046078

Reneker, D. H., & Chun, I. (1996). Nanometre diameter fibres of polymer, produced by electrospinning. *Nanotechnology*, *7*(3), 216–223. doi:10.1088/0957-4484/7/3/009

Reneker, D. H., & Yarin, A. L. (2008). Electrospinning jets and polymer nanofibers. *Polymer*, *49*(10), 2837–2425. doi:10.1016/j.polymer.2008.02.002

Rezvani, M. A., Shojaei, A. F., & Zonoz, F. M. (2014). Anatase titania-vanadium polyphosphomolybdate as an efficient and reusable nano catalyst for the desulphurization of gas oil. *Journal of the Serbian Chemical Society*, *79*(9), 1099–1110. doi:10.2298/JSC131026015A

Rezvani, M. A., Shojaie, A. F., & Loghmani, M. H. (2012). Synthesis and characterization of novel nanocomposite, anatase sandwich type polyoxometalate, as a reusable and green nano catalyst in oxidation desulfurization of simulated gas oil. *Catalysis Communications*, *25*(0), 36–40. doi:10.1016/j.catcom.2012.04.007

Rhule, J. T., Hill, C. L., Judd, D. A., & Schinazi, R. F. (1998). Polyoxometalates in Medicine. *Chemical Reviews*, *98*(1), 327–358. doi:10.1021/cr960396q PMID:11851509

Ribeiro, S., Granadeiro, C. M., Silva, P., Almeida Paz, F. A., de Biani, F. F., Cunha-Silva, L., & Balula, S. S. (2013). An efficient oxidative desulfurization process using terbium-polyoxometalate@MIL-101(Cr). *Catalysis Science & Technology, 3*(9), 2404-2414.

Ribeiro, S., Barbosa, A. D. S., Gomes, A. C., Pillinger, M., Goncalves, I. S., Cunha-Silva, L., & Balula, S. S. (2013). Catalytic oxidative desulfurization systems based on Keggin phosphotungstate and metal-organic framework MIL-101. *Fuel Processing Technology*, *116*, 350–357. doi:10.1016/j.fuproc.2013.07.011

Richard, I. M. (1996). Principles of adsorption and reaction on solid surfaces. J. Wiley & Sons.

Ritter, H. L., & Drake, L. C. (1945). Pore size distribution in porous materials. *Industrial & Engineering Chemistry Fundamentals*, *17*, 782–787.

Rivera, J. L., Navarro-Santos, P., Guerra-Gonzalez, R., & Lima, E. (2014). Interaction of refractory dibenzothiophenes and polymerizable structures. *International Journal of Polymer Science*, *2014*, 1–11. doi:10.1155/2014/103945

Robeus, E., Roquerol, J., Rodriguez-Reinoso, F., Sing, K. S.W., & Under, K. K. (n.d.). *Characterization of porous solids III.* Elsevier Amsterdam.

Rodrigues, L. R., & Teixeira, J. A. (2010). *Biomedical and therapeutic applications of biosurfactants*. New York: Springer.

Rodrigues, L., Banat, I. M., Teixeira, J., & Oliveira, R. (2006). Biosurfactants: Potential applications in medicine. *The Journal of Antimicrobial Chemotherapy*, *57*(4), 609–618. doi:10.1093/jac/dkl024 PMID:16469849

Ronkainen, N. J., Halsall, H. B., & Heineman, W. R. (2010). Electrochemical biosensors. *Chemical Society Reviews*, *39*(5), 1747–1763. doi:10.1039/b714449k PMID:20419217

Roop, C. B. (2005). Activated Carbon Adsorption. Taylor and Francis Group.

Roquerol, F., Roquerol, J., Sing, K. S. W., Llewellyn, P., & Maurin, G. (2014). *Adsorption by powders and porous solids: Principles* (2nd ed.). Elsevier, UK: Methodology and Applications.

Rouse, J. D., Sabatini, D. A., Suflita, J. M., & Harwell, J. H. (1994). Influence of surfactants on microbial degradation of organic compounds. *Critical Reviews in Environmental Science and Technology*, *24*(4), 325–370. doi:10.1080/10643389409388471

Rozanska, X., van Santen, R. A., Hutschka, F., & Hafner, J. (2003). A periodic density functional theory study of thiophenic derivative cracking catalyzed by mordenite. *Journal of Catalysis, 215*(1), 20–29. doi:10.1016/S0021-9517(02)00148-3

Ruiz, M., Sastre, A. M., & Guibal, E. (2000). Comparative adsorption of Cu(II), Zn(II), and Pb(II) ions in aqueous solution on the crosslinked chitosan with epichlorohydrin. *Reactive & Functional Polymers, 45*, 155–173. doi:10.1016/S1381-5148(00)00019-5

Ryu, K., Heo, J., & Yoo, I. K. (2002). Removal of dibenzothiophene and its oxidized product in anhydrous water-immiscible organic solvent by immobilized cytochrome c. *Biotechnology Letters, 24*(2), 143–146. doi:10.1023/A:1013806830105

Ryzhikov, A., Bezverkhyy, I., & Bellat, J. (2008). Reactive adsorption of thiophene on Ni/ZnO: Role of hydrogen pretreatment. *Applied Catalysis B: Environmental, 84*(3-4), 766–772. doi:10.1016/j.apcatb.2008.06.009

Saboktakin, A. (2014). Improvements of reinforced silica aerogel nanocomposites thermal properties for architecture applications. *International Journal of Biological Macromolecules, 72C*, 230–234. PMID:25172161

Sachin, K., & Vimal Chandra Srivastava, R. P. (2011). Studies on adsorptive desulfurization by zirconia based adsorbents. *Fuel, 90*(11), 3209–3216. doi:10.1016/j.fuel.2011.06.029

Sachin, K., & Vimal Chandra Srivastava, R. P. (2012). Oxidative desulfurization by chromium promoted sulfated zirconia. *Fuel Processing Technology, 93*(1), 18–25. doi:10.1016/j.fuproc.2011.08.017

Salavati, H., Tavakkoli, N., & Hosseinpoor, M. (2012). Preparation and characterization of polyphosphotungstate/ZrO 2 nanocomposite and their sonocatalytic and photocatalytic activity under UV light illumination. *Ultrasonics Sonochemistry, 19*(3), 546–553. doi:10.1016/j.ultsonch.2011.09.001 PMID:21986517

Saleh, T.A. (2016). *Applying Nanotechnology to the Desulfurization Process in Petroleum Engineering*. IGI Global.

Salem, A. B. S. H. (1994). Naphtha desulfurization by adsorption. *Industrial & Engineering Chemistry Research, 33*(2), 336–340. doi:10.1021/ie00026a025

Salem, A. B. S. H., & Hamid, H. S. (1997). Removal of sulphur compounds from naphtha solutions using solid adsorbents. *Chemical Engineering & Technology, 20*(5), 342–347. doi:10.1002/ceat.270200511

Sami, M., & Hatch, L. F. (1994). Chemistry of petrochemical processes, Gulf publishing company, Houston, Texas, p11-17.Monticello, D.J. (2000). Biodesulfurization and the upgrading of petroleum distillates. *Current Opinion in Biotechnology, 11*, 540–546.

Sánchez-Segado, S., Lozano, L. J., de Los Ríos, A. P., Hernández-Fernández, F. J., Godínez, C., & Juan, D. (2012). Process design and economic analysis of a hypothetical bioethanol production plant using carob pod as feedstock. *Bioresource Technology, 104*, 324–328. doi:10.1016/j.biortech.2011.10.046 PMID:22100234

Sano, Y. (2004a). Adsorptive removal of sulfur and nitrogen species from a straight run gas oil over activated carbons for its deep hydrodesulfurization. *Applied Catalysis B: Environmental, 49*(4), 219–225. doi:10.1016/j.apcatb.2003.12.007

Sano, Y., Choi, K. K., Korai, Y., & Mochida, I. (2004b). Selection further activation of activated carbons for removal of nitrogen species in gas oil as a pretreatment for its deep hydrodesulfurization. *Energy & Fuels, 18*(3), 644–651. doi:10.1021/ef034080e

Sarda, K. K., Bhandari, A., Pant, K. K., & Jain, S. (2012). Deep desulfurization of diesel fuel by selective adsorption over Ni/Al_2O_3 and Ni/ZSM-5 extrudates. *Fuel, 93*, 86–91. doi:10.1016/j.fuel.2011.10.020

Schilling, B. M., Alvarez, L. M., Wang, D. I. C., & Cooney, C. L. (2002). Continuous desulfurization of dibenzothiophene with *Rhodococcus rhodochrous* IGTS8 (ATCC 53968). *Biotechnology Progress, 18*(6), 1207–1213. doi:10.1021/bp0200144 PMID:12467453

Schneemann, A., Bon, V., Schwedler, I., Senkovska, I., Kaskel, S., & Fischer, R. A. (2014). Flexible metal–organic frameworks. *Chemical Society Reviews, 43*(16), 6062–6096. doi:10.1039/C4CS00101J PMID:24875583

Schulz, H., Böhringer, W., Ousmanov, F., & Waller, P. (1999). Refractory sulfur compounds in gas oils. *Fuel Processing Technology, 61*(1-2), 5–41. doi:10.1016/S0378-3820(99)00028-4

Schulz, H., Böhringer, W., Waller, P., & Ousmanov, F. (1999). Gas oil deep hydrodesulfurization: Refractory compounds and retarded kinetics. *Catalysis Today, 49*(1-3), 87–97. doi:10.1016/S0920-5861(98)00412-X

Scott, K., Jachuck, R. J., & Hall, D. (2001). Crossflow microfiltration of water-in-oil emulsions using corrugated membranes. *Separation and Purification Technology, 22*(1-2), 431–441. doi:10.1016/S1383-5866(00)00180-5

Scott, T. C., & Wham, R. M. (1989). An electrically driven multistage countercurrent solvent extraction device: The emulsion-phase contactor. *Industrial & Engineering Chemistry Research, 28*(1), 94–97. doi:10.1021/ie00085a016

Seah, C. M., Chai, S. P., & Mohamed, A. R. (2014). Mechanisms of graphene growth by chemical vapour deposition on transition metals. *Carbon, 70*, 1–21. doi:10.1016/j.carbon.2013.12.073

Sellergren, B. (2001). *In molecularly Imiprinted polymers: man-made mimics of antibodies and their applications in analytical chemistry.* In B. Sellergren (Ed.), *Techniques and instrumentation in analytical chemistry* (Vol. 23, pp. 113–184). Amsterdam: Elsevier.

Selvavathi, V., Chidambaram, V., Meenakshisundaram, A., Sairam, B., & Sivasankar, B. (2009). Adsorptive desulfurization of diesel on activated carbon and nickel supported systems. *Catalysis Today*, *141*(1-2), 99–102. doi:10.1016/j.cattod.2008.05.009

Setti, L., Lanzarini, G., & Pifferi, P. G. (1997). Whole cell biocatalysis for an oil desulfurization process. *Fuel Processing Technology*, *52*(1-3), 145–153. doi:10.1016/S0378-3820(97)00023-4

Shafi, R., & Hutchings, G. J. (2000). Hydrodesulfurization of hindered dibenzothiophenes:an overview. *Catalysis Today*, *59*(3-4), 423–442. doi:10.1016/S0920-5861(00)00308-4

Shahadat, H. A. M. H., & Bruce, J. (2013). Adsorptive desulfurization of jet and diesel fuels using Ag/TiOx–Al_2O_3 and Ag/TiOx–SiO_2 adsorbents. *Fuel*, *107*, 465–473. doi:10.1016/j.fuel.2012.11.030

Shangguan, J., Zhao, Y., Fan, H., Liang, L., Shen, F., & Miao, M. (2013). Desulfurization behavior of zinc oxide based sorbent modified by the combination of Al_2O_3 and K_2CO_3. *Fuel, 108*, 80-84.

Shan, J. H., Liu, X. Q., Sun, L. B., & Cui, R. (2008). Cu-Ce Bimetal ion-exchanged Y-Zeolites for selective adsorption of thiophenic sulfur. *Energy & Fuels*, *22*(6), 3955–3959. doi:10.1021/ef800296n

Shennan, J. L. (1996). Microbial attack on sulphur-containing hydrocarbons: Implications for the biodesulphurisation of oils and coals. *Journal of Chemical Technology and Biotechnology (Oxford, Oxfordshire)*, *67*(2), 109–123. doi:10.1002/(SICI)1097-4660(199610)67:2<109::AID-JCTB532>3.0.CO;2-Y

Silva, C. R., Silva, M. L. C., Kamida, H. M., Goes-Neto, A., & Koblitz, M. G. B. (2014). Lytic enzyme production optimization using low-cost substrates and its application in the clarification of xanthan gum culture broth. *Food Science & Nutrition*, *2*(4), 299–307. doi:10.1002/fsn3.87 PMID:25473487

Silva, T. P., Paixão, S. M., & Alves, L. (2016). Ability of Gordonia alkanivorans strain 1B for high added value carotenoids production. *RSC Advances*, *6*, 58055–58063.

Silva, T. P., Paixão, S. M., Roseiro, J. C., & Alves, L. (2015). Jerusalem artichoke as low-cost fructose-rich feedstock for fossil fuels desulfurization by a fructophilic bacterium. *Journal of Applied Microbiology*, *118*(3), 609–618. doi:10.1111/jam.12721 PMID:25494982

Silva, T. P., Paixão, S. M., Teixeira, A. V., Roseiro, J. C., & Alves, L. (2013). Optimization of low sulfur carob pulp liquor as carbon source for fossil fuels biodesulfurization. *Journal of Chemical Technology and Biotechnology (Oxford, Oxfordshire)*, *88*(5), 919–923. doi:10.1002/jctb.3921

Singh, A., & Agrawal, M. (2007). Acid rain and its ecological consequences. *Journal of Environmental Biology*, *29*(1), 15. PMID:18831326

Singh, A., Singh, B., & Ward, O. (2012). Potential applications of bioprocess technology in petroleum industry. *Biodegradation*, *23*(6), 865–880. doi:10.100710532-012-9577-2 PMID:22829348

Sing, K. S. W. (1982). Reporting Physiosorption Data Gas/Solid systems. *Pure and Applied Chemistry*, *54*(11), 2201–2218. doi:10.1351/pac198254112201

Sing, K. S. W., Everett, D. H., Haul, R. A., Moscou, L., Pierotti, R. A., Rouquerol, J., & Siemieniewska, T. (1985). Reporting physisorption data for gas/solid systems with special reference to the determination of surface area and porosity. *Pure and Applied Chemistry*, *57*(4), 603–619. doi:10.1351/pac198557040603

Si, X., Cheng, S., Lu, Y., Gao, G., & He, M. Y. (2008). Oxidative desulfurization of model oil over Au/Ti-MWW. *Catalysis Letters*, *122*(3-4), 321–324. doi:10.100710562-007-9380-6

Soleimani, M. M., Bassi, A., & Margaritis, A. (2007). Biodesulfurization of refractory organic sulfur compounds in fossil fuels. *Biotechnology Advances*, *25*(6), 570–596. doi:10.1016/j.biotechadv.2007.07.003 PMID:17716849

Song, C. (1999). Sulfur-resistant noble metal catalysts based on shape-selective exclusion and hydrogen spillover. In *Shape-selective catalysis, chemicals synthesis and hydrocarbon processing* (ACS Symposium Series 738). American Chemical Society.

Song, C., Turaga, U. T., & Ma, X. (2006). Desulfurization. Encyclopedia of Chemical Processing, 651-661. DOI: doi:10.1081/E-ECHP-120007732

Song, C. (1999). Designing sulfur-resistant, noble-metal hydrotreating catalysts. *Chemtech*, *29*(3), 26–30.

Song, C. (2003). An overview of new approaches to deep desulfurization for ultra-clean gasoline, diesel fuel and jet fuel. *Catalysis Today*, *86*(1-4), 211–263. doi:10.1016/S0920-5861(03)00412-7

Song, C. S., & Ma, X. L. (2004). Ultra-deep desulfurization of liquid hydrocarbon fuels: Chemistry and process. *International Journal of Green Energy*, *1*(2), 167–191. doi:10.1081/GE-120038751

Song, C., & Ma, X. (2003). New design approaches to ultra-clean diesel fuels by deep desulfurization and deep dearomatization. *Applied Catalysis B: Environmental*, *41*(1-2), 207–238. doi:10.1016/S0926-3373(02)00212-6

Song, C., & Reddy, K. M. (1999). Mesoporous molecular sieve MCM-41 supported Co-Mo catalyst for hydrodesulfurization of dibenzothiophene in distillate fuels. *Applied Catalysis A, General*, *176*(1), 1–10. doi:10.1016/S0926-860X(98)00230-0

Son, W. K., Youk, J. H., Lee, T. S., & Park, W. H. (2004). The effects of solution properties and polyelectrolyte on electrospinning of ultrafine poly(ethylene oxide) fibers. *Polymer*, *45*(9), 2959–2966. doi:10.1016/j.polymer.2004.03.006

Srivastav, A., & Srivastava, V. C. (2009). Adsorptive desulfurization by activated alumina. *Journal of Hazardous Materials*, *170*(2-3), 1133–1140. doi:10.1016/j.jhazmat.2009.05.088 PMID:19523762

Srivastava, V. C. (2012). An evaluation of desulfurization technologies for sulfur removal from liquid fuels. *RSC Advances*, *2*(3), 759–783. doi:10.1039/C1RA00309G

Stanislaus, A., Marafi, A., & Rana, M. S. (2010). Recent advances in the science and technology of ultra low sulfur diesel (ULSD) production. *Catalysis Today*, *153*(1-2), 1–68. doi:10.1016/j.cattod.2010.05.011

Stellenbosch nanofiber company (SNC). (2015). Retrieved from http://www.sncfibers.com/#!products/c1jm9

Stephanie, N. S., Goubert, R., Xiuling, Z., & Andrzej, W. (2010). Synthesis and characterization of carbon-supported transition metal oxide nanoparticles-Cobalt porphyrin as catalysts for electroreduction of oxygen in acids. *Electrochemistry Communications*, *12*(11), 1457–1461. doi:10.1016/j.elecom.2010.06.004

Stoner, D. L., Wey, J. E., Barrett, K. B., Jolley, J. G., Wright, R. B., & Dugan, P. R. (1990). Modification of water-soluble coal-derived products by dibenzothiophene-degrading microorganisms. *Applied and Environmental Microbiology*, *56*(9), 2667–2676. PMID:16348275

Subhashini, G., & Pant, K. K. (2005). Equilibrium Kinetics and breakthrough studies for adsorption of fluoride on activated alumina. *Separation and Purification Technology*, *42*(3), 265–271. doi:10.1016/j.seppur.2004.09.001

Sughrue, E. L., Khare, G. P., Bertus, B. J., & Johnson, M. M. (2001). *US Patent 6.254.766*. US Patent Office.

Sulfex™ – Desulfurization simplified. (2019). Retrieved from https://www.altpetrol.com/sulfex/

Suman, M., & Khaiwal Ravinda, N. R. (2007). Adsorption of chromium from aqueous solution by activated alumina and activated charcoal. *Bioresource Technology*, *98*(4), 954–957. doi:10.1016/j.biortech.2006.03.018 PMID:16725320

Sundaraman, R., Peduzzi, E., & Song, C. (2009). Catalytic cracking of oxidized sulfur compounds in gas oil stream. *Prepr Pap Am Chem Soc Div Fuel Chem*, *54*(2), 954–955.

Sushree, S. T., & Ashok, M. (2008). Enhanced adsorption capacity of activated alumina by impregnation with alum for removal of As (V) from water. *Chemical Engineering Journal*, *138*(1-3), 179–186. doi:10.1016/j.cej.2007.06.028

Svoboda, K., Lin, W., Hannes, J., Korbee, R., & van den Beek, C. M. (1994). Low temperature flue gas desulfurization by alumina-CaO regenerable sorbents. *Fuel*, *73*(7), 1144–1150. doi:10.1016/0016-2361(94)90251-8

Ta-Chen, L., Chang, J.-S., & Young, C.-C. (2008). Exopolysaccharides produced by *Gordonia alkanivorans* enhance bacterial degradation activity for diesel. *Biotechnology Letters*, *30*(7), 1201–1206. doi:10.100710529-008-9667-8 PMID:18286235

Tajudeen, O. A. (2011). Mophology and composition of nanocrystalline stabilized zirconia using SEM-EDS system. *Leonardo. Journal of Science*, (19): 81–92.

Takada, M., Nomura, N., Okada, H., Nakajima-Kambe, T., Nakahara, T., & Uchiyama, H. (2005). De-repression and comparison of oil-water separation activity of the dibenzothiophene desulfurizing bacterium, *Mycobacterium* sp. G3. *Biotechnology Letters*, *27*(12), 871–874. doi:10.100710529-005-6721-7 PMID:16086250

Talvy, S., Cockx, A., & Line, A. (2005). Global modelling of a gas–liquid–solid airlift reactor. *Chemical Engineering Science*, *60*(22), 5991–6003. doi:10.1016/j.ces.2005.04.067

Tam, P. S., & Kittrell, J. R. (1984). *Process for purifying hydrocarbonaceous oils*. U.S. Patent 4, 485, 007.

Tam, P. S., Kittrell, J. R., & Eldridge, J. W. (1990). Desulfurization of fuel oil by oxidation and extraction. 1. Enhancement of extraction oil yield. *Industrial & Engineering Chemistry Research*, *29*(3), 321–324. doi:10.1021/ie00099a002

Tam, P. S., Kittrell, J. R., & Eldridge, J. W. (1990). Desulfurization of fuel oil by oxidation and extraction. 2. Kinetic modeling of oxidation reaction. *Industrial & Engineering Chemistry Research*, *29*(3), 324–329. doi:10.1021/ie00099a003

Tanabe, K. (1985). Surface and catalytic properties of $Zr0_2$. *Materials Chemistry and Physics*, *13*(3-4), 347–364. doi:10.1016/0254-0584(85)90064-1

Tanaka, Y., Yoshikawa, O., Maruhashi, K., & Kurane, R. (2002). The cbs mutant strain of *Rhodococcus erythropolis* KA2-5-1 expresses high levels of Dsz enzymes in the presence of sulfate. *Archives of Microbiology*, *178*(5), 351–357. doi:10.100700203-002-0466-7 PMID:12375103

Tang, K., & Hong, X. (2014). A characterization of bacterium for fcc diesel oil desulfurization. *Petroleum Science and Technology*, *32*(13), 1545–1552. doi:10.1080/10916466.2012.689795

Tang, M., Zhou, L., Du, M., Lyu, Z., Wen, X.-D., Li, X., & Ge, H. (2015). A novel reactive adsorption desulfurization Ni/MnO adsorbent and its hydrodesulfurization ability compared with Ni/ZnO. *Catalysis Communications*, *61*, 37–40. doi:10.1016/j.catcom.2014.11.005

Tawara, K., Nishimura, T., & Iwanami, H. (2000). Ultra-deep hydrodesulfurization of kerosene for fuel cell system (Part 2) regeneration of sulfurpoisoned nickel catalyst in hydrogen and finding of autoregenerative nickel catalyst. Sekiyu Gakkaishi. *Journal of the Japan Petroleum Institute*, *43*(2), 114–120. doi:10.1627/jpi1958.43.114

Tawara, K., Nishimura, T., Iwanami, H., Nishimoto, T., & Hasuike, T. (2001). New hydrodesulfurization catalyst for petroleum-feed fuel cell vehicles and cogenerations. *Industrial & Engineering Chemistry Research*, *40*(10), 2367–2370. doi:10.1021/ie000453c

Tawara, K., Nishimura, T., Iwanami, H., Nishimoto, T., & Hasuike, T. (2001b). Ultra-deep hydrodesulfurization of kerosene for fuel cell system (Part 3) Development and evaluation of Ni/ZnO catalyst. *Sekiyu Gakkaishi*, *44*(1), 43–51. doi:10.1627/jpi1958.44.43

Tawfik, A. S., Gaddafi, I. D., & Taye, D. S. (2015). Nanocomposites and Hybrid Materials for Adsorptive Desulfurization. Advances in Chemical and Materials Engineering, 129-153.

Tella, A. C., Owalude, S. O., Olatunji, S. J., Oloyede, S. O., Ogunlaja, A. S., & Bourne, S. A. (2018). Synthesis, crystal structure and desulfurization properties of zig-zag 1D coordination polymer of copper(II) containing 4-methoxybenzoic acid ligand. *Journal of Sulfur Chemistry*, *39*(6), 588–606. doi:10.1080/17415993.2018.1489808

Teo, W. E., & Ramakrishna, S. (2005). Electrospun fibre bundle made of aligned nanofibres over two fixed points. *Nanotechnology*, *16*(9), 1878–1884. doi:10.1088/0957-4484/16/9/077

Teo, W. E., & Ramakrishna, S. (2006). A review on electrospinning design and nanofibre assemblies. *Nanotechnology*, *17*(14), 89–106. doi:10.1088/0957-4484/17/14/R01 PMID:19661572

Terrones, M. (2003). Science and technology of twenty-five century: Synthesis, properties and applications. *Annual Review of Materials Research*, *33*, 419. doi:10.1146/annurev.matsci.33.012802.100255

Thanasilp, S., Schwank, J. W., Meeyoo, V., Pengpanich, S., & Hunsom, M. (2013). Preparation of supported POM catalysts for liquid phase oxydehydration of glycerol to acrylic acid. *Journal of Molecular Catalysis A Chemical*, *380*, 49–56. doi:10.1016/j.molcata.2013.09.023

The International Council On Clean Transportation (ICCT). (2011). *Energy Econ. Appl. Optim.*, 1–38.

Theron, S. A., Yarin, A. L., Zussman, E., & Kroll, E. (2005). Multiple jets in electrospinning: Experiment and modelling. *Polymer*, *46*(9), 2889–2899. doi:10.1016/j.polymer.2005.01.054

Thess, A., Smalley, R. E., Roland, L., Nikolaev, P., Dai, H., Petit, P., ... Fischer, J. E. (1996). Crystalline ropes of metallic carbon nanotubes. *Science*, *273*(5274), 483–487. doi:10.1126cience.273.5274.483 PMID:8662534

Thitiwan, N. (2013). Adsorptive desulfurization of dibenzothiophene by sewage sludge-derived activated carbon. *Chemical Engineering Journal*, *228*, 263–271. doi:10.1016/j.cej.2013.04.067

Tian, W.-H., Sun, L.-B., Song, X. L., Liu, X.-Q., Yin, Y., & He, G.-S. (2010). Adsorptive Desulfurization by Copper Species within Confined Space. *Langmuir*, *26*(22), 17398–17404. doi:10.1021/la101856d PMID:20882950

Tony, S. S., & Pant, K. K. (2004). Equilibrium, Kinetics and thermodynamics studies for adsorption of As (III) on activated alumina. *Separation and Purification Technology*, *36*(2), 139–147. doi:10.1016/S1383-5866(03)00209-0

Topsoe, H., Clausen, B. S., & Massoth, F. E. (1996). Hydrotreating catalysis. In *Science and Technology* (p. 310). Berlin: Springer-Verlag.

Torres-García, E., Galano, A., & Rodriguez-Gattorno, G. (2011). Oxidative desulfurization (ODS) of organosulfur compounds catalysed by peroxo metallate complexes of WOx–ZrO_2: Thermochemical, structural, and reactivity indexes analyses. *Journal of Catalysis*, *282*(1), 201–208. doi:10.1016/j.jcat.2011.06.010

Triantafyllidis, K. S., & Deliyanni, E. A. (2014). Desulfurization of diesel fuels: Adsorption of 4, 6-DMDBT on different origin and surface chemistry nanoporous activated carbons. *Chemical Engineering Journal, 236*, 406–414. doi:10.1016/j.cej.2013.09.099

Tsuru, T. (2001). Inorganic porous membranes for liquid phase separation. *Separation and Purification Reviews, 30*(2), 191–220. doi:10.1081/SPM-100108159

Tuli, D., & Kumar, A. (2008). Non-conventional technologies for fuel desulphurization. In *19th World Petroleum Congress*. World Petroleum Congress.

Turk, B. S., & Gupta, R. P. (2001). RTI's TReND Process for deep desulfurization of naphtha, Am. Chem. Soc. *Div. Fuel Chem. Prepr., 46*, 392.

Unger, K. K. (1979). *Porous Silica, its properties and use as support in column chromatography*. Elsevier.

United States Environmental Protection Agency (USEPA). (2006). *Introduction of cleaner-burning diesel fuel enables advanced pollution control for cars, trucks and busses*. EPA420-F-06-064. Retrieved from www.epa.gov/otaq/highway-diesel/regs/420f06064.pdf

United States Environmental Protection Agency (USEPA). (2014b). *EPA sets tier 3 motor vehicle emission and fuel standards, strengthens clean cars program*. EPA-420-F-14-010. Retrieved from www.epa.gov/otaq/documents/tier3/420f14010.pdf

Unites States Environmental Protection Agency (USEPA). (2014). *EPA sets tier 3 motor vehicle emission and fuel standards*. EPA-420-F-14-009. Retrieved from www.epa.gov/otaq/documents/tier3/420f14009.pdf

Unites States Environmental Protection Agency (USEPA). (2014a). *EPA sets tier 3 motor vehicle emission and fuel standards*. EPA-420-F-14-009. Retrieved from www.epa.gov/otaq/documents/tier3/420f14009.pdf

Urum, K., & Pekdemir, T. (2004). Evaluation of biosurfactants for crude oil contaminated soil washing. *Chemosphere, 57*(9), 1139–1150. doi:10.1016/j.chemosphere.2004.07.048 PMID:15504473

Van de Voorde, B., Bueken, B., Denayer, J., & De Vos, D. (2014). Adsorptive separation on metal-organic frameworks in the liquid phase. *Chemical Society Reviews, 43*(16), 5766–5788. doi:10.1039/C4CS00006D PMID:24647892

Vansant, E. F., Vander Voort, P., & Vrancken, K. C. (1995). *Characterization and chemical modification of the silica surface*. Elsevier.

Varesano, A., Carletto, R. A., & Mazzuchetti, G. (2009). Experimental investigations on the electrospinning process. *Journal of Materials Processing Technology, 209*(11), 5178–5185. doi:10.1016/j.jmatprotec.2009.03.003

Vazquez-Duhalt, R., Torres, E., Valderrama, B., & Le Borgne, S. (2002). Will biochemical catalysis impact the petroleum refining industry? *Energy & Fuels, 16*(5), 1239–1250. doi:10.1021/ef020038s

Velu, S., Ma, X., & Song, C. (2003). Selective adsorption for removing sulfur from jet fuel over zeolite based adsorbents. *Industrial & Engineering Chemistry Research*, *42*(21), 5293–5304. doi:10.1021/ie020995p

Venil, C. K., Aruldass, C. A., Dufossé, L., Zakaria, Z. A., & Ahmad, W. A. (2014). Current perspective on bacterial pigments: Emerging sustainable compounds with coloring and biological properties for the industry – an incisive evaluation. *RSC Advances*, *4*, 39523–39529.

Venturiello, C., Alneri, E., & Ricci, M. (1983). A new effective catalytic system for epoxidation of olefins by hydrogen peroxide under phase-transfer conditions. *The Journal of Organic Chemistry*, *48*(21), 3831–3833. doi:10.1021/jo00169a052

Walmsley, R. S., Hlangothi, P., Litwinski, C., Torto, N., Nyokong, T., & Tshentu, Z. R. (2012). Catalytic oxidation of thioanisole using oxovanadium(IV)-functionalized electrospun polybenzimidazole nanofibers. *Journal of Applied Polymer Science*, *127*(6), 4719–4725. doi:10.1002/app.38067

Walmsley, R. S., Ogunlaja, A. S., Coombes, M. J., Chidawanyika, W., Litwinski, C., Torto, N., ... Tshentu, Z. R. (2012). Imidazole-functionalized polymer microspheres and fibers – useful materials for immobilization of oxovanadium(IV) catalysts. *Journal of Materials Chemistry*, *22*(12), 5792–5800. doi:10.1039/c2jm15485d

Wang, W. (2018). *Study on the methanation and desulfurization behaviors on Ni/ZnO* (PhD dissertation). University of Chinese Academy of Sciences.

Wang, B., Zhu, J., & Ma, H. (2009). Desulfurization from thiophene by SO $^{(4)(2-)}$/ZrO_2 catalytic oxidation at room temperature & atmospheric pressure. *Journal of Hazardous Materials*, *164*(1), 256–264. doi:10.1016/j.jhazmat.2008.08.003 PMID:18799257

Wang, D. H., Qian, E. W. H., Amano, H., Okata, K., Ishihara, A., & Kabe, T. (2003). Oxidative desulfurization of fuel oil - Part I. Oxidation of dibenzothiophenes using *tert*-butylhydroperoxide. *Applied Catalysis A, General*, *253*(1), 91–99. doi:10.1016/S0926-860X(03)00528-3

Wang, M. Z., & Ruan, Y. J. (2010). Study on process conditions of adsorption desulfurization reaction of FCC gasoline. *Petroleum Refinery Engineering*, *40*(9), 5–10.

Wang, M., Li, W., Shi, Y., Wang, D., & Feng, H. (2006). Effects of surfactant on biodesulfurization by *Corynebacterium* sp. ZD-1 in the presence of organic phase. *Journal of Zhejiang University. Science A*, *7*(S2), 371–375. doi:10.1631/jzus.2006.AS0371

Wang, P., & Krawiec, S. (1996). Kinetic analyses of desulfurization of dibenzothiophene by *Rhodococcus erythropolis* in batch and fed-batch cultures. *Applied and Environmental Microbiology*, *62*(5), 1670–1675. PMID:16535315

Wang, R., Yu, F., Zhang, G., & Zhao, H. (2010). Performance evaluation of the carbon nanotubes supported Cs2.5H0.5PW12O40 as efficient and recoverable catalyst for the oxidative removal of dibenzothiophene. *Catalysis Today*, *150*(1–2), 37–41. doi:10.1016/j.cattod.2009.10.001

Wang, R., Zhang, G., & Zhao, H. (2010). Polyoxometalate as effective catalyst for the deep desulfurization of diesel oil. *Catalysis Today*, *149*(1–2), 117–121. doi:10.1016/j.cattod.2009.03.011

Wang, W., Li, X., Zhang, Y., Zhang, R., Ge, H., Bi, J., & Tang, M. (2017). Strong metal–support interactions between Ni and ZnO particles and their effect on the methanation performance of Ni/ZnO. *Catalysis Science & Technology*, *7*(19), 4413–4421. doi:10.1039/C7CY01119A

Wang, X., Chen, W., & Song, Y.-F. (2014). Directional Self-Assembly of Exfoliated Layered Europium Hydroxide Nanosheets and Na9EuW10O36·32H2O for Application in Desulfurization. *European Journal of Inorganic Chemistry*, *2014*(17), 2779–2786. doi:10.1002/ejic.201400122

Wang, X.-S., Huang, Y.-B., Lin, Z.-J., & Cao, R. (2014). Phosphotungstic acid encapsulated in the mesocages of amine-functionalized metal-organic frameworks for catalytic oxidative desulfurization. *Dalton Transactions (Cambridge, England)*, *43*(31), 11950–11958. doi:10.1039/C4DT01043D PMID:24967570

Wang, Y. H., Yang, F. H., Yang, R. T., Heinzel, J. M., & Nickens, A. D. (2006). Desulfurization of high sulfur jet fuel by π-Complexation with copper and palladium halide sorbents. *Industrial & Engineering Chemistry Research*, *45*(22), 7649–7655. doi:10.1021/ie060922w

Wang, Y. H., Yang, R. T., & Heinzel, J. M. (2008). Desulfurization of jet fuel by π-complexation with metal halides supported on MCM-41 and SBA-15 mesoporous materials. *Chemical Engineering Science*, *63*, 356–365. doi:10.1016/j.ces.2007.09.002

Wang, Y., & Yang, R. T. (2007). Desulfurization of liquid fuels by adsorption on carbon-based sorbents and ultrasound-assisted sorbent regeneration. *Langmuir*, *23*(7), 3825–3831. doi:10.1021/la063364z PMID:17315903

Wang, Z., Wang, D., Li, Q., Li, W., Tang, H., & Xing, J. (2011). Enhanced biodesulfurization by expression of dibenzothiophene uptake genes in *Rhodococcus erythropolis. World Journal of Microbiology & Biotechnology*, *27*(9), 1965–1970. doi:10.100711274-011-0656-z

Wannatong, L., Sirivat, A., & Supaphol, P. (2004). Effects of solvents on electrospun polymeric fibers: Preliminary study on polystyrene. *Polymer International*, *53*(11), 1851–1859. doi:10.1002/pi.1599

Washburn, E. W. (1921). The dynamics of capillary flow. *Physical Review*, *17*(3), 273–283. doi:10.1103/PhysRev.17.273

Wee, L. H., Bonino, F., Lamberti, C., Bordiga, S., & Martens, J. A. (2014). Cr-MIL-101 encapsulated Keggin phosphotungstic acid as active nanomaterial for catalysing the alcoholysis of styrene oxide. *Green Chemistry*, *16*(3), 1351–1357. doi:10.1039/C3GC41988F

Wen, Y., Wang, G., Wang, Q., Xu, C., & Gao, J. (2012). Regeneration characteristics and kinetics of Ni/ZnO-SiO_2-Al_2O_3 adsorbent for reactive adsorption desulfurization. *Industrial & Engineering Chemistry Research*, *51*(10), 3939–3950. doi:10.1021/ie202730w

Werther, J., Saenger, M., Hartge, E.-U., Ogada, T., & Siagi, Z. (2000). Combustion of agricultural residues. *Progress in Energy and Combustion Science*, *26*(1), 1–27. doi:10.1016/S0360-1285(99)00005-2

Whitehurst, D. D., Isoda, T., & Mochida, I. (1998). Present state of the art and future challenges in the hydrodesulfurization of polyaromatic sulfur compounds. *Advances in Catalysis*, *42*, 345–471. doi:10.1016/S0360-0564(08)60631-8

Wildgoose, G. G., Banks, C. E., & Compton, R. G. (1996). Metal Nanoparticles and Related Materials Supported on Carbon Nanotubes: Methods and Applications. *Small*, *2*(2), 182–193. doi:10.1002mll.200500324 PMID:17193018

Willaert, R. G., & Baron, G. V. (1996). Gel entrapment and micro-encapsulation : Methods, applications and engineering principles. *Reviews in Chemical Engineering*, *12*(1), 1–205. doi:10.1515/REVCE.1996.12.1-2.1

Williams, G., Seger, B., & Kamat, P. V. (2008). TiO_2-graphene nanocomposites. UV-assisted photocatalytic reduction of graphene oxide. *ACS Nano*, *2*(7), 1487–1491. doi:10.1021/nn800251f PMID:19206319

Wondrous World of Carbon Nanotubes. (n.d.). *A Review of Current Carbon Nanotube Technologies*. Eindhoven University of Technology.

Wood, A. E., Sheely, C., & Trusty, A. W. (1925). Corrosion Effect of Naphtha Solutions of Sulfur and Sulfur Compounds. *Industrial & Engineering Chemistry*, *17*(8), 798–802. doi:10.1021/ie50188a008

Wu, B., Kuang, Y., Zhang, X., & Chen, J. (2011). Noble metal nanoparticles /carbon nanotubes nanohybrids: Synthesis and applications. *Nano Today*, *6*(1), 75–90. doi:10.1016/j.nantod.2010.12.008

Wulff, G. (1993). The role of binding-site interactions in the molecular imprinting of polymers. *Trends in Biotechnology*, *11*(3), 85–87. doi:10.1016/0167-7799(93)90056-F PMID:7763512

Wu, S., Lin, J., & Chan, S. I. (1994). Oxidation of dibenzothiophene catalyzed by heme-containing enzymes encapsulated in sol-gel glass. *Applied Biochemistry and Biotechnology*, *47*(1), 11–20. doi:10.1007/BF02788671 PMID:8203869

Xiao, J., Song, C., Ma, X., & Li, Z. (2012). Effects of Aromatics, Diesel Additives, Nitrogen Compounds, and Moisture on Adsorptive Desulfurization of Diesel Fuel over Activated Carbon. *Industrial & Engineering Chemistry Research*, *51*(8), 3436–3443. doi:10.1021/ie202440t

Xiao, J., Wu, L., Wu, Y., Liu, B., Dai, L., Li, Z., ... Xi, H. (2014). Effect of gasoline composition on oxidative desulfurization using a phosphotungstic acid/activated carbon catalyst with hydrogen peroxide. *Applied Energy*, *113*(0), 78–85. doi:10.1016/j.apenergy.2013.06.047

Xie, X., Ju, L., Feng, X., Sun, Y., Zhou, R., Liu, K., ... Jiang, K. (2009). Controlled fabrication of high-quality carbon nanoscrolls from monolayer graphene. *Nano Letters*, *9*(7), 2565–2570. doi:10.1021/nl900677y PMID:19499895

Xie, Y. (2006). Photoelectrochemical reactivity of a hybrid electrode composed of polyoxophosphotungstate encapsulated in titania nanotubes. *Advanced Functional Materials*, *16*(14), 1823–1831. doi:10.1002/adfm.200500695

Xin, C., & Wang, S. (2018). Desulfurization by MOFs as Sorbents for Thiophene Sulfides. *Earth and Environmental Science*, *108*, 042035.

Xinhai, X., Zhang, S., Li, P., & Shen, Y. (2014). Adsorptive desulfurization of liquid Jet-A fuel at ambient conditions with an improved adsorbent for on-board fuel treatment for SOFC applications. *Fuel Processing Technology*, *124*, 140–146. doi:10.1016/j.fuproc.2014.03.001

Xue, H., Lin, H., & Zheng, Y. (2014). Regeneration methods to restore carbon adsorptive capacity of dibenzothiophene and neutral nitrogen heteroaromatic compounds. *Chemical Engineering Journal*, *243*, 315–325. doi:10.1016/j.cej.2013.12.074

Xu, J., Li, H., Wang, S., Luo, F., Liu, Y., Wang, X., & Jiang, Z. (2014). Ultra-deep desulfurization via reactive adsorption on peroxophosphomolybdate/agarose hybrids. *Chemosphere*, *111*(0), 631–637. doi:10.1016/j.chemosphere.2014.05.033 PMID:24997975

Xu, J., Zhao, S., Ji, Y., & Song, Y.-F. (2013). Deep Desulfurization by Amphiphilic Lanthanide-Containing Polyoxometalates in Ionic-Liquid Emulsion Systems under Mild Conditions. *Chemistry (Weinheim an der Bergstrasse, Germany)*, *19*(2), 709–715. doi:10.1002/chem.201202595 PMID:23161825

Yamase, T. (2005). Anti-tumor, -viral, and -bacterial activities of polyoxometalates for realizing an inorganic drug. *Journal of Materials Chemistry*, *15*(45), 4773-4782.

Yamase, T. (1998). Photo- and Electrochromism of Polyoxometalates and Related Materials. *Chemical Reviews*, *98*(1), 307–326. doi:10.1021/cr9604043 PMID:11851508

Yang, F. H., Hernandez-Maldonado, A. J., & Yang, R. T. (2004). Selective adsorptipion of organosulfur compounds from transportation fuels by π-complexation. *Separation Science and Technology*, *39*(8), 1717–1731. doi:10.1081/SS-120035937

Yang, J., Hu, Y., Zhao, D., Wang, S., Lau, P. C. K., & Marison, I. W. (2007). Two-layer continuous-process design for the biodesulfurization of diesel oils under bacterial growth conditions. *Biochemical Engineering Journal*, *37*(2), 212–218. doi:10.1016/j.bej.2007.04.012

Yang, R. T., & Arturo, J. (2003). Desulfurization of Transportation Fuels with Zeolites under Ambient Conditions. *Science*, *301*(5629), 79–81. doi:10.1126cience.1085088 PMID:12843388

Yang, W., Ratinac, K. R., Ringer, S. P., Thordarson, P., Gooding, J. J., & Braet, F. (2010). Carbon nanomaterials in biosensors: Should you use nanotubes or graphene? *Angewandte Chemie International Edition*, *49*(12), 2114–2138. doi:10.1002/anie.200903463 PMID:20187048

Yang, X., Cao, C., Klabunde, K. J., Hohn, K. L., & Erickson, L. E. (2007). Adsorptive desulfurization with xerogel-derived zinc-based nanocrystalline aluminum oxide. *Industrial & Engineering Chemistry Research*, *46*(14), 4819–4823. doi:10.1021/ie0701928

Yang, X., Erickson, L. E., Hohn, K. L., Jeevanandam, P., & Klabunde, K. J. (2006). Sol-Gell Cu-Al_2O_3 Adsorbents for Selective Adsorption of Thiophene out of Hydro carbon. *Industrial & Engineering Chemistry Research*, *45*(18), 6169–6174. doi:10.1021/ie060559t

Yan, X. M., Mei, P., Xiong, L., Gao, L., Yang, Q., & Gong, L. (2013). Mesoporous titania-silica-polyoxometalate nanocomposite materials for catalytic oxidation desulfurization of fuel oil. *Catalysis Science & Technology*, *3*(8), 1985–1992. doi:10.1039/c3cy20732c

Yan, X. M., Mei, Z., Mei, P., & Yang, Q. (2014). Self-assembled HPW/silica-alumina mesoporous nanocomposite as catalysts for oxidative desulfurization of fuel oil. *Journal of Porous Materials*, *21*(5), 729–737. doi:10.100710934-014-9819-2

Yan, X., Su, G., & Xiong, L. (2009). Oxidative desulfurization of diesel oil over Ag-modified mesoporous HPW/SiO2 catalyst. *Journal of Fuel Chemistry and Technology*, *37*(3), 318–323. doi:10.1016/S1872-5813(09)60022-0

Yazu, K., Furuya, T., Miki, K., & Ukegawa, K. (2003). Tungstophosphoric acid-catalysed oxidative desulfurization of light oil with hydrogen peroxide in a light oil/acetic acid biphasic system. *Chemistry Letters*, *32*(10), 920–921. doi:10.1246/cl.2003.920

Yazu, K., Yamamoto, Y., Furuya, T., Miki, K., & Ukegawa, K. (2001). Oxidation of dibenzothiophenes in an organic biphasic system and its application to oxidative desulfurization of light oil. *Energy & Fuels*, *15*(6), 1535–1536. doi:10.1021/ef0101412

Yoshida, H., Shimizu, T., Uchiyama, T., Kohno, H., Homma, Y., & Takeda, S. (2009). Atomic-scaleanalysis on the role of molybdenum in iron-catalyzed carbon nanotube growth. *Nano Letters*, *9*(11), 3810–3815. doi:10.1021/nl9019903 PMID:19731921

Yoshikawa, O., Ishii, Y., Koizumi, K.-I., Ohshiro, T., Izumi, Y., & Maruhashi, K. (2002). Enhancement and stabilization of desulfurization activity of *Rhodococcus erythropolis* KA2-5-1 by feeding ethanol and sulfur components. *Journal of Bioscience and Bioengineering*, *94*(5), 447–452. doi:10.1016/S1389-1723(02)80223-7 PMID:16233332

Young, K. (2002). The adsorption of fluoride ion from aqueous solution by activated alumina. *Water, Air, and Soil Pollution*, *133*(1/4), 349–360. doi:10.1023/A:1012929900113

Yuan, L., Saito, K., Hu, W., & Chen, Z. (2001). Ethylene flame synthesis of well-aligned multi-walled carbon nanotubes. *Chemical Physics Letters*, *346*(1-2), 23–28. doi:10.1016/S0009-2614(01)00959-9

Yuan, X. Y., Zhang, Y. Y., Dong, C. H., & Sheng, J. (2004). Morphology of ultrafine polysulfone fibers prepared by electrospinning. *Polymer International*, *53*(11), 1704–1710. doi:10.1002/pi.1538

Yudasaka, M., Komatsu, T., Ichihashi, T., & Iijima, S. (1997). Single wall carbon nanotube formation by laser ablation using double-targets of carbon and metal. *Chemical Physics Letters*, *278*(1-3), 102–106. doi:10.1016/S0009-2614(97)00952-4

Yu, F. L., & Wang, R. (2013). Deep Oxidative Desulfurization of Dibenzothiophene in Simulated Oil and Real Diesel Using Heteropolyanion-Substituted Hydrotalcite-Like Compounds as Catalysts. *Molecules (Basel, Switzerland)*, *18*(11), 13691–13704. doi:10.3390/molecules181113691 PMID:24284486

Yu, F., Zheng, P. Q., Long, Y. X., Ren, Y. P., Kong, X. J., Long, L. S., ... Zheng, L. S. (2010). Polyoxometalate-based metal-organic frameworks as heterogeneous catalysts for selective oxidation of ethylbenzene. *European Journal of Inorganic Chemistry*, *2010*(28), 4526–4531. doi:10.1002/ejic.201000491

Yuhui, J., Li, A., & Sandra, G. (2009). Amorphous Silica nanohybrids: Synthesis, properties & Applications. *Coordination Chemistry Reviews*, *253*, 2998–3014.

Yujun, W., Zhang, C., Bi, S., & Luo, G. (2010). Preparation of ZnO nanoparticles using the direct precipitation method in a membrane dispersion micro-structured reactor. *Powder Technology*, *202*(1-3), 130–136. doi:10.1016/j.powtec.2010.04.027

Zahedi, P., Rezaeian, I., Ranaei-Siadat, S. O., Jafari, S. H., & Supaphol, P. (2010). A review on wound dressings with an emphasis on electrospun nanofibrous polymeric bandages. *Polymers for Advanced Technologies*, *21*, 77–95.

Zaki, T., & Nermen, H. (2013). Characterization and application of nano-alumina sorbents for desulfurization and dearomatization of Suez crude petrolatum. *Fuel Processing Technology*, *106*, 625–630. doi:10.1016/j.fuproc.2012.09.055

Zaki, T., Riad, M., Saad, L., & Mikhail, S. (2005). Selected oxide materials for sulfur removal. *Chemical Engineering Journal*, *113*(1), 41–46. doi:10.1016/j.cej.2005.08.004

Zannikos, F., Lois, E., & Stournas, S. (1995). Desulfurization of petroleum fractions by oxidation and solvent extraction. *Fuel Processing Technology*, *42*(1), 35–45. doi:10.1016/0378-3820(94)00104-2

Zeki, B. (2013). Food Process Engineering and Technology. In Adsorption and Ion Exchange (2nd ed.). Food Science and Technology.

Zhang, Y., Yang, Y., Han, H., Yang, M., Wang, L., Zhang, Y., Jiang, Z., & Li, C. (2012). Ultra-deep desulfurization via reactive adsorption on Ni/ZnO: The effect of ZnO particle size on the adsorption performance. *Applied Catalysis B: Environmental*, *119– 120*, 13– 19.

Zhang, C. X., Yuan, X. Y., Wu, L. L., Han, Y., & Sheng, J. (2005). Study on morphology of electrospun poly(vinyl alcohol) mats. *European Polymer Journal*, *41*(3), 423–432. doi:10.1016/j.eurpolymj.2004.10.027

Zhang, Q., Huang, J. Q., Qian, W. Z., Zhang, Y. Y., & Wei, F. (2013). The Road for Nanomaterials Industry: A Review of Carbon Nanotube Production, Post-Treatment, and Bulk Applications for Composites and Energy Storage. *Small*, *9*(8), 1237–1265. doi:10.1002mll.201203252 PMID:23580370

Zhang, Q., Yu, Z., Du, P., & Su, C. (2010). Carbon nanomaterials used as conductive additives in lithium ion batteries. *Recent Patents on Nanotechnology*, *4*(2), 100–110. doi:10.2174/187221010791208803 PMID:20415660

Zhang, S., Zhang, Y., Huang, S., Wang, P., & Tian, H. (2012). Mechanistic investigations on the adsorption of thiophene over Zn3NiO4 bimetallic oxide cluster. *Applied Surface Science*, *258*(24), 10148–10153. doi:10.1016/j.apsusc.2012.06.096

Zhang, Y., Lü, H., Wang, L., Zhang, Y., Liu, P., Han, H., ... Li, C. (2010). The oxidation of benzothiophene using the Keggin-type lacunary polytungstophosphate as catalysts in emulsion. *Journal of Molecular Catalysis A Chemical*, *332*(1–2), 59–64. doi:10.1016/j.molcata.2010.08.021

Zhang, Y.-W., Prabhu, P., & Lee, J.-K. (2010). Alginate immobilization of recombinant *Escherichia coli* whole cells harboring L-arabinose isomerase for L-ribulose production. *Bioprocess and Biosystems Engineering*, *33*(6), 741–748. doi:10.100700449-009-0397-7 PMID:19946709

Zhang, Y., Yang, Y., Hana, H., Yang, M., Wang, L., Zhang, Y., ... Li, C. (2012a). Ultra-deep desulfurization via reactive adsorption on Ni/ZnO: The effect of ZnO particle size on the adsorption performance. *Applied Catalysis B: Environmental*, *119-120*, 13–19. doi:10.1016/j.apcatb.2012.02.004

Zhang, Y., Yang, Y., Lin, F., Yang, M., Liu, T., Jiang, Z., & Li, C. (2013). Improvement of adsorptive desulfurization performance of Ni/ZnO adsorbent by doping with Mn additive. *Chinese Journal of Catalysis*, *34*(1), 140–145. doi:10.1016/S1872-2067(11)60513-5

Zhang, Z. Y., Shi, T. B., Jia, C. Z., Ji, W. J., Chen, Y., & He, M. Y. (2008). Adsorptive removal of aromatic organosulfur compounds over the modified Na-Y zeolites. *Applied Catalysis B: Environmental*, *82*(1-2), 1–10. doi:10.1016/j.apcatb.2008.01.006

Zheng, H., Sun, Z., Chen, X., Zhao, Q., Wang, X., & Jiang, Z. (2013). A micro reaction-controlled phase-transfer catalyst for oxidative desulfurization based on polyoxometalate modified silica. *Applied Catalysis A, General*, *467*(0), 26–32. doi:10.1016/j.apcata.2013.06.026

Zhou, G., Sun, W., Jiang, W., & Zhou, H. (2013). Deep desulfurization of hydro-gasoline by adsorption in H_2 atmosphere over Ni/ZnO-SiO_2-Al_2O_3. *Petroleum Processing and Petrochemicals*, *44*(7), 17–21.

Zhou, H., Li, G., Wang, X., Jin, C., & Chen, Y. (2009). Preparation of a kind of mesoporous carbon and its performance in adsorptive desulfurization. *Journal of Natural Gas Chemistry*, *18*(3), 365–368. doi:10.1016/S1003-9953(08)60117-6

Zhou, W. W., Han, Z. Y., Wang, J. Y., Zhang, Y., Jin, Z., Sun, X., ... Li, Y. (2006). Copper catalyzing growth of single walled carbon nanotubes on substrates. *Nano Letters*, *6*(12), 2987–2990. doi:10.1021/nl061871v PMID:17163745

Zhou, X. R., Zhao, C. X., Yang, J. Z., & Zhang, S. F. (2007). Catalytic oxidation of dibenzothiophene using cyclohexanone peroxide. *Energy & Fuels*, *21*(1), 7–10. doi:10.1021/ef060441p

Zhou, Y., Chen, G., Long, Z., & Wang, J. (2014). Recent advances in polyoxometalate-based heterogeneous catalytic materials for liquid-phase organic transformations. *RSC Advances*, *4*(79), 42092–42113. doi:10.1039/C4RA05175K

Zhou, Y., Yang, D., Chen, X., Xu, Q., Lu, F., & Nie, J. (2008). Electrospun water-soluble carboxyethyl chitosan/poly(vinyl alcohol) nanofibrous membrane as potential wound dressing for skin regeneration. *Biomacromolecules*, *9*(1), 349–354. doi:10.1021/bm7009015 PMID:18067266

Zhou, Z., & Zhang, K. (2004). Development situation and prospect of oil fields in China. *Petroleum Exploration And Development*, *31*(1), 84–87.

Zhu, Q.-L., & Xu, Q. (2014). Metal-organic framework composites. *Chemical Society Reviews*, *43*(16), 5468-5512.

Zhu, L., Jia, X., Bian, H., Huo, T., Duan, Z., Xiang, Y., & Xia, D. (2018). Structure and adsorptive desulfurization performance of the composite material MOF-5@AC. *New Journal of Chemistry*, *42*(5), 3840–3850. doi:10.1039/C7NJ04192F

Zhu, W., Huang, W., Li, H., Zhang, M., Jiang, W., Chen, G., & Han, C. (2011). Polyoxometalate-based ionic liquids as catalysts for deep desulfurization of fuels. *Fuel Processing Technology*, *92*(10), 1842–1848. doi:10.1016/j.fuproc.2011.04.030

Zong, X. H., Kim, K., Fang, D. F., Ran, S. F., Hsiao, B. S., & Chu, B. (2002). Structure and process relationship of electrospun bioabsorbable nanofiber membranes. *Polymer*, *43*(16), 4403–4412. doi:10.1016/S0032-3861(02)00275-6

Zubair, Jeon, & Hwa Jhung. (2012). Oxidative desulfurization of benzothiophene and thiophene with WOx/ZrO2 catalysts: Effect of calcination temperature of catalysts. *Journal of Hazardous Materials*, (205–206), 216–221.

Zuo, W. W., Zhu, M. F., Yang, W., Yu, H., Chen, Y. M., & Zhang, Y. (2005). Experimental study on relationship between jet instability and formation of beaded fibers during electrospinning. *Polymer Engineering and Science*, *45*(5), 704–709. doi:10.1002/pen.20304

About the Contributors

Tawfik A. Saleh works, at present, as a faculty member in the Chemistry Department of King Fahd University of Petroleum and Minerals. He has published several papers in refereed journals and conference proceedings, authored and edited several books and book chapters, and he is a member of several local and international scientific societies. He has supervised many graduate students. His expertise includes, but is not limited to, synthesis, design and preparation of materials, nanomaterials, nanocomposites, hybrid materials and their applications. His expertise also includes materials characterization using various spectroscopic and electrochemical techniques.

* * *

Luís Alves is a researcher at the Bioenergy Unit from LNEG. PhD in Microbiology by the Faculty of Sciences - University of Lisbon, Portugal, he has experience in isolation and characterization of novel microorganisms and enzymes, application of fermentation techniques of SSF, SHF and CBP in batch or chemostat for the production of bioproducts (bioethanol, bioH2, lactic acid, carotenoids, biosurfactants, etc.) and biorefinery of fossil fuels. He has supervised several graduate, master and PhD thesis. He has participated in several national and international R&D projects as team member or principal investigator. He published 37 peer-reviewed papers in ISI journals, and presented several communications at national/international scientific meetings.

Bruno Arez has a degree in Cellular Biology and Biotechnology by the Faculty of Sciences- University of Lisbon. He developed a post-graduation project entitled "Production and characterization of a novel yeast extracellular invertase activity towards improved dibenzotiophene biodesulfurization through SSF" in the ambit of the R&D project "Carbon4Desulf", at the Bioenergy unit at LNEG. He has experience in characterization and production of novel enzymes through the use of agroindustrial feedstocks and their application in different bioprocesses through techniques such as SSF and SHF. He published 2 peer-reviewed paper in ISI journals, and presented several communications at national/international scientific meetings.

Salete S. Balula graduated in Technological Chemistry at Univ of Lisbon in 1997 and completed the PhD degree in Chemistry at Univ of Aveiro in 2004. After complete the PhD studies, Balula started the postdoctoral work in the area of polyoxometalates (POMs) and oxometal complexes: synthesis and catalytic applications. Balula shared this postdoctoral period working in the Green Chemistry Centre of the Univ of York, UK, supervised by Prof Clark and in CICECO Laboratory, Univ of Aveiro, supervised by Prof Gonçalves . During this time Balula developed her expertise in material science and sustainable oxidative heterogeneous catalysis. Balula started her independent research career on 2009, as Assistant Investigator in Associated Laboratory REQUIMTE, Univ of Porto. She created a new area of research and currently she is the only researcher in Portugal working in the area of heterogeneous oxidative catalysis based on POMs as active centers. Balula published a total of 60 papers in international peer review journals. As independent researcher, Balula has established important international and national collaborations to develop her career and also to create new lines of work. An example of this is the number papers published by Balula involving authors from international institutions and from other national institutions. Recently, Balula´s research group is developing a new area of expertise based in the desulfurization of diesel using green heterogeneous oxidative catalytic processes. Galp Energia is also an actual prestigious collaboration of Balula to investigate the production of sulfur-free fuels via heterogeneous oxidation.

Tendai Olsen Dembaremba was born in Zimbabwe, where he grew up before moving to South Africa to further his studies at Nelson Mandela University. Having a soft spot for environmental issues, he obtained his Bsc(Honors) and Msc in chemistry at the university, covering research around catalyst development for oxidative desulfurization of crude oil based fuels. He is currently pursuing his PhD under Prof. Tshentu and Dr. Ogunlaja, where he is seeking avenues to improve metal organic frameworks for use in adsorptive separation of sulfur and nitrogen compounds from fuel.

Zhenyu Ge received his Bachelor's degree at School of material science and engineering, Shandong University, China at July 2008. Then he began his research career at the Institute of Coal Chemistry. His research area is now mainly focused on synthesis of novel catalysis materials and their application.

Lichun Huang received his Ph.D. degree in 2011 from Institute of Coal Chemistry, CAS. In the same year, he was appointed member of the Ordos Dalu Institute of Coal Chemical Industry. He was promoted to senior engineer in 2015. The main focus of his work is the heterogeneous catalysis and Fischer–Tropsch synthesis. As of May 2018, He has published more than 10 scientific articles in international journals.

Adeniyi S. Ogunlaja received a B.Sc. (Hons) from University of Ilorin in 2004, and a M.Sc. from the University of Ilorin in 2009. He received his Ph.D. in Chemistry from Rhodes University in 2014. He is a member of the South African Chemical Institute and has received prestigious awards and grants such as the SASOL Post-Graduate Medal (2013) awarded by the South African Chemical Institute (SACI) for innovative research more recently a recipient of the South African National Research Foundation (NRF)-Thuthuka (2018-Rating track). Adeniyi lectures at Nelson Mandela University in Port Elizabeth, South Africa. He currently supervises several postgrads (PhD's and MSc) and has published several peer reviewed journal articles, contributed four book chapters and presented several conference papers. His research interests is in inorganic and analytical chemistry, with focus on oxidative desulfurization (ODS), hydrodesulfurization (HDS), photocatalysis and electrocatalysis.

Susana M. Paixão is a researcher at the Bioenergy Unit from LNEG. PhD in Microbiology by the Faculty of Sciences - University of Lisbon, Portugal, she has expertise in the fields of Environmental Microbiology (bioassays, ecotoxicity, biodegradability) and Bioenergy/Biotechnology (screening and characterization of novel microorganisms/enzymes, bioconversion studies; biomass for bioethanol; biorefinery; development of bioprocesses for bioenergy/bioproducts; experimental design; fermentation techniques such as SHF, SSF and CBP, fed-batch/continuous culture; clean technologies for fossil fuels - biodesulfurization). Experience as team member in several R&D projects, National and International (European) and experience in managing/co-managing R&D projects in Applied Microbiology/Enzymology and in supervising graduate, master and PhD thesis. She published more than 35 peer-reviewed papers, and presented numerous works (oral communications and posters) at national/international scientific meetings.

Susana Ribeiro received her BSc degree and MSc degree in Chemical Engineering from the Department of Chemical Engineering in the School of Engineering – Polytechnic of Porto. Following that, she worked within research fellowships at Institute for Polymer and Composites (IPC) University of Minho - Associated Laboratory I3N and at Department of Chemistry and Biochemistry Faculty of Science of the University of Porto (FCUP) - Associated Laboratory REQUIMTE. Currently, she is a PhD student at Department of Chemistry and Biochemistry Faculty of Science of the University of Porto (FCUP) - Associated Laboratory REQUIMTE. Her research focuses on the development of efficient composite materials to catalyze oxidative desulfurization processes (ODS) in order to reduce the level of sulfur in diesel.

Tiago Silva is a PhD student (SFRH/BD/104977/2014), at the Bioenergy Unit from LNEG, developing a thesis entitled "Zero emissions system for fossil fuel desulfurization using *Gordonia alkanivorans* strain 1B". This thesis is integrated in the R&D project "Greenfuel", of which he is currently part of the team. He has a master's degree in Applied Microbiology by the Faculty of Sciences - University of Lisbon, Portugal, which thesis ("Utilization of agroindustrial materials as alternative carbon sources for the biodesulfurization of fossil fuels by *Gordonia alkanivorans* strain 1B") was performed in the ambit of the R&D project "Carbon4desulf". He has experience in processing agroindustrial feedstoks as well as application of fermentation techniques such as SSF in batch or chemostat for the production of microbial biomass towards application in the biorefinery of fossil fuels and production of bacterial pigments and surfactants. He published 7 peer-reviewed papers in ISI journals, and presented several communications at national/international scientific meetings.

Mingxing Tang studied chemical engineering and technology at the Beijing University of chemical technology (2001-2005) then studied industrial catalysis at Chinese Academy of Sciences and received his Master in 2008. His major research interests are hydrogenation and desulfurization technologies. He is currently an associate professor. As of May 2018, He has published more than 10 scientific articles in various prestigious peer-reviewed journals.

Zenixole R. Tshentu is currently an associated professor of chemistry at the Nelson Mandela University (NMU) in South Africa. He completed his PhD studies in Inorganic Chemistry at NMU in 2005 and he was offered a lectureship position at the Rhodes University Chemistry Department where he remained 7 years until his move back to NMU in 2013 as associated professor. He was appointed as Head of Department of Chemistry at NMU in the period 2016-2018, and as the Director of the School of Biomolecular and Chemical Sciences (2019-2021). He also currently serves as the Executive Secretary of the South African Chemical Institute (SACI) in the period 2017-2021. He has experience in solid state and solution studies of transition metal chemistry as well as using inorganic/organic polymer materials in separation technology and in heterogeneous catalysis. He has been involved in several projects such as designing extractants for the separation of base metals and platinum group metals, designing catalysts and adsorbents for desulfurization and denitrogenation of fuels, designing ligands for stabilization of therapeutic metals in biological systems, and design of nanoparticles for biological targets. He has published more than 80 research articles and communications as well as four book chapters and two patents. He calls his philosophy "the quest for selectivity and specificity in chemical and biological systems". His contribution drives towards beneficiation of earth and secondary materials.

Weixing Wang received his Ph.D. degree at Institute of Coal Chemistry, CAS in 2018. From 2008 until now, he is a faculty member at Southwest Petroleum University in China. His research area is mainly focused on reactive adsorption desulfurization and methanation of syngas.

Index

www.ingramcontent.com/pod-product-compliance
Lightning Source LLC
LaVergne TN
LVHW080600300926
849162LV00005B/1009